Structural Vibration Serviceability and Human Comfort

Structural Vibration Serviceability and Human Comfort

Editor

Haoqi Wang

MDPI

Basel • Beijing • Wuhan • Barcelona • Belgrade • Novi Sad • Cluj • Manchester

Editor
Haoqi Wang
Tongji University
Shanghai, China

Editorial Office
MDPI
St. Alban-Anlage 66
4052 Basel, Switzerland

This is a reprint of articles from the Special Issue published online in the open access journal *Buildings* (ISSN 2075-5309) (available at: https://www.mdpi.com/journal/buildings/special_issues/struct_vibration_comfort).

For citation purposes, cite each article independently as indicated on the article page online and as indicated below:

Lastname, A.A.; Lastname, B.B. Article Title. *Journal Name* **Year**, *Volume Number*, Page Range.

ISBN 978-3-0365-8706-6 (Hbk)
ISBN 978-3-0365-8707-3 (PDF)
doi.org/10.3390/books978-3-0365-8707-3

Contents

About the Editor

Haoqi Wang

Haoqi Wang is currently an associate professor at the College of Civil Engineering, Tongji University, in Shanghai, China. He received his B.E. and M.E. from Tongji University and holds a Ph.D. from the University of Tokyo. His scientific interests include vibration serviceability, structural health monitoring, vehicle-bridge dynamics, and AI techniques in structural engineering.

Preface

This book is aimed at researchers and engineers that are involved or interested in the topic of the vibration serviceability of civil infrastructures. This topic, in recent years, has become more and more popular, owing to the ever-increasing living standards and higher demands for structural serviceability. Human activities, strong winds, heavy machines, and adjacent traffic may cause excessive vibration to structures, which needs to be carefully addressed in engineering practice. The articles in this book present the latest research regarding the above-mentioned problem. We invite the readers to learn about all of these novel findings.

Haoqi Wang
Editor

buildings

MDPI

Article

Dynamic Reliability Analysis of Large-Span Structures under Crowd Bouncing Excitation

Dongjun Zeng, Haoqi Wang * and Jun Chen

College of Civil Engineering, Tongji University, Shanghai 200092, China; 2110005@tongji.edu.cn (D.Z.); cejchen@tongji.edu.cn (J.C.)
* Correspondence: 12wanghaoqi@tongji.edu.cn

Abstract: Bouncing is one of the most common human crowd activities on civil infrastructures such as sports stadiums and concert halls, where the audience tends to make their bodies jump up and down to celebrate or participate in sport and musical events. Dynamic loads are thus generated and exerted on the structures, giving unpleasant structural vibration, which may affect the functionality of the structure or even lead to a panic of the crowd. Although researchers have studied human-induced vibration from many perspectives including load models, calculation methods, criteria for serviceability evaluation, etc., there has been minimal work regarding crowd-induced reliability analysis, mainly because the stochastic feature of the crowd load as well as the mechanism describing the crowd–structure interaction is still not clear. In this paper, a framework to calculate crowd-induced structural vibration that considers the crowd–structure interaction effect is proposed and is validated through an experimental test. The dynamic parameters of the bouncing person in the crowd are adopted from a previous statistical study. The feasibility of a probability density evolution method (PDEM) is proved to be effective to calculate structural stochastic vibration under the bouncing crowd. The dynamic reliability of the structure is thus analyzed based on the stochastic responses. Results show that the consideration of the crowd–structure interaction effect significantly affects the dynamic reliability, which is also dependent on various factors including bouncing frequency, failure criteria, limit threshold, human model parameter distribution, etc. This paper provides a foundation for the performance-based vibration serviceability design of large-span structures.

Keywords: bouncing excitation; crowd–structure interaction; stochastic vibration; dynamic reliability; probability density evolution method

Citation: Zeng, D.; Wang, H.; Chen, J. Dynamic Reliability Analysis of Large-Span Structures under Crowd Bouncing Excitation. *Buildings* **2022**, *12*, 332. https://doi.org/10.3390/buildings12030332

Academic Editor: Giuseppina Uva

Received: 23 January 2022
Accepted: 7 March 2022
Published: 10 March 2022

1. Introduction

In recent decades, large-span structures become more and more popular in structural design due to the requirement from building functionalities as well as from an aesthetic point of view, especially for public facilities including sports stadiums, transport stations, etc. While the developed construction technologies and new materials are capable to ensure safety and durability, large-span structures such as sports stadiums usually suffer from unpleasant vibrations when human crowd activities such as crowd bouncing take place, leading to serviceability problems of the structure [1]. The well-known incident of the Millennium Bridge triggered research on the human-induced structural serviceability problem [2]. Researchers have studied this problem from various perspectives, including the human-induced load models [3,4], structural calculation methods [5,6], comfort criteria [7], and vibration control technologies [8]. The human-induced vibration serviceability problem has nowadays become an important and sometimes dominant issue that must be considered in the design stage of the large-span structures.

Although the research regarding the structural vibration serviceability problem has gained tremendous popularity among many researchers in the past several decades, the dynamic reliability problem of such vibration has rarely been investigated. In most design codes around the world, the serviceability of large-span structures is satisfied according to the allowable response method [9–14], indicating that when the human-induced structural response exceeds a predetermined value, the structure is determined as a failure from a serviceability point of view. However, it is already widely acknowledged that human-induced excitation is highly stochastic, featured by the so-called inter- and intra-subject variability [15]. To fully reflect the stochastic effect, the concept of dynamic reliability needs to be adopted. The failure of the structure should not be based on whether the structural response exceeds the limit value or not. Instead, the dynamic reliability of the structure under the crowd excitation needs to be higher than a predetermined target value.

It is known that the crowd–structure interaction (CSI) has a significant effect on structural vibration, especially when the mass of the crowd is not negligible compared with the mass of the structure, which, unfortunately, is always the case for the light-weight large-span structures [16–19]. To develop the dynamic reliability analysis method for the human-induced structural vibration, an analytical method that quantitatively considers the interaction effect is necessary. Some researchers use a single degree of freedom (DOF) system to represent the crowd and the structure and thus establish a 2 DOF equation of motion of the crowd–structure interaction system [20], which is unable to consider higher vibration modes that may interact with higher harmonics of the crowd load. Nimmen et al. proposed a detailed crowd model as well as a simplified method that makes it possible to evaluate the interaction effect for walking excitation [21,22]. However, there is still no widely accepted solution to this problem.

One more factor that may affect the CSI result is that the dynamic parameters of the human bodies in the crowd are not easily determined. Researchers have found that the human body could be represented by a spring-mass damper (SMD) model with its own mass, stiffness, and damping [5,23]. The values of these dynamic parameters are usually investigated through modal analysis on the empty and the crowd-occupied structure [16,24]. However, when the structure is occupied by an active crowd, the operational modal analysis becomes doubtful because the structure is under forced vibration. Moreover, the dynamic parameters of the human body are highly stochastic, and a large number of tests are necessary to obtain their probability distribution as well as their dependency on the motion frequency. In 2017 and 2019, the particle filter technique is adopted to estimate the dynamic properties of walking [25] and bouncing people [26], giving a reasonable description of the SMD model parameters, which can be used for further analysis of the CSI effect on the structural response calculation.

Due to the huge computational cost for reliability analysis, an efficient technique to calculate stochastic vibration is usually necessary, especially when a refined structural finite-element model is adopted. For such cases, the widely adopted Monte Carlo simulation (MCS), although versatile for stochastic vibration calculation, has its drawback of high computational cost in practical applications [27]. On the other hand, the probability density evolution method (PDEM) is popular for its capability of reducing the computational cost by the selection of limited representative points in the calculation process [28] and has the potential to be used in the dynamic reliability analysis of crowd-induced vibration.

From the above statement, three difficulties that hinder the development of the dynamic reliability for crowd-induced serviceability problems can be roughly seen, including (a) practical mechanism reflecting CSI effect, (b) stochastic feature of human body parameters, and (c) efficient stochastic vibration calculation technique. Therefore, the aim of this research is to develop an efficient framework for dynamic reliability analysis of structures suffering crowd bouncing load from the perspective of vibration serviceability and provide analytical results for the reference of engineering practice. In this paper, the governing dynamic equations with and without CSI are first derived and proposed. Each individual is regarded as a SMD model, and thus the effect of physical properties of each bouncing person is included in the analysis. An experiment quantitatively showing the CSI effect by including physical properties of the human body is carried out for validation and the effect of CSI is further investigated by numerical simulation. PDEM is then used in this study to calculate the stochastic response under crowd excitation efficiently, and the dynamic reliability is thus analyzed. The randomness originating from each bouncing person is included using the probability distribution extracted from a previous study [26]. Finally, the dependency of the dynamic reliability on various factors including the CSI effect, reliability criteria, bouncing frequency, limit threshold, and parameter distribution is investigated for engineering practice.

2. Analytical Model of Crowd–Structure Interaction

2.1. Human–Structure Coupled System

In this section, an analytical model of the crowd–structure interaction is given. The description of this analytical model starts from the SMD model representing single person bouncing on an arbitrary structure, which is illustrated by Figure 1. The equation of motion for single-person bouncing is expressed as Equation (1):

$$m_0\ddot{u}_0 + c_0(\dot{u}_0 - \dot{v}_0) + k_0(u_0 - v_0) = p_0(t) \tag{1}$$

where m_0, c_0, and k_0 are the mass, damping, and stiffness of the human body, u_0 is the displacement of the human system, v_0 is the structural displacement response of the structure at the human excitation point, and $p_0(t)$ is the time history of the biomechanical, which could be defined by Fourier series as:

$$p_0(t) = a_0 + \sum_i (a_i \cos(2\pi f_b t) + b_i \sin(2\pi f_b t)) \tag{2}$$

in which a_0, a_n, b_n are Fourier coefficients, f_b is the motion frequency, and n is the order of the biomechanical force. The nth biomechanical load factor (BLF) is defined by:

$$\mathrm{BLF}_{0n} = \frac{\sqrt{a_n^2 + b_n^2}}{m_0 g} \tag{3}$$

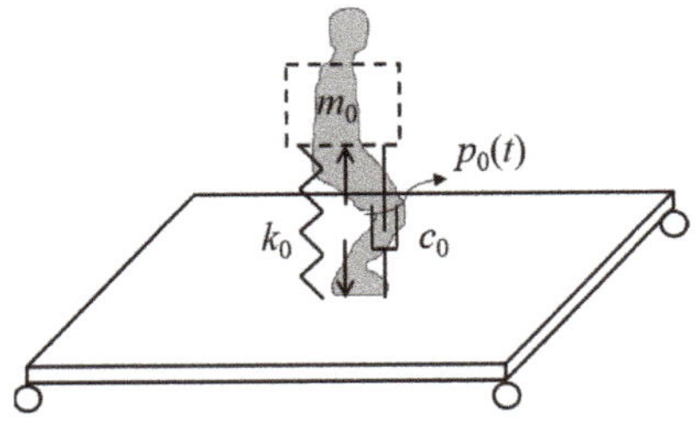

Figure 1. Diagram of analytical model for single-person bouncing.

From the force equilibrium at the surface of the structure, the contact force $F_0(t)$ between the human body and the structure is equal to the human's inertia force, as shown by Equation (4):

$$F_0(t) = -m_0\ddot{u}_0 \tag{4}$$

The equation of motion of the structure is written for each mode according to the modal decomposition principle. For the jth mode of the structure, its equation of motion is:

$$M_{s,j}\ddot{q}_j + C_{s,j}\dot{q}_j + K_{s,j}q_j = F_0(t)\phi_{j0} \tag{5}$$

where $M_{s,j}$, $C_{s,j}$, and $K_{s,j}$ are the mass, damping, and stiffness of the jth decomposed mode, q_j is the jth modal coordinate, and ϕ_{j0} is the jth mode shape value at the point of the bouncing excitation.

Through modal superposition, the physical displacement response at the excitation point is calculated as:

$$v_0 = \sum_j q_j \phi_{j0} \tag{6}$$

It is clearly observed that Equations (1), (4) and (5) are coupled together, showing that the human mechanical system parameters have an influence on the structural response, which in return affects the human body response as well as the contact force.

2.2. Crowd–Structure Coupled System

The above derivation is extended to the case where multiple bouncing people excite the structure, which happens more frequently in engineering practice (e.g., audience celebrating in a grandstand of a sports stadium) and is more likely to give vibration serviceability problems compared to the single-person case. The crowd bouncing case is illustrated in Figure 2.

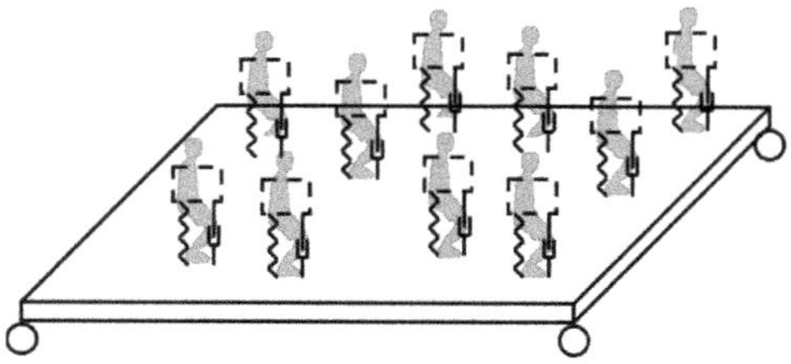

Figure 2. Diagram of analytical model for crowd bouncing.

For the crowd bouncing case, each person in the crowd is represented by a SMD model with a pair of biomechanical forces. For the *i*th person in the bouncing crowd, the equation of motion is indicated by:

$$m_i \ddot{u}_i + c_i(\dot{u}_i - \dot{v}_i) + k_i(u_i - v_i) = p_i(t) \tag{7}$$

in which the symbols share the same definition as in Equation (1).

Considering the multiple excitation points from the crowd, the equation of motion of the *j*th mode of the structure becomes:

$$M_{s,j}\ddot{q}_j + C_{s,j}\dot{q}_j + K_{s,j}q_j = -\sum_i m_i \ddot{u}_i \phi_{ji} \tag{8}$$

where ϕ_{ji} indicates the *j*th mode shape value at the location of the *i*th bouncing person.

From the modal superposition principle, the structural displacement at the location of the *i*th bouncing person is expressed:

$$v_i = \sum_j q_j \phi_{ji} \tag{9}$$

The combination of Equations (7)–(9) governs the crowd–structure coupling system. It is shown that the structural response is affected by all bouncing people in the crowd. In return, the human model parameters of each bouncing person in the crowd have an influence on the structural response.

2.3. Governing Dynamic Equations with and without Interaction Effect

The governing equations expressed by Equations (7)–(9) need to be solved to obtain the structural responses. For the case of *N* people bouncing on a structure decomposed by

Y vibration modes, the total number of degree-of-freedom is $N + Y$. The equations rewritten in the matrix form is expressed by:

$$\begin{bmatrix} \mathbf{m} & \mathbf{O} \\ \mathbf{Z}_\mathrm{m} & \mathbf{M}_\mathrm{s} \end{bmatrix} \begin{bmatrix} \ddot{\mathbf{U}} \\ \ddot{\mathbf{q}} \end{bmatrix} + \begin{bmatrix} \mathbf{c} & \mathbf{Z}_\mathrm{c} \\ \mathbf{O} & \mathbf{C}_\mathrm{s} \end{bmatrix} \begin{bmatrix} \dot{\mathbf{U}} \\ \dot{\mathbf{q}} \end{bmatrix} + \begin{bmatrix} \mathbf{k} & \mathbf{Z}_\mathrm{k} \\ \mathbf{O} & \mathbf{K}_\mathrm{s} \end{bmatrix} \begin{bmatrix} \mathbf{U} \\ \mathbf{q} \end{bmatrix} = \begin{bmatrix} \mathbf{P} \\ \mathbf{O} \end{bmatrix} \quad (10)$$

in which

$$\mathbf{m} = \begin{bmatrix} m_1 & 0 & 0 \\ 0 & \ddots & 0 \\ 0 & 0 & m_N \end{bmatrix}_{N\times N}, \mathbf{c} = \begin{bmatrix} c_1 & 0 & 0 \\ 0 & \ddots & 0 \\ 0 & 0 & c_N \end{bmatrix}_{N\times N}, \mathbf{k} = \begin{bmatrix} k_1 & 0 & 0 \\ 0 & \ddots & 0 \\ 0 & 0 & k_N \end{bmatrix}_{N\times N} \quad (11)$$

$$\mathbf{M}_\mathrm{s} = \begin{bmatrix} M_{\mathrm{s},1} & 0 & 0 \\ 0 & \ddots & 0 \\ 0 & 0 & M_{\mathrm{s},Y} \end{bmatrix}_{Y\times Y}, \mathbf{C}_\mathrm{s} = \begin{bmatrix} C_{\mathrm{s},1} & 0 & 0 \\ 0 & \ddots & 0 \\ 0 & 0 & C_{\mathrm{s},Y} \end{bmatrix}_{Y\times Y}, \mathbf{K}_\mathrm{s} = \begin{bmatrix} K_{\mathrm{s},1} & 0 & 0 \\ 0 & \ddots & 0 \\ 0 & 0 & K_{\mathrm{s},Y} \end{bmatrix}_{Y\times Y} \quad (12)$$

$$\mathbf{Z}_\mathrm{m} = \begin{bmatrix} m_1\phi_{11} & \cdots & m_N\phi_{1N} \\ \vdots & \ddots & \vdots \\ m_1\phi_{Y1} & \cdots & m_N\phi_{YN} \end{bmatrix}, \mathbf{Z}_\mathrm{c} = -\begin{bmatrix} c_1\phi_{11} & \cdots & c_1\phi_{Y1} \\ \vdots & \ddots & \vdots \\ c_N\phi_{1N} & \cdots & c_N\phi_{YN} \end{bmatrix}, \mathbf{Z}_\mathrm{k} = -\begin{bmatrix} k_1\phi_{11} & \cdots & c_k\phi_{Y1} \\ \vdots & \ddots & \vdots \\ k\phi_{1N} & \cdots & k_N\phi_{YN} \end{bmatrix} \quad (13)$$

$$\mathbf{U} = \begin{bmatrix} u_1 & \cdots & u_N \end{bmatrix}^\mathrm{T}, \mathbf{P} = \begin{bmatrix} p_1(t) & \cdots & p_N(t) \end{bmatrix}^\mathrm{T}, \mathbf{q} = \begin{bmatrix} q_1(t) & \cdots & q_Y(t) \end{bmatrix}^\mathrm{T} \quad (14)$$

and **O** indicate a zero matrix. In this work, a three-order Fourier series model is adopted to simulate $p_i(t)$.

From Equations (10)–(14), it is observed that the crowd and the structure are coupled together through the matrices $\mathbf{Z}_\mathrm{c}$ and $\mathbf{Z}_\mathrm{k}$. For the purpose of comparing the structural response with and without the CSI effect, the equation of motion without the coupling term is obtained by replacing them with zero matrices, as shown by Equation (15):

$$\begin{bmatrix} \mathbf{m} & \mathbf{O} \\ \mathbf{Z}_\mathrm{m} & \mathbf{M}_\mathrm{s} \end{bmatrix} \begin{bmatrix} \ddot{\mathbf{U}} \\ \ddot{\mathbf{q}} \end{bmatrix} + \begin{bmatrix} \mathbf{c} & \mathbf{O} \\ \mathbf{O} & \mathbf{C}_\mathrm{s} \end{bmatrix} \begin{bmatrix} \dot{\mathbf{U}} \\ \dot{\mathbf{q}} \end{bmatrix} + \begin{bmatrix} \mathbf{k} & \mathbf{O} \\ \mathbf{O} & \mathbf{K}_\mathrm{s} \end{bmatrix} \begin{bmatrix} \mathbf{U} \\ \mathbf{q} \end{bmatrix} = \begin{bmatrix} \mathbf{P} \\ \mathbf{O} \end{bmatrix} \quad (15)$$

Equations (10) and (15) can be easily solved by the typical numerical algorithm such as the Runge–Kutta algorithm or the Newmark algorithm once the parameters are determined. It is noteworthy that the above analysis applies to all types of structures since only the modal parameters are needed.

3. Numerical and Experimental Test for Structural Response Calculation

In this section, a numerical example is given to calculate the structural acceleration response following the procedure proposed above. The crowd parameter settings and the structural analytical model are briefly introduced and the structural responses with and without the consideration of the CSI effect are compared.

3.1. Analytical Model of a Large-Span Structure

A large-span structure with a size of 10 m × 6 m is adopted as the structural model for the numerical example in this section. The floor is with line supports two long and two short sides. The dynamic properties are acquired through modal tests conducted in advance and are listed in Table 1. The mode shapes of the structure are depicted in Figure 3. Considering crowd bouncing usually results in linear vibration, the modal decomposition method is adopted using the modal parameters in Table 1 for efficiency instead of a finite element model.

Table 1. Dynamic properties of the structural model.

	Mode 1	Mode 2	Mode 3	Mode 4	Mode 5	Mode 6
Modal mass (kg)	8583	2587	9625	2423	2898	4900
Frequency (Hz)	3.500	6.150	6.750	14.120	15.190	18.100
Damping ratio (%)	0.374	0.514	0.614	0.913	0.666	1.497

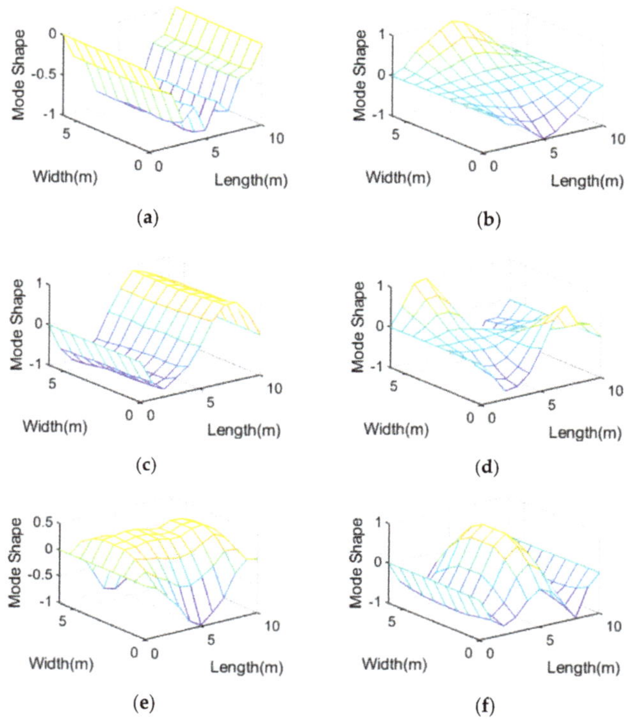

Figure 3. Mode shapes of the structural model: (**a**) Mode 1; (**b**) Mode 2; (**c**) Mode 3; (**d**) Mode 4; (**e**) Mode 5; (**f**) Mode 6.

3.2. Numerical Example Showing CSI Effect

As stated in the introduction section, the randomness of the structural responses originates from the stochastic physical parameters of the crowd. In a previous study, the physical parameters of a bouncing person as well as their probability distribution have been investigated. It is reported that the physical parameters of a bouncing person, including natural frequency, damping ratio, and BLFs, follow a skew-normal distribution [26] defined by its location parameter μ, scale parameter σ, and shape parameter α, as shown by Equation (16):

$$h(x) = \frac{2}{\sigma\sqrt{2\pi}} e^{-\frac{(x-\mu)^2}{2\sigma^2}} \int_{-\infty}^{\alpha\left(\frac{x-\mu}{\sigma}\right)} \frac{1}{\sqrt{2\pi}} e^{-\frac{t^2}{2}} dt \quad (16)$$

where $h(x)$ describes the probability density function (PDF) of parameter x that represents either of the physical parameters of a bouncing person.

Moreover, the mass of crowd follows a normal distribution defined by parameter μ and σ according to Ref [29]. The coefficients of the above distribution when the bouncing frequency f_b equals 1.75 Hz are listed in Table 2.

Table 2. Distribution coefficients of parameters for the analytical model, adapted from Refs. [26,29].

	Natural Frequency (Hz)	Damping Ratio (%)	BLF_1	BLF_2	BLF_3	Mass (kg)
μ	1.73	0.13	0.20	0.15	0.01	62.80
σ	0.37	0.16	0.10	0.13	0.06	10.90
α	−3.52	4.00	2.61	0.75	8.01	——

In this paper, random variables representing the physical parameters of each bouncing person in the crowd are generated following the predetermined probability distribution. In this manner, the crowd is coupled with the structure in the way described in Section 2. Structural responses under the bouncing crowd with six different crowd sizes were calculated. Considering the size of the structure, the number of people in each crowd is decided to be 1, 3, 9, 15, 21, and 27, and their geometrical distribution of this crowd (i.e., the excitation points of the crowd) is illustrated in Figure 4. Because the structure has a fundamental frequency of around 3.500 Hz (See Table 1), a bouncing frequency of 1.750 Hz, which can be easily achieved for the audience activities in a sports stadium or in a concert hall, is analyzed in this numerical example.

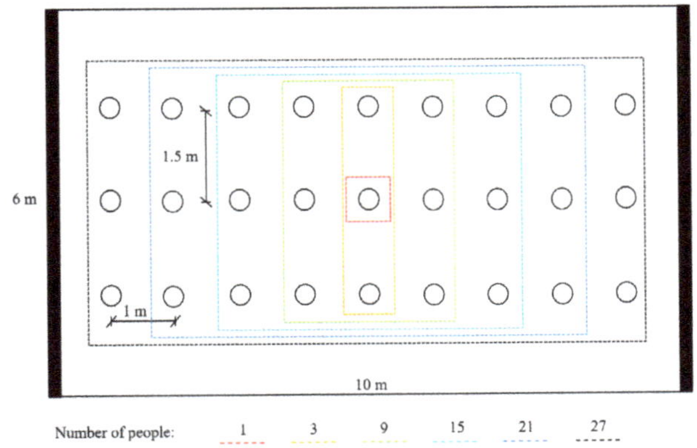

Figure 4. Position of the bouncing people in the crowd in the numerical example.

As a typical illustration of the CSI effect, the structural mid-span acceleration with nine bouncing people is calculated according to Equations (10) and (15), for structural response with and without considering CSI, respectively. The calculated responses are given in Figure 5. Note that the human model parameters of the bouncing crowd are generated following the distribution given in Table 2. It is shown that the response with CSI is smaller than that without CSI, which is explained by the fact that the people in the bouncing crowd act as additional dampers that absorb energy from the structural vibration. A CSI index ε is defined following Equation (17) to show the error if the CSI effect is considered.

$$\varepsilon = \frac{RMS_{withoutCSI} - RMS_{withCSI}}{RMS_{withCSI}} \tag{17}$$

In this calculation, the root mean square (RMS) value of the time history without CSI is calculated to be 4.78 m/s^2, while the response considering CSI is 2.95 m/s^2. Following the above definition, the CSI index is calculated to be 62.03%, clearly showing that large errors may occur if the CSI effect is not properly considered in the structural response prediction at the design stage.

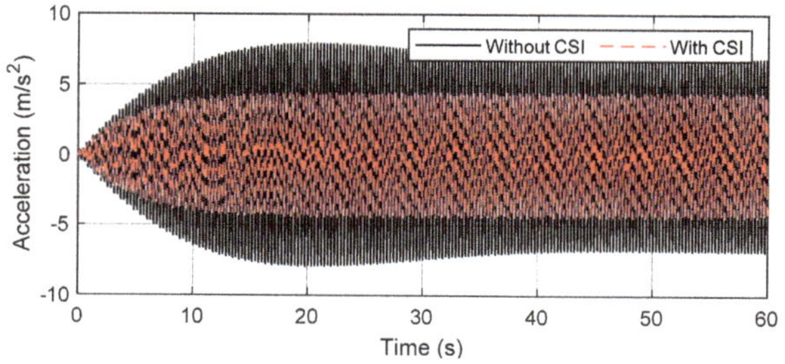

Figure 5. Comparison of structural mid-span acceleration with and without CSI effect.

The CSI index for different numbers of bouncing people is calculated for each crowd size. Because the physical parameters of the people in the bouncing crowd are random variables (See Table 2 for their distribution), the structural responses with and without the CSI effect are also random. The above process is repeated three times to show the increasing trend of the CSI index against the increase in number of people, as shown by Figure 6. With more bouncing people, the CSI effect tends to become larger, mainly because the mass ratio between the crowd and the structure becomes higher. The results show that the error from the negligence of CSI can reach up to 150% for such a large-span structure.

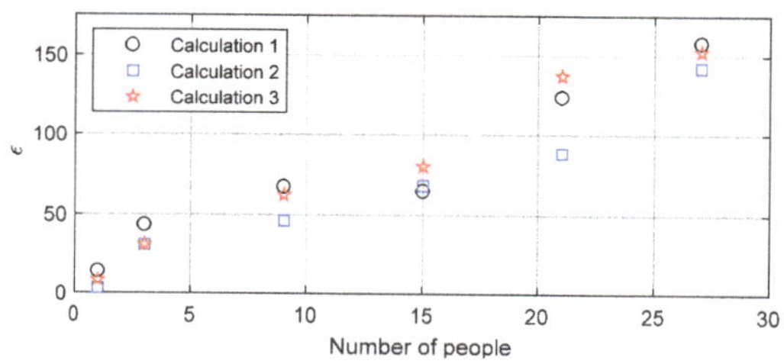

Figure 6. Relation of ε and number of people in a bouncing crowd.

3.3. *Formatting of Mathematical Components*

To validate the procedure of the structural response calculation under crowd bouncing excitation, an experimental test was conducted on a prestressed concrete plate with a weight of 16.5 t. The dynamic properties of this plate have been given in Section 3.1. Accelerometers were attached at the bottom of the plate to capture the structural acceleration. A test participant with a weight of 59.7 kg was asked to bounce at the mid-point of the plate with a bouncing frequency of 1.75 Hz under the guidance of a metronome. The physical parameters of this participant had been identified in advance using an inverse analysis technique described in Ref. [26] and are listed here in Table 3. The overview of the experimental setup is shown in Figure 7.

Table 3. Physical parameters of the bouncing person.

Parameter	Natural Frequency (Hz)	Damping Ratio (%)	BLF_1	BLF_2	BLF_3
Value	1.49	41	0.345	0.267	0.046

The test participant was asked to bounce on the structure three times, each of which lasted around 30 s. The structural mid-span acceleration of Test I was plotted in Figure 8 together with the prediction given by Equations (10) and (15). It was expected that the measured structural acceleration was lower than the prediction because in the real case the test participant could not keep his bouncing frequency as a constant, while in the calculation procedure the bouncing frequency was fixed at 1.75 Hz to assure resonance. However, it is clear that the prediction with the consideration of CSI effect, i.e., responses given by Equation (10), is much closer and sometimes equals to the measured acceleration than the one without CSI effect.

(**a**) (**b**)

Figure 7. Experimental setup: (**a**) Overview of the large-span plate; (**b**) Test participant on the structure.

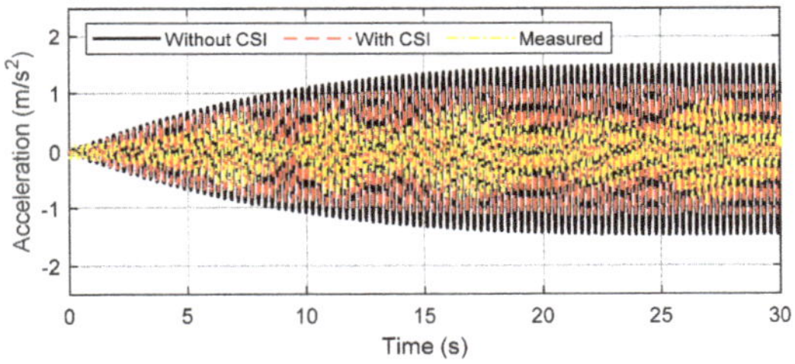

Figure 8. Comparison of predicted and measured mid-span acceleration.

The maximum value of the mid-span acceleration of each bouncing test was listed in Table 4 and compared with the prediction with and without the CSI effect. It is shown that although both predicted values are higher than the measured ones, the prediction with the CSI effect is much closer to the measured values, indicating that the proposed method for structural response calculation is reasonable.

Table 4. Comparison of predicted and measured mid-span RMS acceleration.

Case	No CSI	With CSI	Test 1	Test 2	Test 3
Value (m/s^2)	1.48	1.14	0.93	0.80	0.87

4. PDEM-Based Stochastic Vibration Analysis and Its Verification

4.1. Formatting of Mathematical Components

Because of the variability of the human-induced load, the structural responses under crowd excitation should be considered from the perspective of random vibration. In view of this, the governing equation expressed by Equation (10) is rewritten in the form of Equation (18):

$$\mathbf{M}(\Theta)\ddot{\mathbf{X}}(\Theta,t)+\mathbf{C}(\Theta)\dot{\mathbf{X}}(\Theta,t)+\mathbf{K}(\Theta)\mathbf{X}(\Theta,t)=\mathbf{F}(\Theta,t) \tag{18}$$

where

$$\mathbf{X}=\begin{bmatrix}\mathbf{U} & \mathbf{q}\end{bmatrix}^{\mathrm{T}},\mathbf{F}=\begin{bmatrix}\mathbf{P} & \mathbf{O}\end{bmatrix}^{\mathrm{T}} \tag{19}$$

and **K**, **C**, and **M** represent the matrices in Equation (10) before **X** and its derivatives.

The vector $\mathbf{\Theta}$ in Equation (18) characterizes all random variables involved in the crowd–structure coupling system. For example, if the dynamic parameters of the structural model are considered deterministic, this vector will only contain human model parameters and has the form of Equation (20):

$$\mathbf{\Theta} = \begin{bmatrix} m_1 & \cdots & m_N & c_1 & \cdots & c_N & k_1 & \cdots & k_N & \mathrm{BLF}_{11} & \cdots & \mathrm{BLF}_{1N} & \mathrm{BLF}_{21} & \cdots & \mathrm{BLF}_{2N} & \mathrm{BLF}_{31} & \cdots & \mathrm{BLF}_{3N} \end{bmatrix}^{\mathrm{T}} \quad (20)$$

where BLF_{ij} denotes the ith order coefficient of the bouncing load for the jth person, and the others share the same definition with Equation (11). Therefore, the solution of Equation (18) is the function of $\mathbf{\Theta}$ and t, and is expressed as:

$$\mathbf{X} = \mathbf{H}(\mathbf{\Theta}, t) \quad (21)$$

It is noteworthy that any physical quantities, i.e., acceleration, bending moment, etc., of the structural system could be expressed as a function of the solution $\mathbf{X}$ and/or its derivatives, as shown by Equation (22):

$$\mathbf{Z} = \mathbf{F_Z}(\mathbf{X}) = \mathbf{H_Z}(\mathbf{\Theta}, t) \quad (22)$$

where $\mathbf{Z}$ is the physical quantity of interest of the system and $\mathbf{H_z}$ is a deterministic vector operator that describes the physical mechanism of the system expressed by Equation (18).

According to the density evolution theory [28], the evolutionary joint PDF of $(\mathbf{Z}^{\mathrm{T}}, \mathbf{\Theta}^{\mathrm{T}})^{\mathrm{T}}$, denoted as $p_{\mathbf{Z\Theta}}(\mathbf{z}, \theta, t)$, is governed by Equation (23).

$$\frac{\partial p_{\mathbf{Z\Theta}}(\mathbf{z}, \theta, t)}{\partial t} + \sum_j \dot{Z}_j(\theta, t) \frac{\partial p_{\mathbf{Z\Theta}}(\mathbf{z}, \theta, t)}{\partial z_j} = 0 \quad (23)$$

If only one physical quantity Z is of interest, Equation (23) will reduce to a one-dimensional form with the initial condition as shown by Equations (24) and (25), which is used more often in most situations.

$$\frac{\partial p_{Z\mathbf{\Theta}}(z, \theta, t)}{\partial t} + \dot{Z}(\theta, t) \frac{\partial p_{Z\mathbf{\Theta}}(z, \theta, t)}{\partial z} = 0 \quad (24)$$

$$p_{Z\mathbf{\Theta}}(z, \theta, t_0) = \delta(z - z_0) p_{\mathbf{\Theta}}(\theta) \quad (25)$$

Equation (24) governs the PDF evolution of the PDF of the structural response which evolves with time and is thus denoted as the generalized density evolution equation (GDEE). The purpose to solve the stochastic vibration problem is to obtain the PDF of the response from that of the system parameters. For the crowd-induced structural vibration, the acceleration response is usually much related to the serviceability of the structure in the view of human comfort. Therefore, the physical quantity Z in this study is represented by the structural acceleration $\ddot{v}_i(t)$.

4.2. Procedure to Numerically Solve GDEE

For most engineering problems, the governing equation shown by Equation (24) needs to be solved numerically. In this process, the coefficient of the partial differential equation, $\dot{Z}(\theta, t)$, which can be calculated from Equation (22), should be obtained in the first place. This equation is solved by a point evolution method whose details are found in [27]. The steps using this method are briefly reviewed.

A number of n_{pt} representative points in the distribution domain $\Omega_{\mathbf{\Theta}}$ for vector $\mathbf{\Theta}$ are firstly selected and are denoted as:

$$\theta_q = \begin{pmatrix} \theta_{q,1} & \theta_{q,2} & \cdots & \theta_{q,N_\theta} \end{pmatrix}, q = 1, 2, \cdots, n_{\mathrm{pt}} \quad (26)$$

where $N_{\boldsymbol{\theta}}$ denotes the dimension of vector $\boldsymbol{\Theta}$. The assigned probability corresponding to the qth representative point is:

$$P_q = \int_{\Omega_q} p_{\boldsymbol{\Theta}}(\boldsymbol{\theta})\mathrm{d}\boldsymbol{\theta} \tag{27}$$

where Ω_q is the subdomain for $\Omega_{\boldsymbol{\Theta}}$ that satisfies Equation (28):

$$\begin{cases} \boldsymbol{\Omega}_q \cup \boldsymbol{\Omega}_p = \oslash, \forall p \neq q \\ \cup_{q=1}^{n_{\mathrm{pt}}} \boldsymbol{\Omega}_q = \boldsymbol{\Omega}_{\boldsymbol{\Theta}} \end{cases} \tag{28}$$

In each subdomain Ω_q, it is assumed that the coefficient $\dot{Z}(\boldsymbol{\theta}, t)$ is invariant with respect to $\boldsymbol{\theta}$, as expressed as follows:

$$\dot{Z}(\boldsymbol{\theta}, t) = \dot{Z}(\boldsymbol{\theta}_q, t) \tag{29}$$

The governing equation is transformed to Equation (30) by integrating Equation (24) over Ω_q with respect to $\boldsymbol{\theta}$:

$$\frac{\partial p_q(z,t)}{\partial t} + \dot{Z}(\boldsymbol{\theta}_q, t)\frac{\partial p_q(z,t)}{\partial z} = 0, q = 1, 2, \cdots, n_{\mathrm{pt}} \tag{30}$$

where

$$p_q(z,t) = \int_{\Omega_q} p_{Z\boldsymbol{\Theta}}(z, \boldsymbol{\theta}, t)\mathrm{d}\boldsymbol{\theta} \tag{31}$$

The corresponding initial condition is also transformed to:

$$p_q(z, t_0) = \delta(z - z_0)P_q, q = 1, 2, \cdots, n_{\mathrm{pt}} \tag{32}$$

which could be solved through the finite difference method, giving the numerical solution of $p_q(z, t)$. The PDF of Z at each time instant can be calculated by summation of the PDF of each representative point, as shown by Equation (33):

$$p_Z(z,t) = \int_{\Omega_{\boldsymbol{\Theta}}} p_{Z\boldsymbol{\Theta}}(z, \boldsymbol{\theta}, t)\mathrm{d}\boldsymbol{\theta} = \sum_{q=1}^{n_{\mathrm{pt}}} \int_{\Omega_q} p_{Z\boldsymbol{\Theta}}(z, \boldsymbol{\theta}, t)\mathrm{d}\boldsymbol{\theta} = \sum_{q=1}^{n_{\mathrm{pt}}} p_q(z,t) \tag{33}$$

The PDF expressed in the above equation serves as the basis to evaluate the structural dynamic reliability in further analysis.

4.3. Response Calculation through PDEM and Its Verification

Because the human physical parameters of the bouncing person in a crowd are random parameters, the structural acceleration under such a crowd excitation needs to be considered as a random process. In calculation, the GF-discrepancy method is used for point selection [30]. A difference scheme known as total variation diminishing (TVD) is adopted to solve the GDEE expressed by Equation (24), whose details are found in Ref [31], to obtain the PDF of the structural responses. The mean value and the standard deviation time histories are plotted in Figure 9 for the case described in Section 3.2, to characterize the randomness. As in Figure 9a, the mean time history shares a similar shape with its representative time history shown in Figure 5, while the standard deviation can reach as large as 1.2 $\mathrm{m/s^2}$, indicating that the randomness of the physical parameters in the crowd could not be neglected.

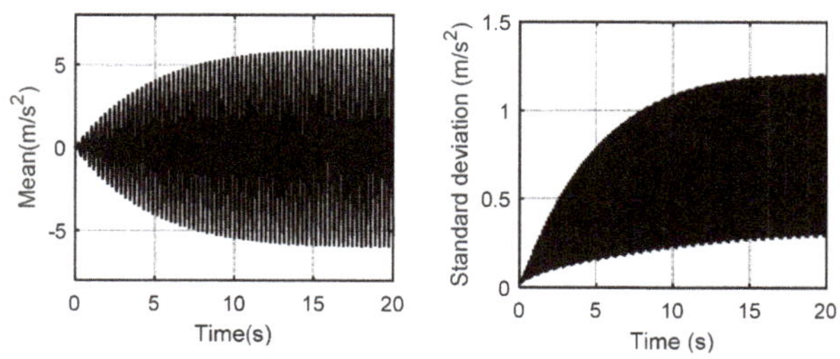

(**a**) (**b**)

Figure 9. The mean and the standard deviation time history of the mid—span acceleration: (**a**) Mean; (**b**) Standard deviation.

Furthermore, the PDF evolution surface of the mid-span acceleration responses is obtained and illustrated in Figure 10. The extreme value of the PDF decreases over time within the first ten seconds, indicating that the structural responses are gradually stimulated, and the variation keeps increasing. After around 15 s, the extreme value becomes nearly constant, mainly because the structural responses have reached the stable stage. The PDF thus starts to evolve with a regulated pattern, as already illustrated in Figures 5 and 8.

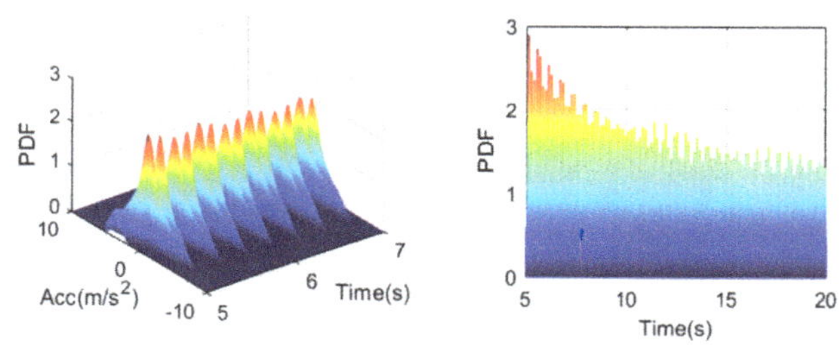

(**a**) (**b**)

Figure 10. Probability density evolution surface of the structural mid-span acceleration: (**a**) Azimuth = −37.5° and elevation = 37.5°; (**b**) Azimuth = 0 and elevation = 0.

The above results are obtained from the PDEM method on the basis of 1000 deterministic analyses, i.e., 1000 points are selected using the GF-discrepancy method. To verify its accuracy, the distribution at some time instants is calculated through Monte-Carlo simulation (MCS) using 100,000 examples. The random distribution is compared in the form of cumulative density function (CDF) as shown by Figure 11. Good consistency is observed from the comparison of CDF from PDEM and MCS, and the results obtained by PDEM are thus demonstrated to be correct.

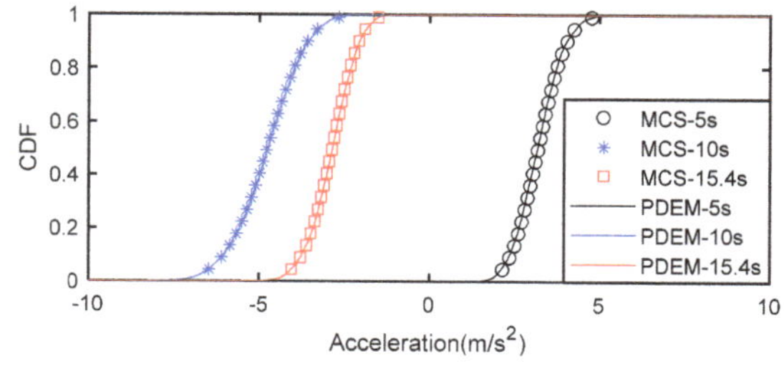

Figure 11. Comparison of CDF of structural mid—span acceleration.

Based on the results of the random vibration analysis above, the dynamic reliability of the structure is evaluated, which will be discussed in the next section.

5. Dynamic Reliability Analysis for Crowd-Induced Structural Vibration

5.1. Failure Criteria

For the calculation of structural dynamic reliability, reasonable failure criteria need to be defined. Because the human comfort on the structure is mainly related to structural acceleration, most design codes require that the maximum acceleration response should not exceed a predefined limit to maintain the vibration serviceability of the structure. Once the acceleration exceeds such a limit, the structure is considered a failure from the perspective of vibration serviceability. If the structural acceleration at the mid-point of the structure is interested, the reliability could be expressed by the most frequently adopted first-passage criterion shown in Equation (34):

$$R(t) = P\{|\ddot{v}_{\text{mid}}(\tau)| \leqslant [\ddot{v}], \tau \in [0, t]\} \quad (34)$$

in which *R*(*t*) is the time-dependent dynamic reliability, $\ddot{v}_{\text{mid}}$ is the acceleration at the mid-point of the structure, and $[\ddot{v}]$ is the predefined limit threshold.

However, the mid-point of the structure may not always be the location where the largest structural acceleration occurs, especially when higher vibration modes are excited. Furthermore, the locations on which people are bouncing or standing are more interested because the structural acceleration at these locations is directly perceived by people. In this manner, it is more appropriate to define the dynamic reliability as the probability that the acceleration responses at a series of selected locations do not exceed the predefined limit threshold, as expressed by Equation (35):

$$R(t) = P\left\{\bigcap_{l=1}^{L} |\ddot{v}_l(\tau)| \leqslant [\ddot{v}], \tau \in [0, t]\right\} \quad (35)$$

where L is the number of locations of interest, and $\ddot{v}_l$ indicates the structural acceleration at the lth location.

5.2. Calculation of Dynamic Reliability

The reliability of a structure is usually defined as the probability that the structure finished its expected function within a certain time period. Li proposed a method to calculate the dynamic reliability by considering the structure as a probability dissipative system in 2020 [32], whose GDEE is in the form of:

$$\frac{\partial p_{U\boldsymbol{\Theta}}(u, \boldsymbol{\theta}, t)}{\partial t} + \dot{U}(\boldsymbol{\theta}, t)\frac{\partial p_{U\boldsymbol{\Theta}}(u, \boldsymbol{\theta}, t)}{\partial u} = -\mathcal{H}[f(U(\boldsymbol{\theta}, t))] \cdot p_{U\boldsymbol{\Theta}}(u, \boldsymbol{\theta}, t) \quad (36)$$

where

$$\mathcal{H}[f(U(\boldsymbol{\theta}, t))] = \begin{cases} 0, f(U(\boldsymbol{\theta}, t)) \in \boldsymbol{\Omega}_{\text{S}} \\ 1, f(U(\boldsymbol{\theta}, t)) \in \boldsymbol{\Omega}_{\text{D}} \end{cases} \quad (37)$$

and u can be the time history of any physical quantity of the system, $f(\cdot)$ is a general function that links u to the structural responses of interest, and $\boldsymbol{\Omega}_{\text{S}}$ and $\boldsymbol{\Omega}_{\text{D}}$ indicate the safety domain and the failure domain of the system, respectively.

$$R(t) = \int_{-\infty}^{\infty} p_U(u, t)\mathrm{d}u \quad (38)$$

5.3. Reliability Analysis Results

In this sub-section, the time-dependent dynamic reliability of the structure in Section 3.3 is analyzed following the procedure described in the previous section. Factors that may affect the reliability are analyzed, including CSI effect, reliability criteria, bouncing frequency, limit threshold, and parameter distribution. Note that the results in this section correspond to the same excitation described in Section 3.2, i.e., a crowd of nine people bouncing at a frequency f_b of 1.75 Hz, unless otherwise specified. Figure 12 exhibits a series of figures that show the relation between dynamic reliability and time. It is observed that all curves decrease with time, in accordance with the absorbing feature of the first-passage criteria.

Figure 12a shows the dependency of the dynamic reliability on the adopted failure criteria, in which the limit threshold equals 5.0 m/s^2 for both situations. If only the mid-point structural acceleration is interested, the reliability becomes stable around 0.21. However, if the structural accelerations at all bouncing locations are considered, the reliability decreases to 0.18. This phenomenon indicates that the mid-point may not always be the location with the largest acceleration response. It is better to consider the acceleration at more locations during vibration serviceability design.

The bouncing frequency is another important factor that highly affects the dynamic reliability of the structure. If the bouncing frequency or its harmonics are close to the fundamental frequencies of the structure, the structural responses tend to become much larger, and the dynamic reliability is thus decreased. In Figure 12b, it is clearly observed that the dynamic reliability significantly fluctuates with the bouncing frequency. When f_b equals 1.75 Hz, whose harmonics give rise to the resonant response of the structure, the dynamic reliability drops to its minimum value, while for other bouncing frequencies, the dynamic reliability becomes much higher.

It is also observed from Figure 12c that if the CSI effect is considered, the dynamic reliability is largely increased, especially for the case when f_b = 1.75 Hz, which is explained by the fact that the CSI mostly affects the resonant vibration. On the other hand, the serviceability problem usually occurs when the structure is under resonant excitation. Therefore, it is important to consider the CSI effect in the vibration serviceability design to avoid under-estimation of the dynamic reliability.

The influence of human parameter distribution is also checked in this study, as shown by Figure 12d. The line styles of the first column of the legend correspond to the results obtained by assuming the human body model parameters follow the distribution in Table 2, i.e., people in the bouncing crowd have independent parameters, while that of the second column assume that all people in the crowd share the same human model parameters, which equal to the mean value of their distributions, respectively. Results show that the ignorance of human model parameter distribution may lead to either an over-estimate or under-estimate of the structural reliability.

It is apparent that the dynamic reliability also changes with the predefined allowable limit. Figure 12e clearly shows the increasing trend of the dynamic reliability against the predefined threshold value.

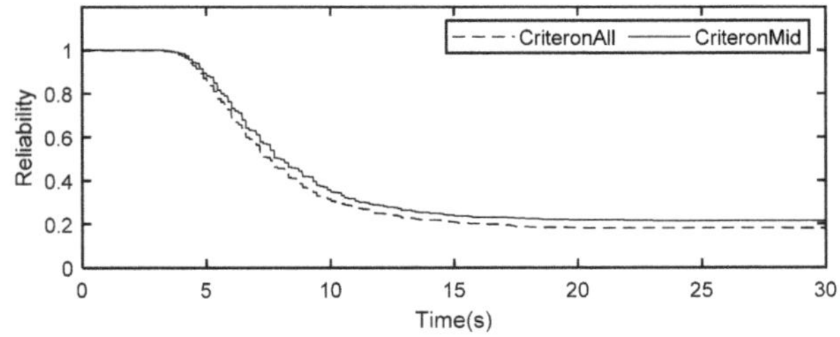

(**a**)

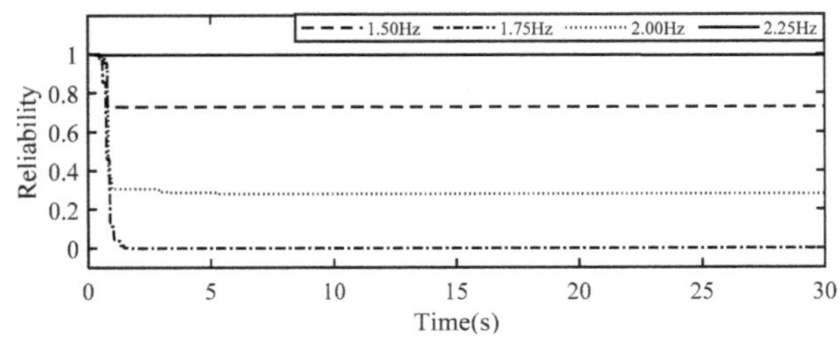

(**b**)

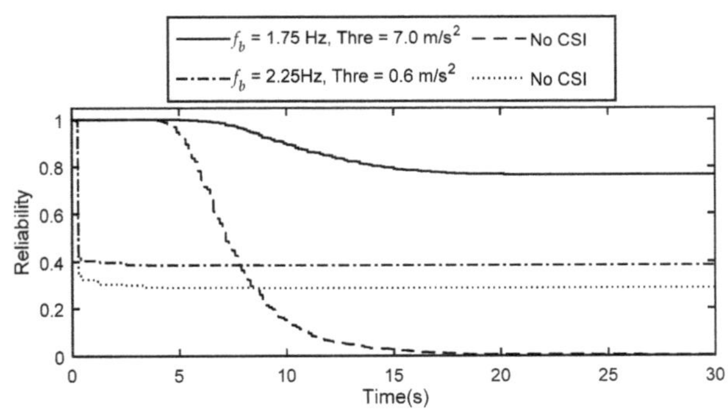

(**c**)

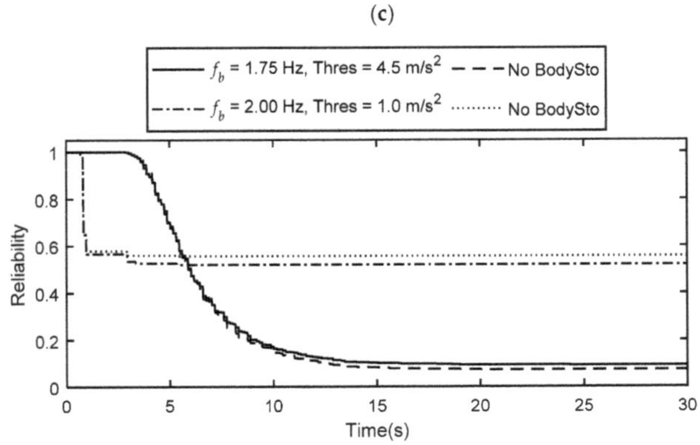

(**d**)

Figure 12. *Cont.*

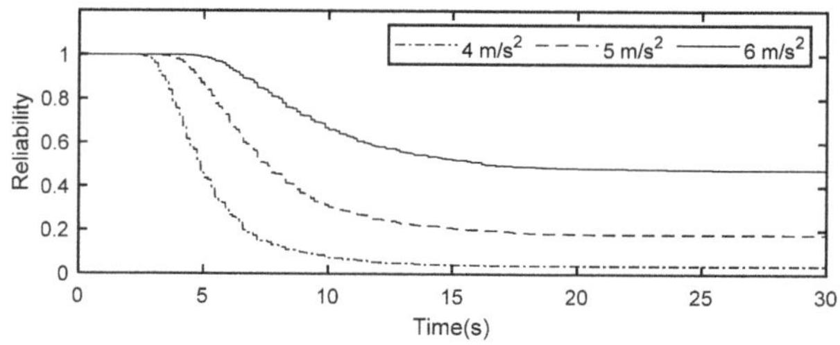

(e)

Figure 12. Reliability analysis for the large-span concrete plate on different parameters: (**a**) Criteria; (**b**) Bouncing frequency; (**c**) CSI effect; (**d**) Human parameter distribution; (**e**) Threshold.

6. Concluding Remarks

The dynamic reliability of a large-span structure is evaluated through PDEM based on the crowd–structure coupled model. The uncertainties of the coupling system come from the random distribution of the model parameters of the human body and the biomechanical forces in the bouncing crowd. The governing equations of motion of the coupled system with and without the CSI effect are given. The features of the random vibration of the coupled system are studied through PDEM and the results lead to the evaluation of dynamic reliability from the perspective of vibration serviceability. Two failure criteria in terms of one-point and multiple-point acceleration passage are adopted for the reliability evaluation, and the affecting factors on the reliability are discussed. The following conclusions are drawn from this study:

(1) The proposed calculation procedure to consider the interaction effect in the crowd–structure coupled system can well predict structural responses and is validated to be reasonable through an experimental test.
(2) The CSI highly affects the structural responses, especially when the crowd size is large.
(3) Through comparison with traditional MCS, the PDEM is tested as capable to conduct human-induced random vibration analysis, which is the foundation of dynamic reliability calculation considering the CSI effect. The PDEM has a great potential when a refined model needs to be adopted or a more complex situation is considered, where MCS becomes unavailable due to its high computational cost.
(4) The dynamic reliability of the large-span structure in terms of vibration serviceability is affected by many factors, including failure criteria, excitation frequency, limit threshold, distribution of human model parameters, and CSI effect.

The current design codes usually require that the structural vibration under human activities does not exceed a predefined limit to maintain the vibration serviceability of the structure. However, because of the uncertainties in the crowd–structure coupling system, this requirement is actually satisfied with a certain degree of probability rather than in a deterministic manner, thus leading to the concept of reliability. This paper presents the procedure of dynamic reliability analysis for large-span structures in terms of vibration serviceability. In future studies, the threshold of vibration perception of each person in the crowd could serve as another random variable in the system. This paper provides a foundation for achieving performance-based vibration serviceability design in the future.

Author Contributions: Conceptualization, D.Z. and H.W.; methodology, D.Z.; validation, H.W.; formal analysis, D.Z.; investigation, D.Z.; writing—original draft preparation, D.Z.; writing—review and editing, H.W. and J.C.; supervision, J.C.; funding acquisition, H.W. and J.C. All authors have read and agreed to the published version of the manuscript.

Funding: This research was funded by National Natural Science Foundation of China, grant number 52008306, 52178151 and Shanghai Sailing Program, grant number 20YF1451300.

Informed Consent Statement: Informed consent was obtained from all subjects involved in the study.

Data Availability Statement: The data presented in this study are available on request from the corresponding author.

Conflicts of Interest: The authors declare no conflict of interest.

References

1. Jones, C.; Reynolds, P.; Pavic, A. Vibration serviceability of stadia structures subjected to dynamic crowd loads: A literature review. *J. Sound Vib.* **2011**, *330*, 1531–1566. [CrossRef]
2. Dallard, P.; Fitzpatrick, A.J.; Flint, A.; Le Bourva, S.; Low, A.; Ridsdill Smith, R.M.; Willford, M. The london millennium footbridge. *Struct. Eng.* **2001**, *79*, 17–33.
3. Duarte, E.; Ji, T. Action of Individual Bouncing on Structures. *ASCE J. Struct. Eng.* **2009**, *135*, 818–827. [CrossRef]
4. Racic, V.; Chen, J. Data-driven generator of stochastic dynamic loading due to people bouncing. *Comput. Struct.* **2015**, *158*, 240–250. [CrossRef]
5. Xiong, J.; Chen, J.; Caprani, C. Spectral analysis of human-structure interaction during crowd jumping. *Appl. Math. Model.* **2020**, *89*, 610–626. [CrossRef]
6. Chen, J.; Ren, J.; Racic, V. Prediction of floor responses to crowd bouncing loads by response reduction factor and spectrum method. *Adv. Struct. Eng.* **2021**, *24*, 1427–1438. [CrossRef]
7. Ebrahimpour, A.; Sack, R.L. A review of vibration serviceability criteria for floor structures. *Comput. Struct.* **2005**, *83*, 2488–2494. [CrossRef]
8. Lievens, K.; Lombaert, G.; De Roeck, G.; Broeck, P.V.D. Robust design of a TMD for the vibration serviceability of a footbridge. *Eng. Struct.* **2016**, *123*, 408–418. [CrossRef]
9. Feldmann, M.; Heinemeyer, C.; Lukic, M. *Human Induced Vibrations of Steel Structures: Design of Footbridges Guideline*; Publication Office of the European Commission: Luxembourg, 2008.
10. Murray, T.M.; Allen, D.E.; Ungar, E.E. *Steel Design Guide Series 11: Floor Vibrations Due to Human Activity*; American Institute of Steel Construction, Inc.: Chicago, IL, USA, 1997.
11. ATC. *ATC Design Guide 1: Minimizing Floor Vibration*; Applied Technology Council: Redwood City, CA, USA, 1999.
12. Frédéric Sébastien Collette, Footbridge Vibrations due to Pedestrian Load. Danish Guidelines and Examples. 2005. Available online: https://www.cnisf.dk/ev/050414/Presentations/Footbridge%20dynamics%20-%20danish%20guidelines%20%26%20examples.pdf (accessed on 27 October 2021).
13. Highways England. *Design Manual for Roads and Bridges: Highway Structures and Bridges Design*; CD 353 Design Criteria for Footbridges, BD 29/17; Highways England: Birmingham, UK, 2017.
14. *JGJ/T 441-2019*; Technical Standard for Human Comfort of the Floor Vibration. Ministry of Housing and Urban-Rural Development of the People's Republic of China: Beijing, China, 2019.
15. Racic, V.; Pavic, A.; Brownjohn, J.M.W. Experimental identification and analytical modelling of human walking forces: Literature review. *J. Sound Vib.* **2009**, *326*, 1–49. [CrossRef]
16. Ellis, B.R.; Ji, T. Human-structure interaction in vertical vibrations. *Proc. ICE Struct. Build.* **1997**, *122*, 1–9. [CrossRef]
17. Salyards, K.A.; Noss, N.C. Experimental Evaluation of the Influence of Human-Structure Interaction for Vibration Serviceability. *J. Perform. Constr. Facil.* **2014**, *28*, 458–465. [CrossRef]
18. Cappellini, A.; Manzoni, S.; Vanali, M. Quantification of Damping Effect of Humans on Lightly Damped Staircases. In *Dynamics of Civil Structures: Proceedings of the 31st IMAC, a Conference on Structural Dynamics*; Society for Experimental Mechanics: Bethel, CT, USA, 2013; pp. 453–460. [CrossRef]
19. Griffin, M.J.; Erdreich, J. Handbook of Human Vibration. *J. Acoust. Soc. Am.* **1991**, *90*, 2213. [CrossRef]
20. Sachse, R.; Pavic, A.; Reynolds, P. Parametric study of modal properties of damped two-degree-of-freedom crowd–structure dynamic systems. *J. Sound Vib.* **2004**, *274*, 461–480. [CrossRef]
21. Van Nimmen, K.; Lombaert, G.; De Roeck, G.; Van den Broeck, P. The impact of vertical human-structure interaction on the response of footbridges to pedestrian excitation. *J. Sound Vib.* **2017**, *402*, 104–121. [CrossRef]
22. Van Nimmen, K.; Pavic, A.; Broeck, P.V.D. A simplified method to account for vertical human-structure interaction. *Structures* **2021**, *32*, 2004–2019. [CrossRef]
23. Dougill, J.W.; Wright, J.R.; Parkhouse, J.G.; Harrison, R.E. Human structure interaction during rhythmic bobbing. *Struct. Eng.* **2006**, *84*, 32–39.
24. Zheng, X.; Brownjohn, J.M.W. Modeling and simulation of human-floor system under vertical vibration. *Proc. SPIE Smart Struct. Syst.* **2001**, *4327*, 513–521. [CrossRef]
25. Wang, H.; Chen, J.; Brownjohn, J.M. Parameter identification of pedestrian's spring-mass-damper model by ground reaction force records through a particle filter approach. *J. Sound Vib.* **2017**, *411*, 409–421. [CrossRef]
26. Wang, H.; Chen, J.; Nagayama, T. Parameter identification of spring-mass-damper model for bouncing people. *J. Sound Vib.* **2019**, *456*, 13–29. [CrossRef]
27. Song, Y.; Basu, B.; Zhang, Z.; Sørensen, J.D.; Li, J.; Chen, J. Dynamic reliability analysis of a floating offshore wind turbine under wind-wave joint excitations via probability density evolution method. *Renew. Energy* **2021**, *168*, 991–1014. [CrossRef]

28. Li, J.; Chen, J. The principle of preservation of probability and the generalized density evolution equation. *Struct. Saf.* **2008**, *30*, 65–77. [CrossRef]
29. Jia, Y.; Yang, N.; Bai, F.; Lyu, Z. Dynamic reliability analysis of structures under stochastic human-induced loads. *J. Vib. Eng.* **2020**, *33*, 509–516. (In Chinese) [CrossRef]
30. Chen, J.; Yang, J.; Li, J. A GF-discrepancy for point selection in stochastic seismic response analysis of structures with uncertain parameters. *Struct. Saf.* **2016**, *59*, 20–31. [CrossRef]
31. Chen, J.-B.; Li, J. Dynamic response and reliability analysis of non-linear stochastic structures. *Probab. Eng. Mech.* **2005**, *20*, 33–44. [CrossRef]
32. Li, J. A PDEM-based perspective to engineering reliability: From structures to lifeline networks. *Front. Struct. Civ. Eng.* **2020**, *14*, 1056–1065. [CrossRef]

buildings MDPI

Article

Sensitivity Analysis for Pedestrian-Induced Vibration in Footbridges

Xiaojun Wei [1,*], Jingwei Zhang [1], Hao Zhou [1] and Stana Živanović [2]

1 School of Civil Engineering, Central South University, Changsha 410075, China; zhangjingwei2019@csu.edu.cn (J.Z.); hao.zhou@csu.edu.cn (H.Z.)
2 College of Engineering, Mathematics and Physical Sciences, University of Exeter, Exeter EX4 4QF, UK; s.zivanovic@exeter.ac.uk
* Correspondence: xiaojun.wei@csu.edu.cn

Abstract: This paper aims to provide a novel insight into the influence of uncertainties in system- and pedestrian-induced load parameters on the vibration response of footbridges. The study begins with a sensitivity analysis for the vertical vibration response of a representative footbridge to two loading cases: a single pedestrian and a crowd. Two methods are utilized: the Sobol'-based global sensitivity analysis method and the local sensitivity analysis method. Uncertainties in all model parameters (which include bridge and human body dynamics in a walking posture, as well as dynamic force generated by humans) are considered in stochastic response estimation. Parametric analysis is then performed to investigate the influence of the variation of the mean values of the bridge modal mass, damping ratio, and natural frequency on the results of global and local sensitivity analysis. Systematic comparison of the results of global and local sensitivity analysis is performed to identify their similarities and differences. It has been found that the sensitive parameters and their importance ranking strongly depend on bridge modal properties and loading scenarios (i.e., a single pedestrian or a crowd crossing). The damping ratio and natural frequency of the human body are found to be the only two insensitive parameters. Therefore, they could be treated as deterministic parameters in the stochastic estimation of human-induced vibration. Global sensitivity analysis is recommended as a choice for the sensitivity analysis of pedestrian-induced vibration of footbridges as it leads to more reliable results, owing to the advantage of characterizing model sensitivity over the entire input spaces.

Keywords: global sensitivity analysis; local sensitivity analysis; Sobol' method; pedestrian-induced vibration; footbridge; uncertainty

Citation: Wei, X.; Zhang, J.; Zhou, H.; Živanović, S. Sensitivity Analysis for Pedestrian-Induced Vibration in Footbridges. *Buildings* **2022**, *12*, 883. https://doi.org/10.3390/buildings12070883

Academic Editor: Weixin Ren

Received: 8 April 2022
Accepted: 20 June 2022
Published: 22 June 2022

1. Introduction

The application of the lightweight high-strength construction materials enabled the design of slender and aesthetically remarkable footbridges [1,2]. Such structures are prone to excessive vibration under the excitation of walking pedestrian(s) [3,4]. Therefore, vibration serviceability assessment plays a key role in the design of modern footbridges [5].

For the vibration serviceability assessment of a slender footbridge with well-separated vibration modes, design standards routinely require the estimation of the vertical resonant response of the footbridge induced by an average pedestrian traversing the bridge. In such an analysis, the bridge is modeled, in the modal domain, as a single degree of freedom (SDoF) system representing a relevant mode. The walking force is assumed to be a harmonic force having the frequency that matches the natural frequency of the relevant vibration mode to create resonance. More advanced design guidelines also require estimation of the vibration response induced by a crowd [6,7]. In this case, the crowd-induced response is calculated by multiplying the resonant response to an average pedestrian by a factor that is a function of the number of pedestrians present on the bridge at any one time and, in some cases, the damping ratio of the bridge. The human–structure interaction (HSI) is usually excluded from the analysis, apart from a basic consideration that pedestrians contribute to the modal mass of the structure.

The dynamic properties and load parameters of a human–structure system are assumed to be deterministic in the analysis above. This leads to a deterministic estimation of vibration response. However, there are inherent uncertainties in load parameters that are due to the inter- and intra-subject variabilities of the pedestrian action. In addition, there unavoidably exist uncertainties in dynamic properties of the structure owing to the inherent uncertainties in the construction process of the structure. As a consequence, the resultant vibration response is difficult to predict accurately. Recently, probability-based human-induced vibration estimation in footbridges by taking into account uncertainties in load parameters has drawn increasing attention [8–12], as it has become clear that stochastic, rather than deterministic, assessment of human-induced vibration response is more suitable for vibration serviceability assessment.

In the last four decades, great efforts have been devoted to experimental characterization of the uncertainties in walking-force parameters [1,13–25]. The mean value and coefficient of variation (COV) for step frequency, step length, and pedestrian weight obtained by various researchers are summarized in Table 1. The dynamic load factor was found to correlate with step frequency in the references [13,26], while in the study [5] the dynamic load factor was assumed as an independent variable following the normal distribution with a mean value of 0.35 and a COV of 20%. The normal distribution parameters expressed using the two parameters will be presented in short form hereafter, in this case being N (0.35, 20%).

Table 1. Stochastic properties of load parameters.

Parameter	Mean Value	COV
Step frequency [13–21,25]	1.77–2.20 Hz	6–14%
Step length [15,21–24]	0.65–0.75 m	1–11%
Pedestrian weight [1,21]	640–744 N	1–21%

Accurate estimation of human-induced vibration in footbridges may also require taking HSI into account [4,27,28]. In the modeling of HSI, the human body in a walking posture is usually modeled as a SDoF mass-damper-spring system [4,29–32]. Several experiments have been conducted to determine the dynamic properties of a human body in a walking posture [33–36]. The results show that the natural frequency and damping ratio vary significantly from person to person and between different studies, ranging from 1.25 Hz to 3 Hz and from 28% to 70%, respectively. In addition, some studies found that the human body's natural frequency and damping ratio are either related to step frequency [33,34] or walking speed [36]. Rather than by fixed values or intervals, the human body's natural frequency measured by Shahabpoor et al. [37] was described by a normal distribution N (2.86 Hz, 12%). They also concluded that the damping ratio follows the normal distribution N (29.5%, 16%). Jiménez-Alonso and Sáez [38] also described the measured natural frequency and damping ratio by normal distributions: N (2.76 Hz, 6%) and N (47%, 6%), respectively. In addition, an inverted pendulum model [39,40] and bipedal model [41,42] were also employed to account for HSI, which are not considered in the paper.

Given that there are inherent uncertainties in the construction process of the structure, the natural frequencies, modal mass (corresponding to modal shapes normalized to the largest component equal to unity), and damping ratios of a bridge are also random parameters. The COVs of the fundamental natural frequency and modal mass are usually no more than 10% [5], while the COV for the damping ratio could be relatively large, ranging from 4% to 25% [43]. Note that experiments for stochastic characterization of uncertain parameters in a human–structure system usually require a significant number of test subjects, a large number of trials, and expensive test facilities, making them time consuming and expensive.

Recently, increasing attention has been paid to the sensitivity of human-induced vibration response of a footbridge to uncertainties in the system and load parameters. Sensitivity analysis aims to study how variations in model outputs may be attributed to model inputs. It informs selection of influential parameters required for analyzing uncertainty propagation in stochastic response estimation, as well as for experimental characterization of randomness. This analysis therefore results in significant savings in computational effort in stochastic response estimation, as well as in reducing the cost of experimental characterization. Wei et al. [44] used a polynomial chaos-expansion-based global sensitivity analysis (GSA) method to determine the influence of uncertainties in the mechanical properties of fiber-reinforced polymer (FRP) components on the vibration response of an FRP footbridge exposed to a single pedestrian crossing. They found that the influence of the uncertainties could be neglected on the investigated footbridge. Van Nimmen et al. [45] focused on the influence of variations in a footbridge's natural frequency on the estimated response caused by a crowd crossing. The vibration response was estimated according to design guides SETRA [6] and HIVOSS [46], in which the walking force is treated as a deterministic, uniformly distributed load. It was found that for bridges with a frequency of no more than the second harmonic of the walking force, a COV of 10% for natural frequency may lead to a big scatter of estimated peak acceleration responses, e.g., ranging from 1.4 m/s^2 to 5.2 m/s^2. Caprani et al. [47] evaluated the stochastic response of a virtual footbridge with dynamic properties treated as deterministic parameters while a single pedestrian loading was modeled as a stochastic force model. The uncertainties in step frequency, step length, and weight of pedestrian were considered. They found the estimated footbridge acceleration response is most sensitive to the step frequency. Similarly, Pedersen and Frier [8] studied the sensitivity of footbridge response to step frequency, walking speed, pedestrian weight, and dynamic load factor characterizing a single pedestrian and concluded that the step frequency is the most influential parameter. Shahabpoor et al. [29] examined the impact of the mass, natural frequency, and damping ratio of the human body and a pedestrian's walking speed, as well as the pedestrians' arrival rate on the multi-pedestrian-induced footbridge response experienced by walking pedestrians rather than at a fixed location. The HSI was considered by modeling each pedestrian as an SDoF system. The results showed that the footbridge acceleration response is most sensitive to a human body's frequency in cases when this frequency is close to the bridge frequency. The sensitivity of the response to the other four parameters was significantly lower. More specifically, a 30% variation in one of these four parameters changed the response up to only 10%, whereas a 10% variation in a human body's frequency can achieve the same effect.

The existing literature provides an incomplete picture of the sensitive parameters, i.e., parameters that strongly influence footbridge vibration response. This is mainly due to the fact that only partial uncertainties were examined, i.e., either uncertainties in the parameters of the structure, or human-induced dynamic force, or human body dynamics. Another possible reason is the choice of the local sensitivity analysis (LSA) in the existing research. LSA is a one-at-a-time technique, evaluating variations in model outputs with respect to one input, with the remaining parameters fixed [48]. For an explicit input–output mapping relationship, the local sensitivity of the output may be defined as the partial derivative of the output with respect to one input of interest. A local sensitivity index is only valid in the vicinity of the base point where it is evaluated and may vary with the location of the base point [49]. Hence, LSA provides only a limited insight into model sensitivity and may lead to inconsistent sensitivity analysis results. A third possible reason is that the results from sensitivity analysis may vary with the input spaces of uncertain parameters. In existing literature, different stochastic distributions or intervals were employed for the same uncertain parameters by different researchers.

The contribution of this paper is to provide a more conclusive insight into the sensitivity of a pedestrian-induced vibration response. More specifically, stochastic vibration responses of a representative footbridge induced by a single pedestrian or a crowd walking are first estimated. Uncertainties in the dynamic properties of the structure and a human body, as well as parameters of human-induced load, are all included, which belong to aleatory uncertainty. The epistemic uncertainty, such as the uncertainty in distribution types of parameters, are out of the scope of the paper. The HSI is taken into account by modeling the human body as an SDoF system in conjunction with a vertical ground reaction force crossing the bridge at a constant speed. The crowd is assumed to constitute spatially unrestricted pedestrian traffic (characterized by a relatively low traffic density in which each person is free to choose a walking speed independently from the other pedestrians). GSA is then carried out using the Sobol' method to investigate the influence of uncertainties in the system and load parameters. In this analysis, all parameters are varied simultaneously over the specified input space. As a result, the contributions of each individual parameter and the interactions between parameters to model output are evaluated simultaneously. GSA provides more reliable results than LSA by characterizing the response sensitivity across the entire input space. Then, parametric analysis is conducted to study the influence of the input spaces of bridge modal mass, damping ratio, and natural frequency on the results of GSA. In addition, LSAs are also performed in all cases so that the similarities and differences between GSA and LSA can be observed and evaluated.

The layout of this article is as follows. This introductory section is followed by Section 2, which briefly introduces the Sobol' method. Section 3 presents the estimation methods for the vertical vibration response of a footbridge induced by a single pedestrian and a crowd. In Section 4, sensitivity analysis is carried out to reveal the global and local sensitivity of the acceleration response of a representative footbridge to the two load cases. Parametric analysis is performed in Section 5 to investigate how the variations in the input spaces of modal mass, damping ratio, and natural frequency of the footbridge affect the results of GSA and LSA. The discussion is presented in Section 6. The concluding remarks are given in Section 7.

2. The Sobol' Method

This section presents a brief introduction of the Sobol' method. A detailed description of the method can be found elsewhere [50].

2.1. Formulation of the Sobol' Method

The model under investigation may be described by the following function:

$$Y = f(\boldsymbol{x}) \tag{1}$$

where Y is a scalar model output and $\boldsymbol{x} = (x_1, x_2, \ldots, x_k)$ is a vector of k model inputs.

It is assumed that $f(\boldsymbol{x})$ is square integrable. In addition, the input parameters are assumed to be independent, random, and within the unit hypercube (i.e., $\boldsymbol{x} \in \boldsymbol{I}^k$ and $\boldsymbol{I}$ is the unit interval [0, 1]). This incurs no loss of generality because any space of a random variable can be transformed onto this unit hypercube.

The right-hand side of Equation (1) can be decomposed into the form of ANOVA-representation [51]:

$$f(\boldsymbol{x}) = f_0 + \sum_i f_i(x_i) + \sum_{i<j} f_{ij}(x_i, x_j) + \ldots + f_{12\ldots k}(x_1, x_2, \ldots, x_k) \tag{2}$$

if $\int_0^1 f_{i_1,i_2,\ldots,i_s}(x_{i_1}, x_{i_2}, \ldots, x_{i_s})\mathrm{d}x_w = 0$ for $w = i_1, i_2 \cdots, i_s$ and $1 \le i_1 < \cdots < i_s \le k$. It follows that the summands in Equation (2) are orthogonal and can be expressed as the integrals of $f(\boldsymbol{x})$. The physical meanings of the summands are: f_0 represents the average of the model $f(\boldsymbol{x})$, $f_i(x_i)$ describes the main effect of input x_i, $f_{ij}(x_i, x_j)$ represents the effect of the interaction between x_i and x_j, and so on.

Squaring Equation (2) and integrating over I^k, we can deduce that the variance of model output can be decomposed into fractions which are contributed by the sets of inputs:

$$D = \sum_i D_i + \sum_{i<j} D_{ij} + \ldots + D_{12\ldots k} \tag{3}$$

where D is the variance of model output. D_i is the variance corresponding to input x_i, D_{ij} is the variance contributed by both inputs x_i and x_j, and so on.

The Sobol' indices can be defined as:

$$S_i = \frac{D_i}{D}, \; S_{ij} = \frac{D_{ij}}{D}, \cdots, \; S_{ij\cdots k} = \frac{D_{ij\cdots k}}{D} \tag{4}$$

where S_i is the first-order sensitivity index (FSI) that measures the contribution of x_i alone to the output variance D. S_{ij} is the second-order sensitivity index that measures the contribution of the terms involving the interaction between x_i and x_j to the total variance D, and so on. Each sensitivity index is in the range from 0 to 1. The sum of all of sensitivity indices is equal to 1.

In addition, a total sensitivity index (TSI), measuring the contribution of the input x_i and all the possible joint terms between x_i and all the other inputs, may be given by:

$$S_i^T = \frac{D - D_{-i}}{D} \tag{5}$$

where S_i^T is TSI for input x_i. D_{-i} represents the total variance uncorrelated to input x_i. Usually, the larger the TSI of an input, the more influence it has on the model output. In addition, the difference between the FSI and TSI of a parameter is accounted by its interaction with other parameters.

Let $\boldsymbol{y} = (x_{l_1}, \ldots, x_{l_m})$, $1 \le l_1 < \cdots < l_m \le k$, be an arbitrary subset of $\boldsymbol{x}$ and $\boldsymbol{z}$ is the complementary set. Thus, $\boldsymbol{x} = (\boldsymbol{y}, \boldsymbol{z})$. Let $L = (l_1, \cdots, l_m)$. The variance corresponding to input $\boldsymbol{y}$ can be defined as [50]:

$$D_{\boldsymbol{y}} = \sum_{s=1}^{m} \sum_{(i_1 < \cdots < i_s) \in L} D_{i, \cdots, i_s} \tag{6}$$

Similarly, $D_{\boldsymbol{z}}$ can be defined. The FSI $S_{\boldsymbol{y}}$ and TSI $S_{\boldsymbol{y}}^T$ of subset $\boldsymbol{y}$ then can be expressed as:

$$S_{\boldsymbol{y}} = \frac{D_{\boldsymbol{y}}}{D} \tag{7}$$

$$S_{\boldsymbol{y}}^T = \frac{D - D_{\boldsymbol{z}}}{D} \tag{8}$$

Equations (7) and (8) become Equation (4) (the first equation) and Equation (5), respectively, when $\boldsymbol{y} = x_i$.

2.2. Implementation of the Sobol' Method

The variances D, $D_{\boldsymbol{y}}$, and $D_{\boldsymbol{z}}$ can be calculated using a Monte Carlo algorithm as [50]:

$$D = [\frac{1}{N} \sum_{d=1}^{N} f^2(\boldsymbol{y}_d, \boldsymbol{z}_d)] - \left[\frac{1}{N} \sum_{d=1}^{N} f(\boldsymbol{y}_d, \boldsymbol{z}_d)\right]^2 \tag{9}$$

$$D_{\boldsymbol{y}} = [\frac{1}{N} \sum_{d=1}^{N} f(\boldsymbol{y}_d, \boldsymbol{z}_d) f(\boldsymbol{y}_d, \boldsymbol{z}'_d)] - \left[\frac{1}{N} \sum_{d=1}^{N} f(\boldsymbol{y}_d, \boldsymbol{z}_d)\right]^2 \tag{10}$$

$$D_{\boldsymbol{z}} = [\frac{1}{N} \sum_{d=1}^{N} f(\boldsymbol{y}_d, \boldsymbol{z}_d) f(\boldsymbol{y}'_d, \boldsymbol{z}_d)] - \left[\frac{1}{N} \sum_{d=1}^{N} f(\boldsymbol{y}_d, \boldsymbol{z}_d)\right]^2 \tag{11}$$

where N is the number of Monte Carlo trails, which should be large enough to ensure convergence. To perform Equations (10) and (11), $\boldsymbol{y}$ ($\boldsymbol{z}$) needs to be sampled twice for each simulation, such that $\boldsymbol{y}_d$ and $\boldsymbol{y}'_d$ ($\boldsymbol{z}_d$ and $\boldsymbol{z}'_d$) are independent from each other. $\boldsymbol{y}_d$ represents the sample of $\boldsymbol{y}$ in the first sampling for the dth simulation and $\boldsymbol{y}'_d$ represents the sample of $\boldsymbol{y}$ in the second sampling for the dth simulation. Similarly, $\boldsymbol{z}_d$ and $\boldsymbol{z}'_d$ can be defined.

3. Pedestrian-Induced Vibration Analysis for Footbridges

This section describes the approaches for estimating the vibration response of a footbridge induced by a single pedestrian and a crowd. The footbridge is modeled as an SDoF system associated with a relevant vibration mode, and the influence of a pedestrian walking on a footbridge is modeled as another SDoF system in conjunction with a ground reaction force crossing the bridge at a constant speed.

3.1. Estimation of the Vibration Response Induced by a Single Pedestrian

The equations of motion of a footbridge with a single pedestrian crossing may be written as [4]:

$$\mathbf{M}_s(t)\ddot{\mathbf{q}}_s(t) + \mathbf{C}_s(t)\dot{\mathbf{q}}_s(t) + \mathbf{K}_s(t)\mathbf{q}_s(t) = \mathbf{P}_s(t) \tag{12}$$

where:

$$\mathbf{M}_s(t) = \begin{bmatrix} 1 & \frac{m_p\Phi(vt)}{m_b} \\ 0 & m_p \end{bmatrix}, \mathbf{C}_s(t) = \begin{bmatrix} 4\pi f_b\xi_b & 0 \\ -4\pi m_p f_p \xi_p \Phi(vt) & 4\pi m_p f_p \xi_p \end{bmatrix}$$

$$\mathbf{K}_s(t) = \begin{bmatrix} 4\pi^2 f_b^2 & 0 \\ -4\pi^2 f_p^2 m_p \Phi(vt) & 4\pi^2 f_p^2 m_p \end{bmatrix}, \mathbf{P}_s(t) = \begin{pmatrix} F(t)\Phi(vt)/m_b \\ 0 \end{pmatrix}, \mathbf{q}_s(t) = \begin{pmatrix} x_b(t) \\ x_p(t) \end{pmatrix}.$$

m_b, ξ_b, and f_b are the modal mass (kg), damping ratio (%), and natural frequency (Hz) of the footbridge, respectively, while m_p, ξ_p, and f_p are the mass (kg), damping ratio (%) and natural frequency (Hz) of the pedestrian, respectively. $x_b(t)$ and $x_p(t)$ are the modal displacements of the footbridge and pedestrian, respectively. $F(t) = W\alpha\cos(2\pi f_s t)$ is the walking force. W, α, and f_s are the weight (N), dynamic load factor, and step frequency (Hz) of the pedestrian, respectively. $\Phi(vt)$ is the mode shape of the bridge and v is the walking speed, which is the product of step frequency f_s and step length l_s. $(\ddot{\cdot})$ and $(\dot{\cdot})$ are the second-order and first-order derivatives of () with respect to time t.

3.2. Estimation of the Vibration Response Induced by a Crowd

Consider a crowd of r pedestrians crossing a footbridge in one direction. Pedestrian arrival is assumed to be a Poisson process. This means that the time lag (τ) between pedestrians follows an exponential distribution [52]. The equations of motion of a footbridge with a crowd crossing are [4]:

$$\mathbf{M}_c(t)\ddot{\mathbf{q}}_c(t) + \mathbf{C}_c(t)\dot{\mathbf{q}}_c(t) + \mathbf{K}_c(t)\mathbf{q}_c(t) = \mathbf{P}_c(t) \tag{13}$$

where:

$$\mathbf{M}_c(t) = \begin{bmatrix} 1 & \mathbf{M}_{12}(t) \\ \mathbf{0}_{r\times 1} & \mathbf{M}_p \end{bmatrix}, \mathbf{C}_c(t) = \begin{bmatrix} 4\pi f_b\xi_b & \mathbf{0}_{1\times r} \\ \mathbf{C}_{21}(t) & \mathbf{C}_p \end{bmatrix}$$

$$\mathbf{K}_c(t) = \begin{bmatrix} 4\pi^2 f_b^2 & \mathbf{0}_{1\times r} \\ \mathbf{K}_{21}(t) & \mathbf{K}_p \end{bmatrix}, \mathbf{P}_c(t) = \begin{pmatrix} \frac{1}{m_b}\sum_{n=1}^{r} F_n(t)\hat{\Phi}(v_n t) \\ \mathbf{0}_{r\times 1} \end{pmatrix}$$

$$\mathbf{q}_c(t) = \begin{bmatrix} x_b(t) & x_{p_1}(t) & \cdots & x_{p_r}(t) \end{bmatrix}^{\mathrm{T}}.$$

The submatrices in Equation (13) are defined as follows:

$$\mathbf{M}_{12}(t) = \begin{bmatrix} \frac{m_{p,1}}{m_b}\hat{\Phi}(v_1 t) & \frac{m_{p,2}}{m_b}\hat{\Phi}(v_2 t) & \cdots & \frac{m_{p,r}}{m_b}\hat{\Phi}(v_r t) \end{bmatrix},$$

$$\mathbf{C}_{21}(t) = \begin{bmatrix} -c_{p,1}\hat{\Phi}(v_1 t) & -c_{p,2}\hat{\Phi}(v_2 t) & \cdots & -c_{p,r}\hat{\Phi}(v_r t) \end{bmatrix}^{\mathrm{T}},$$

$$\mathbf{K}_{21}(t) = \left[-k_{p,1}\hat{\Phi}(v_1 t) \quad -k_{p,2}\hat{\Phi}(v_2 t) \quad \cdots \quad -k_{p,r}\hat{\Phi}(v_r t)\right]^{\mathrm{T}},$$
$$\mathbf{M}_p = \mathrm{diag}(m_{p,1},\ m_{p,2}, \ldots, m_{p,r}),$$
$$\mathbf{C}_p = \mathrm{diag}(4\pi m_{p,1} f_{p,1}\xi_{p,1},\ 4\pi m_{p,2} f_{p,2}\xi_{p,2}, \ldots, 4\pi m_{p,r} f_{p,r}\xi_{p,r}),$$
$$\mathbf{K}_p = \mathrm{diag}\left(4\pi^2 f_{p,1}^2 m_{p,1},\ 4\pi^2 f_{p,2}^2 m_{p,2}, \ldots, 4\pi^2 f_{p,r}^2 m_{p,r}\right),$$
$$\hat{\Phi}(v_n t) = \delta_n(t)\Phi(v_n t), \delta_n(t) = \begin{cases} 1,\ t_{on,n} \le t \le t_{off,n} \\ 0, t < t_{on,n} \text{ or } t > t_{off,n} \end{cases}$$
$$F_n(t) = W_n \alpha_n \cos(2\pi f_{s,n} t + \varphi_n)$$

where $m_{p,n}$, $\xi_{p,n}$, $f_{p,n}$, v_n, $x_{p,n}(t)$, W_n, α_n, $f_{s,n}$, and φ_n, $n = 1, 2, \cdots, r$, are the mass, damping ratio, natural frequency, walking speed, modal displacement, weight, dynamic load factor, step frequency, and phase angle for the n-th pedestrian, respectively. The arrival time of the n-th pedestrian on the bridge can be expressed as $t_{on,n} = \sum_{i=1}^{n} \tau_i$. The departure time of the n-th pedestrian off the bridge can be expressed as $t_{off,n} = t_{on,n} + L/(f_{s,n} l_{s,n})$. L, $l_{s,n}$ and τ_i are the bridge length, step length of the n-th pedestrian, and time lag between the i-th pedestrian and the $(i-1)$-th pedestrian, respectively. The definitions of the other variables have been given in Section 3.1.

4. Sensitivity Analysis of the Vibration Response of a Representative Footbridge

This section presents the GSA and LSA of the vibration response of a representative footbridge induced by a single pedestrian and a crowd. The representative footbridge is an idealized bridge, having a span of 50 m and a deck width of 2 m, which is typically employed in numerical examples in the literature related to human-induced vibration [4,5].

4.1. Representative Values for System and Load Parameters

The first vertical mode of the bridge is considered only. Thus, the mode shape is a half-sine function. According to the database of measured dynamic properties of 138 footbridges summarized by Wei et al. [53], the mean value for the modal mass, damping ratio, and fundamental frequency of the representative footbridge can be assumed to take values of 40,000 kg (based on a representative mass per deck area of 800 kg/m^2), 1% and 2 Hz, respectively. The COVs for the modal mass and fundamental frequency of the footbridge are both assumed to be 10%, while the COV for the damping ratio is doubled because of the usually larger uncertainty in damping identification.

The pedestrian-generated dynamic load from Section 3 is a harmonic function characterized by force amplitude (α times W), step frequency f_s, and step length l_s in the case of a single pedestrian. There are two additional parameters, phase angle φ and time-lag τ, in the case of a crowd. The dynamic model for a walking pedestrian is characterized by the mass m_p (W divided by gravitational acceleration), natural frequency f_p, and damping ratio ξ_p. The mean and COV values for step frequency, step length, weight, damping ratio, and natural frequency of a single pedestrian are assigned the average measured values reported in the literature and summarized in Section 1. They are listed in Table 2. The pedestrian density is set to be 0.3 pedestrians/m^2 of deck area, which corresponds to free (i.e., spatially unrestricted) walking [54]. All the parameters characterizing dynamic properties and dynamic load are assumed to follow the normal distribution, except the phase angle and time lag between pedestrians. The phase angle is assumed to obey the uniform distribution between 0 and 2 π [4]. The time lag between pedestrians follows an exponential distribution with a mean value of 1.2 s, which is the reciprocal of mean arrival rate of 0.84 pedestrians/s. Such an arrival rate corresponds to an average pedestrian density of 0.3 pedestrians/m^2 for a crowd walking, on average, at a walking speed of 1.4 m/s (which is equal to the mean step frequency of 2 Hz multiplied by the mean step length of 0.7 m). In addition, the mean value and COV of the dynamic load factor are in agreement with those provided by Tubino et al. [5], in which the dynamic load factor was assumed to independent of step frequency. All parameters in Table 2 are assumed to be mutually independent random variables. Since the fundamental frequency of the bridge is in the range of the step frequency of a normal-walking pedestrian, only the first-order harmonic of the walking force is considered when estimating human-induced vibration. In the case of a single pedestrian, the input parameters for sensitivity analysis are the step frequency f_s, step length l_s, dynamic load factor α, weight W, damping ratio ξ_p, and natural frequency f_p of the pedestrian, and the fundamental frequency f_b, damping ratio ξ_b, and modal mass m_b of the footbridge. For the case of a crowd, there are two additional inputs: phase angle φ and time lag τ. In both cases, the model output for sensitivity analysis is the maximum of one-second root-mean-squared (RMS) acceleration at the mid-span of the footbridge. Based on the input and output parameters mentioned above, the sensitivity analysis of the human-induced footbridge response can be carried out using the Sobol' method and the local sensitivity analysis method.

Table 2. Statistical characteristics of system and load parameters.

Parameter	Distribution Type	Mean Value μ	Standard Deviation σ	COV
f_s (Hz)	Normal	2.0	0.20	10%
l_s (m)	Normal	0.70	0.04	6%
α	Normal	0.35	0.07	20%
W (N)	Normal	700	70	10%
ξ_p (%)	Normal	38.25	4.21	11%
f_p (Hz)	Normal	2.81	0.25	9%
φ	Uniform $[0, 2\pi]$ [1]	-	-	-
τ (s)	Exponential	1.2	-	-
f_b (Hz)	Normal	2.00	0.20	10%
ξ_b (%)	Normal	1.00	0.20	20%
m_b (kg)	Normal	40000	4000	10%

[1]: The range for uniform distribution.

4.2. Vibration Response Induced by a Single Pedestrian

The vibration response of the footbridge induced by a single pedestrian is estimated by solving Equation (12) using the Newmark-β numerical integration method [55]. Due to the step frequency and step length of pedestrian are random variables, the time to cross a 50 m long bridge is different for each pedestrian. Therefore, the time duration for each simulation is different, which is set to be 110% of the time for crossing the bridge to ensure the time history experiencing peak response and then decaying to a low response level at the end [4]. Monte Carlo simulations based on low-discrepancy sequences are conducted. It is found that the results of the sensitivity analysis are almost unchanged with a simulation size larger than 5000. Therefore, 10,000 Monte Carlo simulations are performed for reliable results.

4.2.1. Global Sensitivity Analysis

In this paper, the aim of the sensitivity analysis is to identify and rank the parameters that influence the vibration response most. Both FSI and TSI from GSA could be employed for distinguishing between sensitive and insensitive parameters, but a literature search does not provide consistent thresholds for either. For example, Tang et al. [56], Coppitters et al. [57], and Hsieh et al. [58] considered parameters of an FSI below 0.01, 0.02, and 0.05, respectively, to be insensitive. By contrast, Zhang et al. [59] and Chan et al. [60] employed a TSI with threshold values of 0.05 and 0.3, respectively. Although FSIs are frequently utilized for parameter selection and ranking, they are insufficient to characterize parameter importance when there are strong parameter interactions [50]. Hence, the selection and ranking of sensitive parameters in this paper are decided using a TSI threshold value of 0.07. However, both FSI and TSI will be calculated and presented.

The FSIs and TSIs of the maximum of one-second RMS acceleration at the mid-span to all the parameters are calculated using the Sobol' method described in Section 2. The results from GSA are depicted in Figure 1a, where S_1 and S_T denote the FSI and TSI, respectively. According to the TSIs, the bridge's natural frequency is the most important parameter, followed closely by step frequency and then dynamic load factor, while the TSI for other parameters is less than 0.07 and is of little importance. The interactions between the uncertain parameters can be detected according to the difference between TSI and FSI, owing to the fact that TSI can measure the contributions of an uncertain parameter, as well as of the interactions between this parameter and all other parameters to the variation of vibration response, but FSI measures the contribution of this parameter only. It can be found that the TSIs for a bridge's natural frequency and step frequency are significantly larger than their first-order counterparts. This indicates a strong interaction between these two parameters, which is accounted for by the high probability of occurrence of resonance with a step frequency close to the bridge's natural frequency.

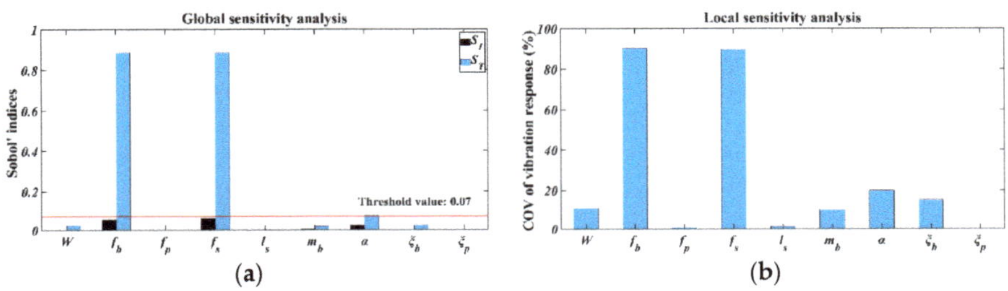

Figure 1. The results from GSA and LSA of the vibration response of a representative footbridge induced by a single pedestrian: (**a**) the Sobol' indices for system and load parameters; and (**b**) the COV of vibration response when one of the system and load parameters varies.

4.2.2. Local Sensitivity Analysis

In this section, LSA for vibration response of the representative footbridge under the excitation of a single pedestrian is carried out. Specifically, one input parameter varies at a time according to its statistical characteristics and the other inputs are kept at their nominal values. The corresponding COV for the vibration response of the bridge, estimated using the Monte Carlo simulation method, is then calculated. Note that the LSA can only produce qualitative results [48], i.e., ranking the input parameters. In other words, it is impossible to establish a quantitative criterion to distinguish important from unimportant parameters. For the sake of comparing the results of GSA and LSA, a top few parameters ranked by LSA are considered as the sensitive parameters as much as those given by GSA.

Figure 1b presents the results of LSA for vibration response induced by a single pedestrian. By comparing with Figure 1a, it can be found that the result of LSA is similar to the TSIs, rather than the FSIs, from the GSA. This is mainly due to the fact that for an uncertain parameter, both the TSI and the result of LSA simultaneously measure the contributions of this parameter as well as of the interactions between this parameter and all other parameters to the variation of vibration response, while the FSI only measure the contribution of this single parameter. The results of LSA show that the bridge natural frequency is the most important parameter, followed by step frequency and then dynamic load factor. This agrees with the conclusion from GSA. However, LSA cannot identify the interactions between the uncertain parameters as the same as the GSA.

Note that a 95th percentile of the peak of one-second RMS acceleration of the 10,000 cases for each analysis in the GSA and LSA in Section 4.2 is in the range of 0.16 m/s^2 to 0.23 m/s^2.

4.3. Vibration Response Induced by a Crowd

The vibration response of the footbridge induced by a crowd is computed using the same numerical integration method to solve Equation (13). To ensure a long-enough duration of steady-state pedestrian flow, i.e., the period after the departure time of the first pedestrian off the bridge and before the arrival time of the last pedestrian on the bridge, the number of people in a crowd is set to be 150. In this case, the maximum of one-second RMS acceleration exhibits no obvious variation during this period. Due to the randomness in the pedestrian step frequency, step length and time-lag between pedestrians, the time required for a crowd to cross the bridge is different each time. Therefore, the time duration for each simulation is set to be the time for a crowd to cross the bridge completely. Again, 10,000 Monte Carlo simulations are conducted.

4.3.1. Global Sensitivity Analysis

The FSIs and TSIs for the system and load parameters are shown in Figure 2a. The sensitive parameters, in descending order (of TSI), are the bridge natural frequency, step frequency, phase angle, time lag, step length, and bridge modal mass. Hereafter, the sensitive parameters are listed in descending order by default. Compared with the case of a single pedestrian, the dynamic load factor loses on significance. In addition, there are big differences between the TSIs and FSIs for the bridge's natural frequency, step frequency, phase angle, and time lag because of the interactions between these four parameters. It should be noted that for a system with nonlinearity like the one in this paper, TSI is recommended to measure the sensitivity of uncertain parameters instead of FSI. This is because the FSI cannot evaluate interactions from nonlinear effects, which may lead to the misleading result.

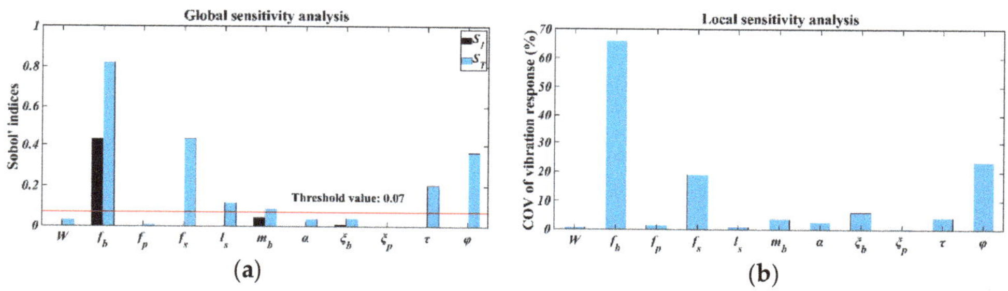

Figure 2. The results from GSA and LSA of the vibration response of a representative footbridge induced by a crowd: (**a**) the Sobol' indices for system and load parameters; and (**b**) the COV of vibration response when one of the system and load parameters varies.

4.3.2. Local Sensitivity Analysis

The results from LSA are shown in Figure 2b. The most influential parameter is the bridge's natural frequency, followed by phase-angle and then step frequency. The bridge damping ratio, time lag, and bridge modal mass are the least sensitive parameters. The other parameters are of little influence. Comparison of the results in Figure 2a,b shows that LSA and GSA differ in ranking the parameters. This inconsistency may be attributed to the fact that a local sensitivity index is estimated only in the vicinity of the nominal point, leading to the incomplete investigation of the full input space, while GSA produces an estimation of model sensitivity over the full input space including the nominal point. As a consequence, GSA, rather than LSA, is recommended as a choice for the sensitivity analysis of the pedestrian-induced vibration of footbridges. This is owing to the fact that the results of LSA may not include the effects of interactions between uncertain parameters in the unexplored input space, while GSA does not have this problem and therefore leads to more reliable results.

Note that a 95th percentile of the peak of one-second RMS acceleration of the 10,000 cases for each analysis in the GSA and LSA in Section 4.3 is in the range of 0.61 m/s^2 to 2.64 m/s^2.

5. Parametric Analysis of the Input Spaces of Modal Mass, Damping Ratio, and Natural Frequency

The GSA and LSA presented in Section 4 are for a particular footbridge. To provide more robust insight, the sensitivity analysis for different footbridges is performed in this section. It includes both GSA and LSA for the vibration response of footbridges with different mean values of modal mass, damping ratio, and natural frequency.

5.1. Natural Frequency

The footbridges with a natural frequency up to 3 Hz may be strongly excited by the first-order harmonic of a pedestrian's walking force. Thus, six footbridges with different mean natural frequencies from 0.5 Hz to 3 Hz (with a step size of 0.5 Hz) are considered in the parametric analysis. The natural frequency of these six footbridges is described by normal distribution: N (0.5–3 HZ, 10%). The statistical characteristics for the other parameters are taken from Table 2. These six footbridges are referred to as Bridge Series A.

5.1.1. Vibration Response Induced by a Single Pedestrian

Global Sensitivity Analysis

The generated TSIs for Bridge Series A are presented in Figure 3a, leading to the following observations:

- For the bridges with the mean natural frequency of 0.5 Hz, the vibration responses are most sensitive to dynamic load factor, followed by pedestrian weight and then bridge modal mass. For the bridges with the mean natural frequency of 1 Hz, there are two additional sensitive parameters step frequency and bridge natural frequency.
- For the bridges with the mean natural frequency no smaller than 1.5 Hz, the variations in step frequency and bridge natural frequency have most significant impact on the vibration response. The dynamic load factor becomes the third sensitive parameter for the bridges with the mean natural frequency of 2 Hz.

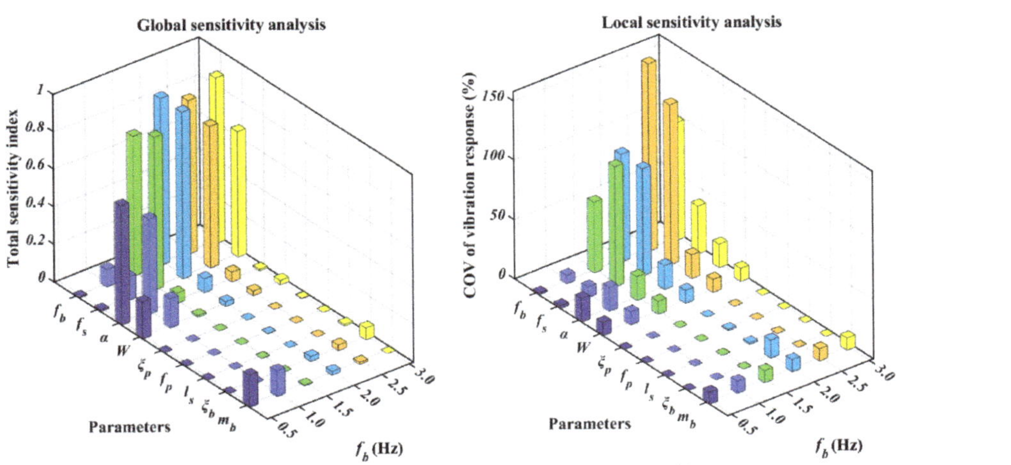

(a) (b)

Figure 3. The results from GSAs and LSAs of the vibration response of Bridge Series A with a single pedestrian crossing: (**a**) the TSIs for system and load parameters; and (**b**) the COV of vibration response when one of the system and load parameters varies.

Local Sensitivity Analysis

The results of LSAs for Bridge Series A are presented in Figure 3b. The sensitivity of parameters and their ranking agree with those from GSA.

The results from GSA and LSA suggest that the parameters that influence vibration response most strongly depend on the natural frequency of the bridge, or more precisely, its closeness to the step frequency and therefore likelihood of exciting the resonance.

Note that a 95th percentile of the peak of one-second RMS acceleration of the 10,000 cases for each analysis in the GSA and LSA in Section 5.1.1 is in the range of 0.01 m/s^2 to 0.23 m/s^2.

5.1.2. Vibration Response Induced by a Crowd

Global Sensitivity Analysis

The TSIs from the GSAs for Bridge Series A are presented in Figure 4a, which suggests that:

- For the bridges with mean natural frequency of 0.5 Hz, the step frequency and phase angle are the first two contributors to the response's variation, followed by the time lag and then the bridge modal mass. The step length and dynamic load factor are the remaining two sensitive parameters.
- For the bridges with the mean natural frequency of 1 Hz, an additional sensitive parameter is the bridge natural frequency.
- For the bridges with the mean natural frequencies of 1.5 Hz and 2.5 Hz, the most sensitive parameter is the bridge frequency, followed by step frequency and then phase angle. For the bridges with the mean natural frequency at 2 Hz, there are three additional sensitive parameters: time lag, step length, and bridge modal mass. For the bridges with the mean natural frequency of 3 Hz, only the first two parameters are sensitive.

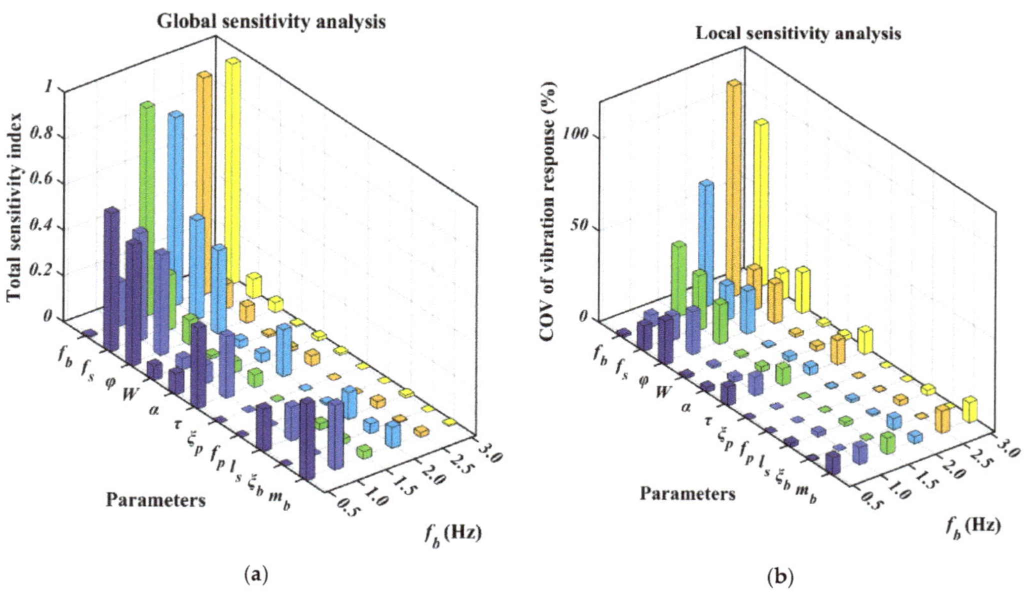

Figure 4. The results from GSAs and LSAs of the vibration response of Bridge Series A induced by a crowd: (**a**) the TSIs for system and load parameters; and (**b**) the COV of vibration response when one of the system and load parameters varies.

Local Sensitivity Analysis

The results of LSAs for Bridge Series A induced by a crowd are presented in Figure 4b.

- For the bridges with the mean natural frequency of 0.5 Hz, there are six sensitive parameters, i.e., the phase angle, step frequency, time lag, bridge modal mass, dynamic load factor, and step length. Apart from these six parameters, an additional parameter, the bridge natural frequency, becomes sensitive for the bridges with the mean natural frequency of 1 Hz.

- For the bridges with the mean natural frequency of 1.5 Hz and 2.5 Hz, the most sensitive parameter is the bridge natural frequency, followed by step frequency and then phase angle. When the bridges have the mean natural frequency of 3 Hz, the step frequency becomes insensitive.
- For the bridges with the mean natural frequency of 2 Hz, the variation in bridge natural frequency impacts most the vibration response, followed by phase angle and then step frequency. The bridge damping ratio, time lag, and bridge modal mass are the last three sensitive parameters.

By comparing Figure 4a,b, it is found that GSA and LSA generate the same sensitive parameters and their ranking for the bridges with the mean natural frequency of 1.5 Hz and 2.5 Hz, different sensitive parameters when the mean natural frequency is taken to be 2 Hz and 3 Hz, and the same parameters but different ranking when the mean natural frequency no larger than 1 Hz. Again, both the results from GSA and LSA vary with the input space of bridge natural frequency.

Note that a 95th percentile of the peak of one-second RMS acceleration of the 10,000 cases for each analysis in the GSA and LSA in Section 5.1.2 is in the range of 0.04 m/s^2 to 2.64 m/s^2.

5.2. Modal Mass

Similarly, parametric analysis is carried out to investigate the robustness of results of sensitivity analysis to the variation of the input space of modal mass. According to [53], the physical mass per square meter of deck area for a footbridge ranges from 52 kg/m^2 to 1955 kg/m^2. The footbridge with low modal mass is prone to vibration problems. Here, seven footbridges, the mean physical mass density ranging from 50 kg/m^2 to 1600 kg/m^2, are analyzed. Their mean modal mass is taken to be 2500 kg, 5000 kg, 7500 kg, 10,000 kg, 20,000 kg, 40,000 kg, and 80,000 kg, respectively. The modal mass of these seven footbridges is described by normal distribution: N (5000–80,000 kg, 10%). The statistical characteristics of the remaining parameters are taken from Table 2. These seven footbridges are designated as Bridge Series B.

5.2.1. Vibration Response Induced by a Single Pedestrian

Global Sensitivity Analysis

Figure 5a shows the TSIs from the GSAs for Bridge Series B, which indicate that the bridge frequency is always the main contributor, with its TSI slightly larger than that of step frequency. The dynamic load factor is the third sensitive parameter with a TSI of about 0.07.

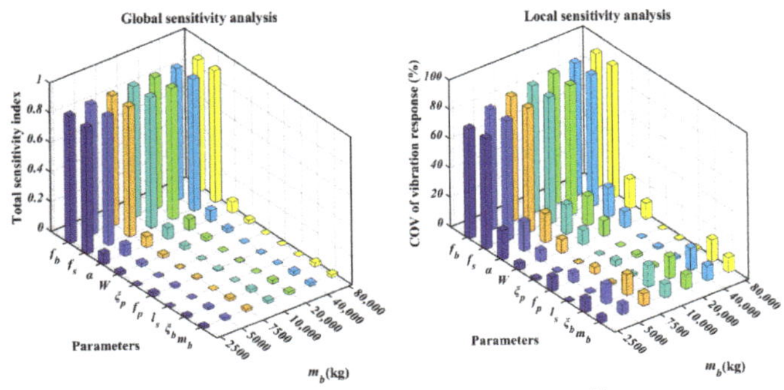

(a) (b)

Figure 5. The results from GSAs and LSAs of the vibration response of Bridge Series B with a single pedestrian crossing: (**a**) the TSIs for system and load parameters; and (**b**) the COV of vibration response when one of the system and load parameters varies.

Local Sensitivity Analysis

The results of LSAs are shown in Figure 5b. The influential parameters are the bridge frequency, step frequency, and dynamic load factor, and they are independent from modal mass.

The comparison between Figure 5a,b indicates that LSA and GSA produce consistent results in terms of sensitive parameters and their ranking. In addition, it can be concluded that the variation of the input space of modal mass will not affect the results of GSA and LSA under the excitation of a single pedestrian.

Note that a 95th percentile of the peak of one-second RMS acceleration of the 10,000 cases for each analysis in the GSA and LSA in Section 5.2.1 is in the range of 0.08 m/s^2 to 2.17 m/s^2.

5.2.2. Vibration Response Induced by a Crowd

Global Sensitivity Analysis

The TSIs from GSAs for Bridge Series B are obtained and presented in Figure 6a. For all seven bridges, there are six influential parameters, i.e., the bridge natural frequency, step frequency, phase angle, time lag, step length, and bridge modal mass. For the bridges with the mean modal mass no larger than 10,000 kg, there is an additional sensitive parameter, the dynamic load factor.

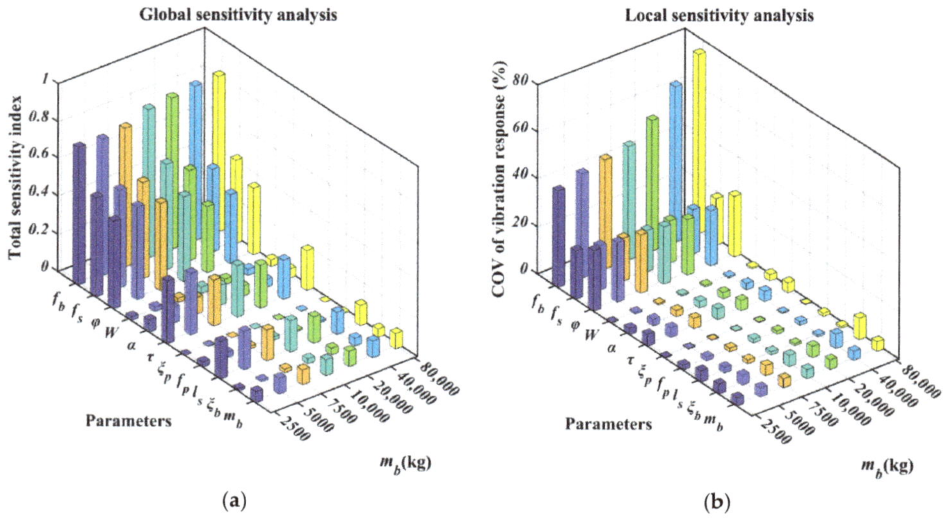

Figure 6. The results from GSAs and LSAs of the vibration response of Bridge Series B induced by a crowd: (**a**) the TSIs for system and load parameters; and (**b**) the COV of vibration response when one of the system and load parameters varies.

Local Sensitivity Analysis

For the scenario of a crowd, the results from LSA are presented in Figure 6b. For the bridges with the mean modal mass no smaller than 20,000 kg, there are six influential parameters, which are the bridge's natural frequency, phase angle, step frequency, bridge damping ratio, time lag, and bridge modal mass. For the bridges with the mean modal mass of 2500 kg, there is an additional sensitive parameter, the step length, and for the bridges with the mean modal mass ranging from 5000 kg to 10,000 kg, the seventh sensitive parameter is the dynamic load factor.

Compared with the results of GSA, LSA may lead to different influential parameters. Based the analysis above, it can be concluded that the results of GSA and LSA may vary with the input space of modal mass.

Note that a 95th percentile of the peak of one-second RMS acceleration of the 10,000 cases for each analysis in the GSA and LSA in Section 5.2.2 is in the range of 0.35 m/s^2 to 8.4 m/s^2.

5.3. Damping Ratio

This section examines the influence of the variation of the input space of the damping ratio on the results of sensitivity analysis. The footbridge with a low damping ratio, usually no more than 2%, is prone to vibration problems. Here, nine footbridges are analyzed. Their mean damping ratios are taken to be 0.2%, 0.3%, 0.4%, 0.5%, 0.75%, 1%, 1.25%, 1.5%, and 2%, respectively. The damping ratio of these nine footbridges is described by normal distribution: N (0.2–2%, 20%). The statistical characteristics of the remaining parameters are taken from Table 2. These nine footbridges are denoted as Bridge Series C.

5.3.1. Vibration Response Induced by a Single Pedestrian

Global Sensitivity Analysis

The generated TSIs for Bridge Series C are displayed in Figure 7a. For all nine bridges, there are two sensitive parameters, among which the step frequency is the most important with the TSI slightly larger than that of the bridge's natural frequency. The dynamic load factor becomes the third sensitive parameter for the bridges with the mean damping ratio no smaller than 1%.

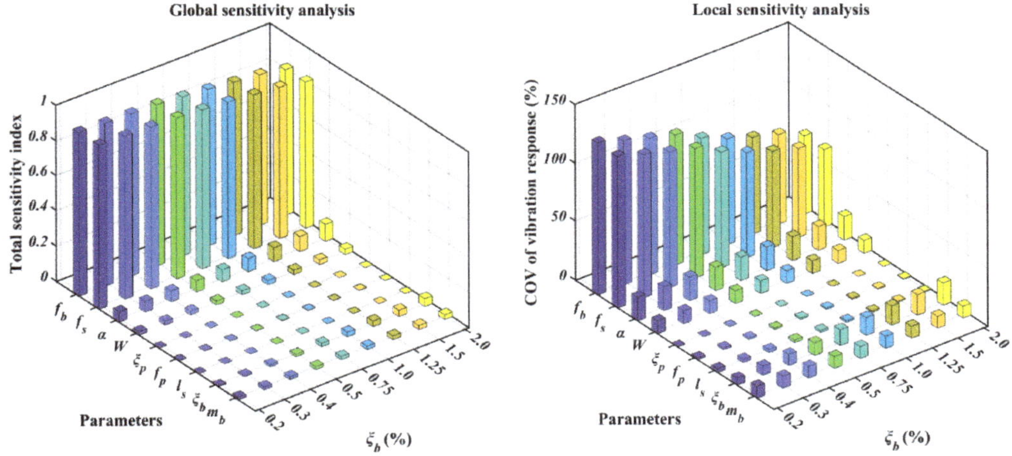

(a) (b)

Figure 7. The results from GSAs and LSAs of the vibration response of Bridge Series C with a single pedestrian crossing: (**a**) the TSIs for system and load parameters; and (**b**) the COV of vibration response when one of the system and load parameters varies.

Local Sensitivity Analysis

The LSAs for Bridge Series C in the single pedestrian scenario are presented in Figure 7b. For all nine bridges, the vibration responses are most sensitive to bridge natural frequency and step frequency. For the bridges with the mean damping ratio no smaller than 1%, there is an additional sensitive parameter, the dynamic load factor.

The comparison between the results in Figure 7a,b indicates that GSA and LSA produce the same sensitive parameters but different rankings for all nine bridges. It can be found that the results from GSA and LSA may change with the input space of the damping ratio.

Note that a 95th percentile of the peak of one-second RMS acceleration of the 10,000 cases for each analysis in the GSA and LSA in Section 5.3.1 is in the range of 0.1 m/s^2 to 0.5 m/s^2.

5.3.2. Vibration Response Induced by a Crowd

Global Sensitivity Analysis

The TSIs from GSAs for Bridge Series C are shown in Figure 8a. For all nine bridges, the sensitive parameters are the bridge natural frequency, step frequency, phase angle, time lag, step length, and bridge modal mass. For the bridges with the mean damping ratio of 2%, there is an additional sensitive parameter, the bridge damping ratio.

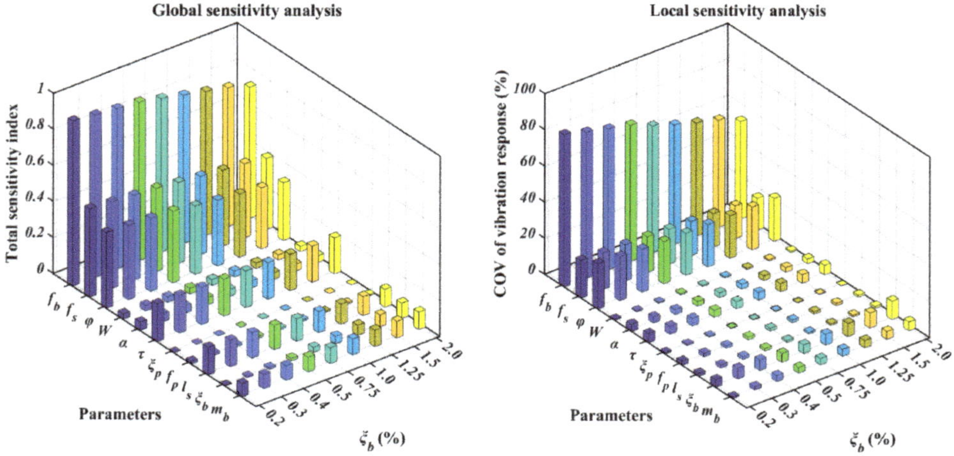

(a) (b)

Figure 8. The results from GSAs and LSAs of the vibration response of Bridge Series C induced by a crowd: (**a**) the TSIs for system and load parameters; and (**b**) the COV of vibration response when one of the system and load parameters varies.

Local Sensitivity Analysis

The results of LSAs for Bridge Series C caused by the excitation of a crowd are shown in Figure 8b, which shows:

- When the bridges have the mean damping ratio no bigger than 0.3%, there are six sensitive parameters, i.e., the bridge natural frequency, phase angle, step frequency, bridge damping ratio, time lag, and dynamic load factor.
- For the bridges with the mean damping ratio ranging from 0.4% to 1.5%, the first five sensitive parameters are the same as those for the bridges with the mean value of damping ratio no bigger than 0.3%, as well as the ranking, but the sixth sensitive parameter is replaced by the bridge modal mass. Apart from these six parameters, an additional parameter, the dynamic load factor, becomes sensitive for the bridges with the mean damping ratio of 2%.

In summary, the results of GSA and LSA sensitive to the input space of damping ratio. Compared with GSA, LSA produces the different sensitive parameters for each damping ratio case.

Note that a 95th percentile of the peak of one-second RMS acceleration of the 10,000 cases for each analysis in the GSA and LSA in Section 5.3.2 is in the range of 0.47 m/s^2 to 4.31 m/s^2.

6. Discussion

The results of GSA and LSA for the vibration responses of the representative bridge, Bridge Series A, Bridge Series B, and Bridge Series C to the two load cases are summarized in Table 3. In the table, S_x^L denotes the local sensitivity index of parameter x and S_x^T denotes the TSI obtained from the Sobol' method. It can be concluded that:

- In the case of a single pedestrian crossing, both GSA and LSA select the same sensitive parameters and produce the same ranking for the representative bridge, Bridge Series A and Bridge Series B. However, for the bridges in Bridge Series C, the same sensitive parameters are selected but ordered differently.
- In the case of the excitation of a crowd, both GSA and LSA select the same influential parameters and their ranking for the bridges in Bridge Series A with the mean natural frequency of 1.5 Hz and 2.5 Hz. The same sensitive parameters but ordered differently are generated for bridges in Bridge Series A with the mean natural frequency no larger than 1 Hz. However, for the representative bridge, bridges in Bridge Series A with the mean natural frequency of 2 Hz and 3 Hz, Bridge Series B and Bridge Series C, GSA and LSA generate different sensitive parameters.
- The results of GSA and LSA for all kinds of bridges in the two load cases are generally sensitive to the variation of input space of modal mass, natural frequency, and damping ratio of the bridge. The exception is the results of GSA and LSA for Bridge Series B under the excitation of a single pedestrian.
- In addition, when the pedestrian step frequency is close to the bridge frequency, the TSIs of important parameters in all cases are found to be much greater than their first-order counterparts, which indicates strong interaction between important parameters. This agrees with the fact that large human-induced vibration response only occurs when these two frequencies are close to each other. It is strong interaction between these two frequencies that leads to significant discrepancies between the TSIs and FSIs.
- The damping ratio and natural frequency of a human body are found to be the only two insensitive parameters in all the cases. Additional numerical results, which are not presented in the paper, show that the human body's frequency may become sensitive only in the case where a single pedestrian crosses a bridge with a mean modal mass no larger than 350 kg. However, there are rarely real footbridges with modal masses lower than 350 kg. Therefore, there is no need to quantify the uncertainty in these two parameters in practice.
- Sensitive parameters from both the GSA and LSA may vary with the load case. For example, for a footbridge in Bridge Series A with the mean natural frequency of 1.5 Hz, the step frequency is the most influential parameter in the case of a single pedestrian crossing, while the bridge's natural frequency becomes the most influential parameter in the case of a crowd crossing. This is explained by the fact that the model outputs for sensitivity analysis are different between the situations of a single pedestrian and a crowd.

The analysis presented in this paper indicates why there is inconsistency in the results of sensitivity analysis for the pedestrian-induced vibrations of footbridges in existing literature. These reasons are given as follows.

- The results of sensitivity analysis may vary with the input space of parameters. In other words, for different footbridges and different load cases, different sensitive parameters may be produced by GSA or LSA. Therefore, the analysis in the paper is limited to the parameter space covered in the paper.
- The inclusion of partial uncertain parameters in sensitivity analysis may lead to a misleading conclusion. Take the representative footbridge considered in Section 4 for example: additional analysis shows that the vibration response to the two load cases is most sensitive to step frequency if uncertainties in the step frequency, step length, and pedestrian weight are considered only. However, the bridge's natural frequency becomes most influential if all uncertain parameters are considered, as shown in Section 4.
- The results of LSA may be inconsistent with those of GSA. This is attributed to the fact that LSA evaluates the sensitivity in the vicinity of a particular point and provides only a limited view of model sensitivity, which may exclude the effects of interactions between uncertain parameters in the unexplored input space and lead to misleading results. However, GSA provides more reliable results by characterizing the sensitivity

across all input spaces. Therefore, GSA is suggested for the sensitive analysis of pedestrian-induced vibration of footbridges.

Table 3. The results from GSA and LSA of vibration responses of bridges.

Bridge	Load Case	Sensitivity Analysis Results	
		Local Sensitivity Analysis	**Global Sensitivity Analysis**
Representative footbridge	Single	$S^L_{f_b} > S^L_{f_s} > S^L_{\alpha}$	$S^T_{f_b} > S^T_{f_s} > S^T_{\alpha}$
	Crowd	$S^L_{f_b} > S^L_{\varphi} > S^L_{f_s} > S^L_{\xi_b} > S^L_{\tau} > S^L_{m_b}$	$S^T_{f_b} > S^T_{f_s} > S^T_{\varphi} > S^T_{\tau} > S^T_{l_s} > S^T_{m_b}$
Bridge Series A	Single	$f_b = 0.5$ Hz, $S^L_{\alpha} > S^L_{w} > S^L_{m_b}$; $f_b = 1.0$ Hz, $S^L_{\alpha} > S^L_{w} > S^L_{m_b} > S^L_{f_s} > S^L_{f_b}$; $f_b = 1.5$ Hz, $S^L_{f_s} > S^L_{f_b}$; $f_b = 2.0$ Hz, $S^L_{f_b} > S^L_{f_s} > S^L_{\alpha}$; $2.5 \leq f_b \leq 3.0$Hz, $S^L_{f_b} > S^L_{f_s}$;	$f_b = 0.5$ Hz, $S^T_{\alpha} > S^T_{w} > S^T_{m_b}$; $f_b = 1.0$Hz, $S^T_{\alpha} > S^T_{w} > S^T_{m_b} > S^T_{f_s} > S^T_{f_b}$; $f_b = 1.5$ Hz, $S^T_{f_s} > S^T_{f_b}$; $f_b = 2.0$ Hz, $S^T_{f_b} > S^T_{f_s} > S^T_{\alpha}$; $2.5 \leq f_b \leq 3.0$ Hz, $S^T_{f_b} > S^T_{f_s}$;
	Crowd	$f_b = 0.5$ Hz, $S^L_{\varphi} > S^L_{f_s} > S^L_{\tau} > S^L_{m_b} > S^L_{\alpha} > S^L_{l_s}$; $f_b = 1.0$ Hz, $S^L_{\varphi} > S^L_{f_s} > S^L_{\tau} > S^L_{m_b} > S^L_{f_b} >$ $S^L_{\alpha} > S^L_{l_s}$; $f_b = 1.5$ Hz, $S^L_{f_b} > S^L_{f_s} > S^L_{\varphi}$; $f_b = 2.0$ Hz, $S^L_{f_b} > S^L_{\varphi} > S^L_{f_s} > S^L_{\xi_b} > S^L_{\tau} > S^L_{m_b}$; $f_b = 2.5$ Hz, $S^L_{f_b} > S^L_{f_s} > S^L_{\varphi}$; $f_b = 3.0$ Hz, $S^L_{f_b} > S^L_{\varphi}$;	$f_b = 0.5$ Hz, $S^T_{f_s} > S^T_{\varphi} > S^T_{\tau} > S^T_{m_b} > S^T_{l_s} > S^T_{\alpha}$; $f_b = 1.0$ Hz, $S^T_{f_s} > S^T_{\varphi} > S^T_{m_b} > S^T_{\tau} > S^T_{f_b} > S^T_{l_s} > S^T_{\alpha}$; $f_b = 1.5$ Hz, $S^T_{f_b} > S^T_{f_s} > S^T_{\varphi}$; $f_b = 2.0$ Hz, $S^T_{f_b} > S^T_{f_s} > S^T_{\varphi} > S^T_{\tau} > S^T_{l_s} > S^T_{m_b}$; $f_b = 2.5$ Hz, $S^T_{f_b} > S^T_{f_s} > S^T_{\varphi}$; $f_b = 3.0$ Hz, $S^T_{f_b} > S^T_{f_s}$;
Bridge Series B	Single	$S^L_{f_b} > S^L_{f_s} > S^L_{\alpha}$	$S^T_{f_b} > S^T_{f_s} > S^T_{\alpha}$
	Crowd	$m_b = 2500$ kg, $S^L_{f_b} > S^L_{\varphi} > S^L_{f_s} > S^L_{\tau} > S^L_{l_S} >$ $S^L_{\xi_b} > S^L_{m_b}$; $5000 \text{ kg} \leq m_b \leq 7500$ kg, $S^L_{f_b} > S^L_{\varphi} > S^L_{f_s} >$ $S^L_{\tau} > S^L_{\xi_b} > S^L_{m_b} > S^L_{\alpha}$; $m_b = 10,000$ kg, $S^L_{f_b} > S^L_{\varphi} > S^L_{f_s} > S^L_{\xi_b} > S^L_{\tau} >$ $S^L_{m_b} > S^L_{\alpha}$; $m_b \geq 20,000$ kg, $S^L_{f_b} > S^L_{\varphi} > S^L_{f_s} > S^L_{\xi_b} > S^L_{\tau} >$ $S^L_{m_b}$;	$m_b \leq 10,000$ kg, $S^T_{f_b} > S^T_{f_s} > S^T_{\varphi} > S^T_{\tau} > S^T_{l_s} > S^T_{\alpha} >$ $S^T_{m_b}$; $m_b \geq 20,000$ kg, $S^T_{f_b} > S^T_{f_s} > S^T_{\varphi} > S^T_{\tau} > S^T_{l_s} > S^T_{m_b}$;
Bridge Series C	Single	$\xi_b \leq 0.75\%$, $S^L_{f_b} > S^L_{f_s}$; $\xi_b \geq 1.0\%$, $S^L_{f_b} > S^L_{f_s} > S^L_{\alpha}$;	$\xi_b \leq 0.75\%$, $S^T_{f_s} > S^T_{f_b}$; $\xi_b \geq 1.0\%$, $S^T_{f_s} > S^T_{f_b} > S^T_{\alpha}$;
	Crowd	$\xi_b \leq 0.3\%$, $S^L_{f_b} > S^L_{\varphi} > S^L_{f_s} > S^L_{\xi_b} > S^L_{\tau} > S^L_{\alpha}$; $0.4\% \leq \xi_b \leq 1.5\%$, $S^L_{f_b} > S^L_{\varphi} > S^L_{f_s} > S^L_{\xi_b} >$ $S^L_{\tau} > S^L_{m_b}$; $\xi_b \geq 2.0\%$, $S^L_{f_b} > S^L_{\varphi} > S^L_{f_s} > S^L_{\xi_b} > S^L_{\tau} > S^L_{m_b} >$ S^L_{α};	$\xi_b \leq 1.5\%$, $S^T_{f_b} > S^T_{f_s} > S^T_{\varphi} > S^T_{\tau} > S^T_{l_s} > S^T_{m_b}$; $\xi_b \geq 2.0\%$, $S^T_{f_b} > S^T_{f_s} > S^T_{\varphi} > S^T_{\tau} > S^T_{l_s} > S^T_{m_b} > S^T_{\xi_b}$;

7. Conclusions

This paper investigates the impact of uncertainty in system and load parameters on vibration responses induced by walking pedestrian(s) using the approaches of LSA and Sobol'-based GSA. First, the LSA and GSA are conducted for the vibration responses of a representative bridge under two load cases, i.e., a single pedestrian and a crowd. Parametric analysis is then carried out to study the sensitivity of the results of LSA and GSA to the input spaces of the modal mass, natural frequency, and damping ratio of a footbridge. It is concluded that in the two load cases, GSA and LSA may produce consistent results in terms of sensitive parameters and their ranking, different sensitive parameters, or the same sensitive parameters but ordered differently, depending upon the bridge's modal properties. The results of GSA and LSA for all kinds of bridges in the two load cases are generally sensitive to the variation of input space of modal mass, natural frequency, and damping ratio of the bridge. The exception is the result for Bridge Series B under the excitation of a single pedestrian. The damping ratio and natural frequency of human body are found to be the only two insensitive parameters. Hence, there is no need to quantify experimentally or numerically the uncertainty in these two parameters in vibration response analysis for footbridges. Compared with the LSA, GSA provides more reliable results by characterizing the sensitivity across entire input spaces; therefore, GSA is suggested for the sensitive analysis of the pedestrian-induced vibration of footbridges. This study provides a comprehensive and conclusive insight into the sensitivity of a pedestrian-induced vibration response and explains why there are inconsistent results in existing literature. It is beneficial for selecting sensitive parameters required for analyzing uncertainty propagation in stochastic response assessment, as well as for the experimental characterization of randomness.

Author Contributions: Conceptualization, X.W.; methodology, X.W.; software, J.Z.; validation, X.W. and J.Z.; investigation, J.Z.; writing—original draft preparation, J.Z.; writing—review and editing, X.W., H.Z. and S.Ž.; funding acquisition, X.W. All authors have read and agreed to the published version of the manuscript.

Funding: The authors acknowledge the support of the National Natural Science Foundation of China (grant number 52178181) and the Fundamental Research Funds for the Central Universities, China (grant number 2020CX035).

Institutional Review Board Statement: Not applicable.

Informed Consent Statement: Not applicable.

Data Availability Statement: Data are contained within the article. Additional supporting data presented in this study are available upon request from the corresponding authors.

Conflicts of Interest: The authors declare no conflict of interest.

References

1. García-Diéguez, M.; Zapico-Valle, J. Sensitivity of the vertical response of footbridges to the frequency variability of crossing pedestrians. *Vibration* **2018**, *1*, 290–311. [CrossRef]
2. Hawryszków, P.; Biliszczuk, J. Vibration serviceability of footbridges made of the sustainable and eco structural material: Glued-laminated wood. *Materials* **2022**, *15*, 1529. [CrossRef] [PubMed]
3. Lievens, K.; Lombaert, G.; Van Nimmen, K.; De Roeck, G.; Van den Broeck, P. Robust vibration serviceability assessment of footbridges subjected to pedestrian excitation: Strategy and applications. *Eng. Struct.* **2018**, *171*, 236–246. [CrossRef]
4. Caprani, C.C.; Ahmadi, E. Formulation of human–structure interaction system models for vertical vibration. *J. Sound Vib.* **2016**, *377*, 346–367. [CrossRef]
5. Tubino, F.; Pagnini, L.; Piccardo, G. Uncertainty propagation in the serviceability assessment of footbridges. *Struct. Infrastruct. Eng.* **2020**, *16*, 123–137. [CrossRef]
6. Sétra. *Technical Guide Footbridges: Assessment of Vibrational Behaviour of Footbridges under Pedestrian Loading*; Association Franaise de Génie Civil (AFGC): Paris, France, 2006.
7. BSI. *UK National Annex to Eurocode 1: Actions on Structures—Part 2: Traffic Loads on Bridges*; NA to BS EN 1991-2: 2003; British Standards Institution: London, UK, 2008.
8. Pedersen, L.; Frier, C. Sensitivity of footbridge vibrations to stochastic walking parameters. *J. Sound Vib.* **2010**, *329*, 2683–2701. [CrossRef]
9. Živanović, S.; Pavić, A.; Reynolds, P. Probability-based prediction of multi-mode vibration response to walking excitation. *Eng. Struct.* **2007**, *29*, 942–954. [CrossRef]
10. Demartino, C.; Avossa, A.M.; Ricciardelli, F. Deterministic and probabilistic serviceability assessment of footbridge vibrations due to a single walker crossing. *Shock Vib.* **2018**, *2018*, 1917629. [CrossRef]
11. Wei, X.; Liu, J.-C.; Bi, S. Uncertainty quantification and propagation of crowd behaviour effects on pedestrian-induced vibrations of footbridges. *Mech. Syst. Signal Process.* **2022**, *167*, 108557. [CrossRef]
12. Avossa, A.M.; Demartino, C.; Ricciardelli, F. Probability distribution of footbridge peak acceleration to single and multiple crossing walkers. *Procedia Eng.* **2017**, *199*, 2766–2771. [CrossRef]
13. Kerr, S.; Bishop, N. Human induced loading on flexible staircases. *Eng. Struct.* **2001**, *23*, 37–45. [CrossRef]
14. Sahnaci, C.; Kasperski, M. Random loads induced by walking. In Proceedings of the Sixth European Conference on Structural Dynamics, Paris, France, 4–7 September 2005.
15. Kasperski, M.; Sahnaci, C. Serviceability of pedestrian structures. In Proceedings of the 25th Conference and Exposition on Structural Dynamics, Orlando, FL, USA, 19–22 February 2007.
16. Kramer, H.; Kebe, H. Man-induced structural vibrations. *Der Bauing.* **1980**, *54*, 195–199.
17. Ricciardelli, F.; Briatico, C.; Ingólfsson, E.T.; Georgakis, C. Experimental validation and calibration of pedestrian loading models for footbridges. In Proceedings of the Experimental Vibration Analysis for Civil Engineering Structures, Porto, Portugal, 24–26 October 2007.
18. Živanović, S.; Pavić, A.; Ingólfsson, E.T. Modeling spatially unrestricted pedestrian traffic on footbridges. *J. Struct.* **2010**, *136*, 1296–1308. [CrossRef]

19. Sahnaci, C.; Kasperski, M. Simulation of random pedestrian flow. In Proceedings of the 8th International Conference on Structural Dynamics, Leuven, Belgium, 4–6 July 2011.
20. Ji, T.; Pachi, A. Frequency and velocity of people walking. *Struct. Eng.* **2005**, *84*, 36–40.
21. Niu, S.; Wang, C.; Gao, S. A novel crowd random loads model for pedestrians walking on footbridge. *Arch. Appl. Mech.* **2016**, *86*, 1369–1381. [CrossRef]
22. Sekiya, N.; Nagasaki, H.; Ito, H.; Furuna, T. Optimal walking in terms of variability in step length. *J. Orthop. Sports Phys. Ther.* **1997**, *26*, 266–272. [CrossRef]
23. Barela, A.M.; Duarte, M. Biomechanical characteristics of elderly individuals walking on land and in water. *J. Electromyogr. Kinesiol.* **2008**, *18*, 446–454. [CrossRef]
24. Živanović, S. Benchmark footbridge for vibration serviceability assessment under the vertical component of pedestrian load. *J. Struct.* **2012**, *138*, 1193–1202. [CrossRef]
25. Matsumoto, Y. *Dynamic Design of Footbridges*; IABSE: Zurich, Switzerland, 1978.
26. Ellis, B. On the response of long-span floors to walking loads generated by individuals and crowds. *Struct. Eng.* **2000**, *78*, 17–25.
27. Ahmadi, E.; Caprani, C.; Živanović, S.; Heidarpour, A. Assessment of human-structure interaction on a lively lightweight GFRP footbridge. *Eng. Struct.* **2019**, *199*, 109687. [CrossRef]
28. Hawryszków, P.; Pimentel, R.; Silva, R.; Silva, F. Vertical vibrations of footbridges due to group loading: Effect of pedestrian–structure interaction. *Appl. Sci.* **2021**, *11*, 1355. [CrossRef]
29. Shahabpoor, E.; Pavić, A.; Racić, V. Structural vibration serviceability: New design framework featuring human-structure interaction. *Eng. Struct.* **2017**, *136*, 295–311. [CrossRef]
30. Tubino, F. Probabilistic assessment of the dynamic interaction between multiple pedestrians and vertical vibrations of footbridges. *J. Sound Vib.* **2018**, *417*, 80–96. [CrossRef]
31. Van Nimmen, K.; Lombaert, G.; De Roeck, G.; Van den Broeck, P. The impact of vertical human-structure interaction on the response of footbridges to pedestrian excitation. *J. Sound Vib.* **2017**, *402*, 104–121. [CrossRef]
32. Gallegos-Calderón, C.; Naranjo-Pérez, J.; Díaz, I.M.; Goicolea, J.M. Identification of a human-structure interaction model on an ultra-lightweight FRP footbridge. *Appl. Sci.* **2021**, *11*, 6654. [CrossRef]
33. Lou, J.; Zhang, M.; Chen, J. Identification of stiffness, damping and biological force of SMD model for human walking. In *Dynamics of Civil Structures*; Springer: Berlin/Heidelberg, Germany, 2015; Volume 2, pp. 331–337.
34. Wang, H.; Chen, J.; Brownjohn, J.M. Parameter identification of pedestrian's spring-mass-damper model by ground reaction force records through a particle filter approach. *J. Sound Vib.* **2017**, *411*, 409–421. [CrossRef]
35. Shahabpoor, E.; Pavic, A.; Racic, V. Identification of mass–spring–damper model of walking humans. *Structures* **2016**, *5*, 233–246. [CrossRef]
36. Bertos, G.A.; Childress, D.S.; Gard, S.A. The vertical mechanical impedance of the locomotor system during human walking with applications in rehabilitation. In Proceedings of the 9th International Conference on Rehabilitation Robotics, Chicago, IL, USA, 28 June–1 July 2005.
37. Shahabpoor, E.; Pavić, A.; Racić, V. Identification of walking human model using agent-based modelling. *Mech. Syst. Signal Process.* **2018**, *103*, 352–367. [CrossRef]
38. Jiménez-Alonso, J.; Sáez, A. A direct pedestrian-structure interaction model to characterize the human induced vibrations on slender footbridges. *Inf. Constr.* **2014**, *66*, 1–9. [CrossRef]
39. Lin, B.; Zhang, Q.; Fan, F.; Shen, S. A damped bipedal inverted pendulum for human–structure interaction analysis. *Appl. Math. Model.* **2020**, *87*, 606–624. [CrossRef]
40. Yang, H.; Wu, B.; Li, J.; Bao, Y.; Xu, G. A spring-loaded inverted pendulum model for analysis of human-structure interaction on vibrating surfaces. *J. Sound Vib.* **2022**, *522*, 116727. [CrossRef]
41. Liang, H.; Zhang, Z.; Wei, P. Theoretical derivation and parameters analysis of a human-structure interaction system with the bipedal walking model. *J. Math.* **2021**, *2021*, 6683083. [CrossRef]
42. Gao, Y.-A.; Yang, Q.-S.; Dong, Y. A three-dimensional pedestrian–structure interaction model for general applications. *Int. J. Struct. Stab. Dyn.* **2018**, *18*, 1850107. [CrossRef]
43. Alampalli, S. Effects of testing, analysis, damage, and environment on modal parameters. *Mech. Syst. Signal Process.* **2000**, *14*, 63–74. [CrossRef]
44. Wei, X.; Wan, H.-P.; Russell, J.; Živanović, S.; He, X. Influence of mechanical uncertainties on dynamic responses of a full-scale all-FRP footbridge. *Compos. Struct.* **2019**, *223*, 110964. [CrossRef]
45. Van Nimmen, K.; Lombaert, G.; De Roeck, G.; Van den Broeck, P. Vibration serviceability of footbridges: Evaluation of the current codes of practice. *Eng. Struct.* **2014**, *59*, 448–461. [CrossRef]
46. Feldmann, M.; Heinemeyer, C.; Lukic, M.; Caetano, E.; Cunha, Á.; Goldack, A.; Keil, A.; Schlaich, M.; Hicks, S.; Smith, A.; et al. *Human-Induced Vibration of Steel Structures*; HIVOSS Project; European Commission: Brussels, Belgium, 2008.
47. Caprani, C.C.; Keogh, J.; Archbold, P.; Fanning, P. Enhancement factors for the vertical response of footbridges subjected to stochastic crowd loading. *Comput. Struct.* **2012**, *102–103*, 87–96. [CrossRef]
48. Haaker, M.; Verheijen, P.J. Local and global sensitivity analysis for a reactor design with parameter uncertainty. *Chem. Eng. Res. Des.* **2004**, *82*, 591–598. [CrossRef]

49. Razavi, S.; Gupta, H.V. What do we mean by sensitivity analysis? The need for comprehensive characterization of "global" sensitivity in earth and environmental systems models. *Water Resour.* **2015**, *51*, 3070–3092. [CrossRef]
50. Sobol, I.M. Global sensitivity indices for nonlinear mathematical models and their Monte Carlo estimates. *Math. Comput. Simul.* **2001**, *55*, 271–280. [CrossRef]
51. Archer, G.; Saltelli, A.; Sobol, I. Sensitivity measures, ANOVA-like techniques and the use of bootstrap. *J. Stat. Comput. Simul.* **1997**, *58*, 99–120. [CrossRef]
52. Piccardo, G.; Tubino, F. Equivalent spectral model and maximum dynamic response for the serviceability analysis of footbridges. *Eng. Struct.* **2012**, *40*, 445–456. [CrossRef]
53. Wei, X.; Russell, J.; Živanović, S.; Mottram, J.T. Measured dynamic properties for FRP footbridges and their critical comparison against structures made of conventional construction materials. *Compos. Struct.* **2019**, *223*, 110956. [CrossRef]
54. Schlaich, M. Guidelines for the Design of Footbridges, Guide to Good Practice-Bulletin 32. In Proceedings of the Footbridge 2005 (Second International Congress), Venice, Italy, 6–8 December 2005.
55. Subbaraj, K.; Dokainish, M. A survey of direct time-integration methods in computational structural dynamics—II. Implicit methods. *Comput. Struct.* **1989**, *32*, 1387–1401. [CrossRef]
56. Tang, Y.; Reed, P.; Van Werkhoven, K.; Wagener, T. Advancing the identification and evaluation of distributed rainfall-runoff models using global sensitivity analysis. *Water Resour.* **2007**, *43*, W06415. [CrossRef]
57. Coppitters, D.; De Paepe, W.; Contino, F. Surrogate-assisted robust design optimization and global sensitivity analysis of a directly coupled photovoltaic-electrolyzer system under techno-economic uncertainty. *Appl. Energy* **2019**, *248*, 310–320. [CrossRef]
58. Hsieh, N.-H.; Reisfeld, B.; Bois, F.Y.; Chiu, W.A. Applying a global sensitivity analysis workflow to improve the computational efficiencies in physiologically-based pharmacokinetic modeling. *Front. Pharmacol.* **2018**, *9*, 588. [CrossRef]
59. Zhang, X.-Y.; Trame, M.; Lesko, L.; Schmidt, S. Sobol sensitivity analysis: A tool to guide the development and evaluation of systems pharmacology models. *CPT Pharmacomet. Syst. Pharmacol.* **2015**, *4*, 69–79. [CrossRef]
60. Chan, K.; Saltelli, A.; Tarantola, S. Sensitivity analysis of model output: Variance-based methods make the difference. In Proceedings of the 29th Conference on Winter Simulation, Atlanta, GA, USA, 7–10 December 1997.

buildings

MDPI

Article

Vibration Serviceability Assessment of a Historic Suspension Footbridge

Elyas Bayat [1], Angelo Milone [2], Federica Tubino [1,*] and Fiammetta Venuti [3]

1 Department of Civil, Chemical and Environmental Engineering, University of Genoa, 16415 Genova, Italy; elyas.bayat@edu.unige.it
2 Direzione Generale del Genio Militare di La Spezia, 19121 La Spezia, Italy; a.milone3@marina.difesa.it
3 Department of Architecture and Design, Politecnico di Torino, 10125 Torino, Italy; fiammetta.venuti@polito.it
* Correspondence: federica.tubino@unige.it

Abstract: Experimental and numerical studies for the structural and vibration serviceability assessment of a historic suspension footbridge adopting non-invasive surveys and low-cost equipment are presented. Field surveys have been carried out to determine geometric properties, ambient vibration tests have been performed to estimate the dynamic properties, and the dynamic response of the footbridge under the action of a single crossing pedestrian has been recorded. Based on field surveys, a 3D Finite Element model was built and was then calibrated against ambient vibration test results. The experimentally-measured maximum acceleration under the action of one crossing pedestrian is compared with the ones obtained numerically and analytically. Furthermore, vibration serviceability assessment under multi-pedestrian loading is carried out, adopting the simplified procedure recommended by a recent guideline. Results show that low-cost non-invasive dynamic testing is suitable to correctly identify the footbridge vertical natural frequencies and mode shapes, including higher-order ones, and to draw considerations about the state of degradation of the structure. Moreover, the level of vibration under the action of a single pedestrian can be estimated with sufficient accuracy using a simplified loading model, provided that the modal damping ratio is properly tuned.

Keywords: dynamic identification; pedestrian excitation; suspension footbridge; vibration serviceability

Citation: Bayat, E.; Milone, A.; Tubino, F.; Venuti, F. Vibration Serviceability Assessment of a Historic Suspension Footbridge. *Buildings* **2022**, *12*, 732. https://doi.org/10.3390/buildings12060732

Academic Editors: Jun Chen and Haoqi Wang

Received: 29 April 2022
Accepted: 26 May 2022
Published: 28 May 2022

1. Introduction

During the 19th Century and the beginning of the 20th Century, a great number of suspended bridges were built in Europe, especially in Italy, Switzerland, and France [1]. The majority of these bridges were destroyed during the 20th Century and sometimes reconstructed, mainly because they were no longer adequate to support the new traffic loadings. Fortunately, some of them still exist in their original configuration, but very often they are closed to traffic, due to the uncertain conditions concerning their current structural properties and serviceability for the current loading scenarios. These surviving bridges are an important part of our architectural heritage, due to their elegance and construction technique. The preservation of these historic bridges implies the need to investigate their structural behavior to assess their safety and serviceability under the expected traffic conditions. Actually, the extreme lightness and flexibility of these kinds of structures, combined with very low structural damping, make them highly prone to human-induced vibrations [2].

Figure 1 shows some examples of suspended bridges in the north of Italy. The Morca bridge (Figure 1b) is the only one allowing for light vehicular traffic: it was retrofitted in 2003 and an extensive program of non-destructive tests and analytical investigations on its dynamic behavior was carried out [3]. The Ramello footbridge (Figure 2), built in 1954 and located in the countryside of La Spezia in Italy, is another example of a historic suspension footbridge. The footbridge was employed for pedestrian crossing and light vehicles until

2019 and then closed, like many Italian bridges that were not considered safe after the collapse of the Morandi Bridge in Genova.

Figure 1. (**a**) Footbridge "Gaietta", Millesimo, Italy (photo by F. Venuti); (**b**) Morca bridge, Italy; (**c**) Footbridge of Vocca Island, Italy; (**d**) Footbridge at Millesimo, Italy (photo by F. Venuti).

Figure 2. General views of the Ramello footbridge: (**a**) lateral view and (**b**) longitudinal view.

The aim of this study is to provide a structural and vibration serviceability assessment of the Ramello footbridge, using non-invasive surveys and low-cost equipment, that could assist public administration in the preservation of the structure. In order to avoid invasive interventions, the structural assessment is carried out in an indirect way by the modal testing of the footbridge and the comparison of the obtained modal properties with those derived from a Finite Element (FE) model built according to the nominal properties of the materials.

Modal testing can be carried out based on controlled input that is measured and used in the identification process (Experimental Modal Analysis, EMA), or on ambient vibration tests where only the response is measured and the force is due to environmental excitation

(Operational Modal Analysis, OMA) [4]. A brief review of modal testing methods for bridges can be found, e.g., in [5]. With the exception of modern non-contact methods developed mainly for laboratory tests (e.g., [6]), EMA usually involves the excitation of the structure through contact methods. It has been applied for the modal identification of footbridges using hammer or shaker excitation (e.g., [7,8]). However, OMA testing techniques have now become attractive, due to their relatively low cost, speed of implementation, and the recent improvements in recording equipment and computational methods (e.g., [9–14]). The low amplitude of vibrations in operational conditions requires very sensitive, low-noise sensors and a high-performance measurement chain [4]. OMA testing techniques are based on the assumption that the excitation is a stationary random process with approximately white noise characteristics. Possible extensions to non-stationary long-term vibration monitoring have been proposed (e.g., [15]). Modal parameters can then be extracted, adopting frequency or time-domain methods [4]: the most commonly adopted are Peak Picking (PP) (e.g., [3,9]), Frequency Domain Decomposition (FDD) ([16], e.g., [3,13,14,17]), and Stochastic Subspace Identification (SSI) ([18], e.g., [8,11,12,14]). The identification of the structural modal parameters can also be employed in the framework of damage detection, since variations in the structural physical properties reflect variations of the modal parameters [19,20]. In particular, mode shapes are more sensitive to damage than natural frequencies, and recent research is focused on the detection of damage using damping [19].

Vibration serviceability assessment requires the evaluation of the level of vibration due to multi-pedestrian traffic, which calls for a probabilistic model of the loading (see e.g., [21–27]). However, current guidelines provide simplified equivalent loading conditions [12], e.g., an equivalent, uniformly-distributed resonant loading is suggested by SÉTRA [28].

The experimental campaign on Ramello Footbridge includes a field survey of the footbridge geometry and element properties, ambient vibration tests, and live load tests of a single pedestrian crossing. The aim is to obtain the information necessary to build a reliable FE model, to get measurements for the dynamic identification of the footbridge, and to have a preliminary assessment of the level of vibration induced by pedestrians. Notwithstanding that the low amplitude of vibrations in operational conditions requires very sensitive, low-noise sensors, ambient vibrations are measured through low-cost equipment and the reliability of such measurement equipment is assessed. A comparison of the different techniques for modal parameters extraction is carried out in order to select the most appropriate one. Then, the FE model, built according to the field survey of the footbridge geometry, is validated against experimentally determined mode shapes and natural frequencies, and the correspondence of experimental and numerical modal properties is used as an indication of the structural health.

The paper develops through the following sections. Section 2 provides a detailed description of the footbridge geometry and element cross-sections estimated through field surveys. Section 3 describes the experimental campaign conducted in 2019 and 2021 and the subsequent identification of the footbridge dynamic properties by operational modal analysis. In Section 4, a 3D FE model of the footbridge is developed and calibrated based on the results obtained from ambient vibration tests. In Section 5, numerical simulations of the footbridge's dynamic response under single pedestrian loading are carried out and a comparison with experimental results is performed. Moreover, the vibration serviceability of the footbridge is investigated based on SÉTRA guidelines. Finally, conclusions are outlined in Section 6.

2. Description of the Footbridge

The Ramello suspension footbridge (Figures 2 and 3) was built in 1954 in the countryside of La Spezia in Italy. Due to the lack of drawing details, the geometric properties of the footbridge and its elements (Figures 4 and 5) were obtained from field surveys. The footbridge has a span of length $L = 90$ m and a width of 2.28 m. Two main suspended cables, with a sag of 7.05 m, connect two reinforced concrete pylons from one side of the river to the opposite side. Each suspension cable is made of three individual spiral strands, with a nominal diameter of 40 mm. The main cables are anchored into the ground at distances of 7.9 m and 7.2 m from the left and right pylons, respectively. The bridge deck is supported by I-shaped transversal steel beams with a step of one meter (Figure 3a), which are suspended to the main cables by means of 89 hangers. The latter are made of rolled steel and have a C-shaped cross-section. The transversal steel floor beams have I-shaped cross-sections and a length of 2.4 m. They support two longitudinal rolled I-shaped steel beams, located at 1.78 m of distance, and three longitudinal timber planks, above which transverse timber planks form the floor (Figure 3a). Both transverse and longitudinal planks have a square cross-section with a side of around 80 mm.

Figure 3. Details of the bridge deck (**a**), connections between the hanger and deck transversal beam (**b**) and between the hanger and main cable (**c**).

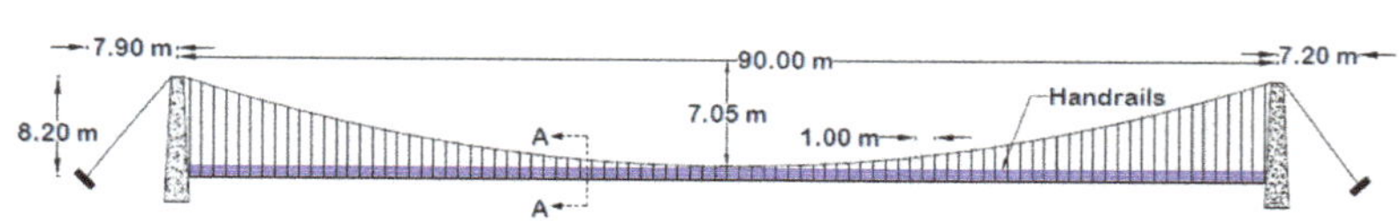

Figure 4. Elevation of the footbridge.

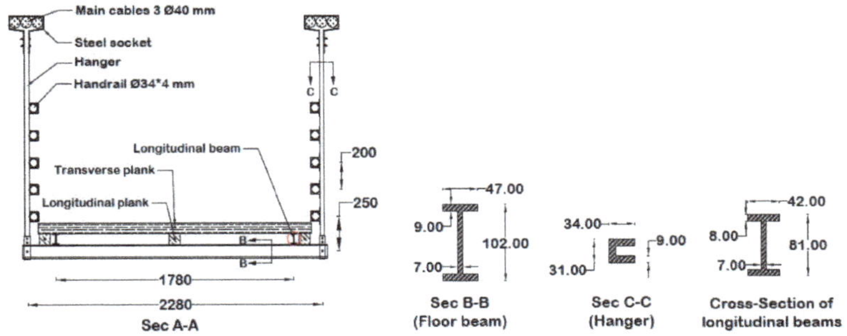

Figure 5. Cross-sections and dimensions of the footbridge structural elements (in mm).

Hangers are connected to floor beams by bolted connections (Figure 3b) and to main cables by means of steel sockets (Figure 3c), which keep the three strands in contact with each other, avoiding any relative displacement. Welded connections link the bottom flanges of the longitudinal beams to the top flanges of the transversal floor beams. Handrails are made of steel circular hollow sections (Figure 2b) and linked to the hangers with joints that restrain the vertical relative translational movement between the handrails and the hangers while enabling horizontal movement. Surface corrosion can be observed widely over the main structural elements (Figure 3b,c).

It is worth noting that the tension force of cables was not measured during the field surveys; therefore, it is considered an unknown parameter that is estimated based on the Finite Element (FE) model of the footbridge that will be discussed in the next sections.

3. Experimental Campaigns and Operational Modal Analysis

This section describes the experimental campaigns carried out in 2019 and 2021 with the aim of identifying the footbridge's dynamic properties and measuring the footbridge's dynamic response under the dynamic excitation of a single pedestrian.

3.1. Ambient Vibration Testing

Four ambient vibration tests with different measurement setup arrangements were conducted on the footbridge to detect its dynamic characteristics. Figure 6a shows the positions of accelerometers in the 4 setups. Suspended footbridges are characterized by extremely low natural frequencies, so the modes at risk of human-induced vibrations are usually higher modes, whose shapes are difficult to be correctly identified unless a great number of sensors are installed. Setup 1 was implemented in 2019, while the other three measurement setups were adopted in 2021 to identify more accurately the mode shapes of the footbridge. The low-cost equipment used for the tests included a 14-channel Labjack U6 data acquisition system with 10 TE Connectivity 4030 signal conditioned MEMS DC triaxial accelerometers that are able to measure accelerations in ± 6 g range with a sensitivity of 333 mV/g and a nominal 0–200 Hz bandwidth. Figure 6b shows the detail of the accelerometers' mounting: they were bolted to an aluminum profile, which was rigidly connected through tight bands at the bottom of the hanger as close as possible to the joint with the transversal beam.

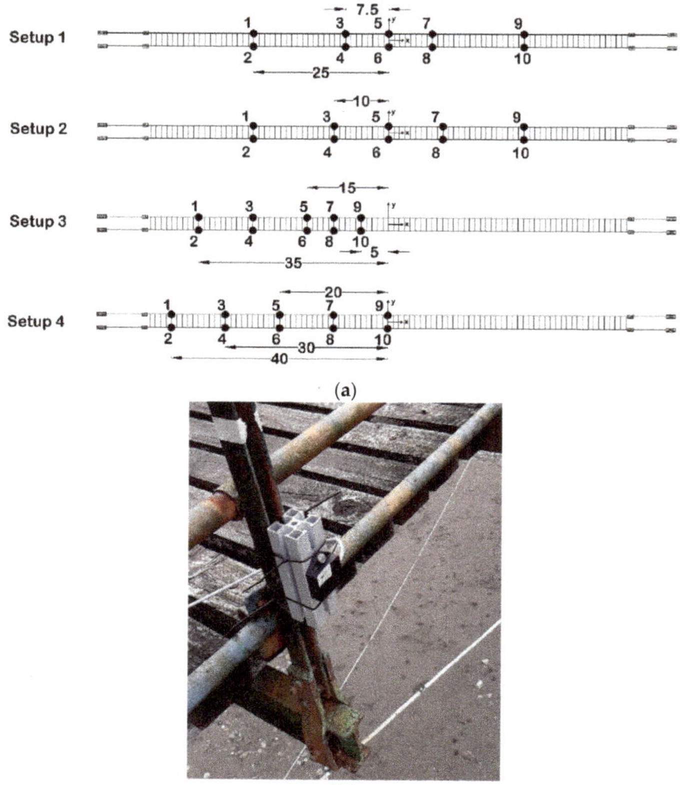

(a)

(b)

Figure 6. Layout and location of accelerometers along the bridge deck for different measurement setups (in m) (**a**) and detail of the accelerometers' mounting (**b**).

The dynamic response of the footbridge was measured at 10 different locations in each measurement setup: vertical acceleration was measured at all the locations, while transverse accelerations were only at four locations (locations 1, 3, 7, and 9). In each measurement setup, the ambient vibrations were simultaneously recorded for one hour with a sampling frequency of 500 Hz.

3.2. Modal Identification of the Footbridge

The data processing and modal identification were performed by using MACEC, a MATLAB-based toolbox for the experimental modal analysis of structures [29]. The identification of the modal parameters was carried out by adopting the SSI, PP, and FDD techniques.

Figure 7, referred to as Setup 1, plots the stabilization diagram derived from the SSI technique (a), the averaged normalized power spectral density function for the identification based on the PP technique (b), and the singular values for FDD (c). In Figure 7a, the power spectral density functions of the measured accelerations are superimposed on the stabilization diagrams in order to verify that the identified poles correspond to the peaks in the acceleration spectrum. It can be deduced that most of the stable poles in Figure 7a correspond to peaks in the power spectral density function, also detected in Figure 7b,c.

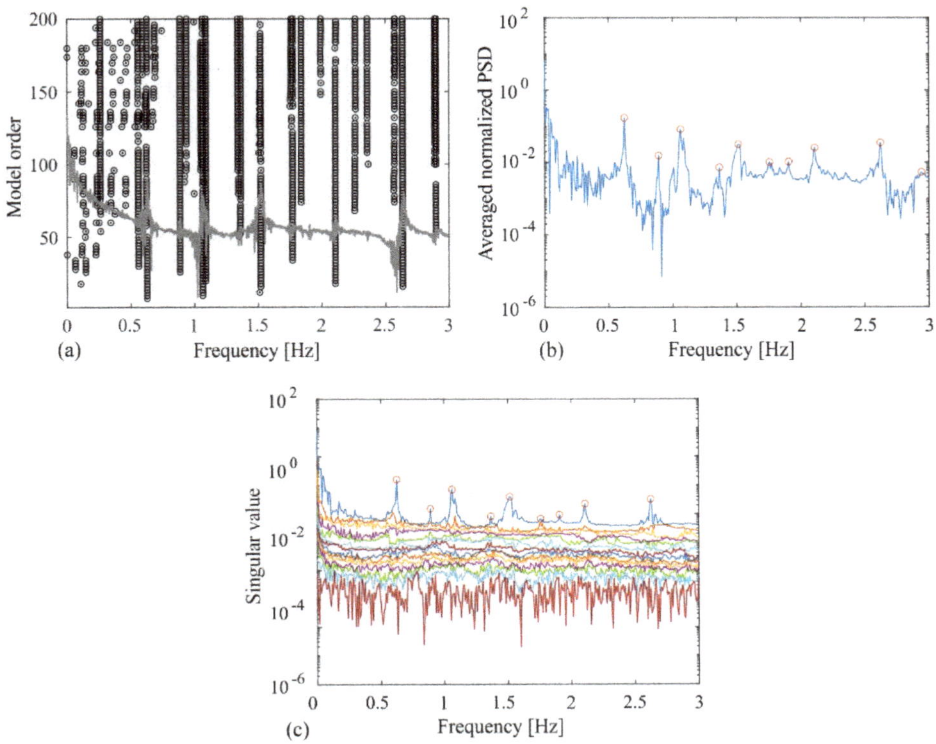

Figure 7. Setup 1: Stabilization diagram (**a**), averaged power spectral density functions (**b**), singular values (**c**).

Table 1 reports the natural frequencies identified through the three techniques, together with a synthetic description of the corresponding mode shape (L = Lateral, V = Vertical, T = Torsional; S = Symmetric, A = Asymmetric). It can be deduced that the natural frequencies identified with the PP and FDD techniques are coincident and they are generally in very good agreement with the ones identified by SSI. However, some of the vibration modes are identified only through the SSI technique. This circumstance is due to the weak ambient excitation and to the significant noise measured by the accelerometers, which does not allow for the clear identification of the peaks corresponding to some of the vibration modes through frequency-domain methods. The results in Table 1 show that the SSI technique is the one that allows the identification of the greater number of modes. Furthermore, frequency-domain methods require long time histories for the reliable identification of the modal damping ratios. Thus, in the following, results obtained by adopting SSI for all the tested setups are reported.

Table 1. Setup 1: Modal parameters of the footbridge obtained from the different identification techniques.

Mode	Mode Shape	Frequency SSI [Hz]	Frequency PP [Hz]	Frequency FDD [Hz]
1	LS	0.260	-	-
2	VA	0.557	-	-
3	VS	0.626	0.625	0.625
4	VA	0.887	0.890	0.890
5	TS	1.064	1.062	1.062
6	VA	1.360	1.367	1.367
7	VS	1.519	1.515	1.515
8	TS	1.768	1.757	1.757
9	VA	1.838	1.905	1.905
10	VS	2.106	2.108	2.108
11	TS	2.259	-	-
12	VA	2.356	-	-
13	VS	2.633	2.624	2.624

Figure 8 shows the stabilization diagrams extracted from Setups 2–4.

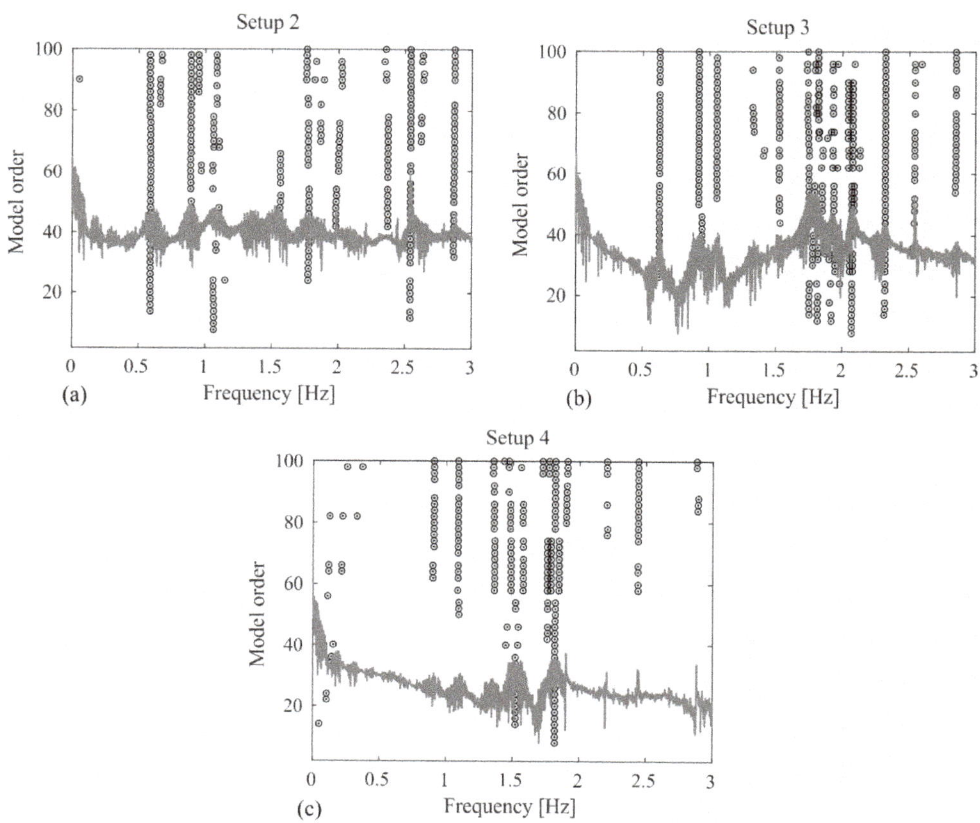

Figure 8. Stabilization diagrams of Setups 2 (**a**), 3 (**b**) and 4 (**c**) obtained from the SSI technique and power spectral density functions of the measured accelerations (in grey).

The identified dynamic characteristics of the footbridge are summarized in Table 2. The natural frequencies and damping ratios are obtained by combining and averaging the results extracted from different measurement setups. The detected modal shapes consist of one lateral bending mode, two torsional modes, and ten vertical bending modes that have symmetric or asymmetric shapes. The results show that the identified modes with frequencies greater than 1.3 Hz are quite stable and reliable since those modes were found in most of the ambient vibration measurements. There are four vertical bending modes (modes 9–12) whose natural frequencies fall within the frequency range of dynamic loading induced by the first walking harmonic [22]. Therefore, the resonant condition could occur for the footbridge due to walking pedestrians. The identified damping ratios of the modes at resonance risk for each Setup are reported in Table 3. It is worth noting that the damping ratios identified from Setups 2 to 4 are generally higher than those from Setup 1 and quite high with respect to the characteristic damping ratios of steel footbridges (around 0.5% [30]). In the Morca suspended footbridge [3], the identified damping ratios were quite high (2.73–7.69%) as well, and this unusual result was explained as due to energy dissipation in the connections between structural elements of the deck. Very large coefficients of variation of bridge damping ratios obtained from different experiments were also reported in [31], where it was observed that operating conditions affect the modal properties and, in particular, damping ratios are very sensitive to test and analysis methods. Therefore, damping average values are not considered fully reliable and will be verified and tuned within the simulation of the footbridge dynamic response (Section 5).

Table 2. Average identified modal parameters using the SSI technique.

Mode	Mode Shape	Setup	Frequency [Hz]	Damping [%]
1	LS	1	0.259	2.051
2	VA	1	0.56	2.462
3	VS	1,2,3	0.612	3.041
4	VA	1,2,4	0.894	3.018
5	TA	3	0.922	3.600
6	TS	1,2,3	1.062	2.500
7	VS	4	1.089	3.024
8	VA	3,4	1.345	2.681
9	VS	1,2,3,4	1.531	2.157
10	VA	1,2,3,4	1.803	1.732
11	VS	1,2,3,4	1.980	1.585
12	VA	1,2,3,4	2.311	0.766
13	VS	1,2,3,4	2.523	0.177

Table 3. Identified damping ratios [%] for each setup and average values for the modes at resonance risk.

Frequency [Hz]	Setup 1 [%]	Setup 2 [%]	Setup 3 [%]	Setup 4 [%]	Average [%]
1.531	1.167	3.248	1.344	2.869	2.157
1.803	0.78	2.588	1.504	2.058	1.732
1.98	0.501	2.661	1.448	1.732	1.585
2.311	0.551	1.602	0.617	0.294	0.766

3.3. Measurements of Dynamic Response Due to a Single Pedestrian Crossing

In the experimental campaign in 2021, in addition to ambient vibration tests, three single walking load tests were carried out on the footbridge. Based on the results obtained from Setup 1 in 2019, a pedestrian, with a weight of 780 N, crossed the footbridge with his step frequency synchronized by a metronome to the frequencies of 1.5, 1.75, and 2.05 Hz, respectively. These frequencies do not exactly match the natural frequencies of the footbridge, which were correctly identified after the 2021 survey, but are very close to the resonant condition.

The layout of accelerometers in Setup 4 was used to record the acceleration of the footbridge deck in the lateral (L) and vertical (V) directions. As an example, the vertical acceleration response $\ddot{q}(t)$ and its corresponding Fourier spectrum $\tilde{\ddot{q}}(f)$ measured by accelerometer 2 (V2) during the passage of the pedestrian with a step frequency of 1.75 Hz are shown in Figure 9. The Fourier spectrum for this case shows that the dynamic response of the footbridge is dominated by the mode with the natural frequency of 1.803 Hz. Moreover, to exclude the effects of noises on the peak accelerations, the responses were lowpass filtered and peak accelerations were extracted from the filtered signal.

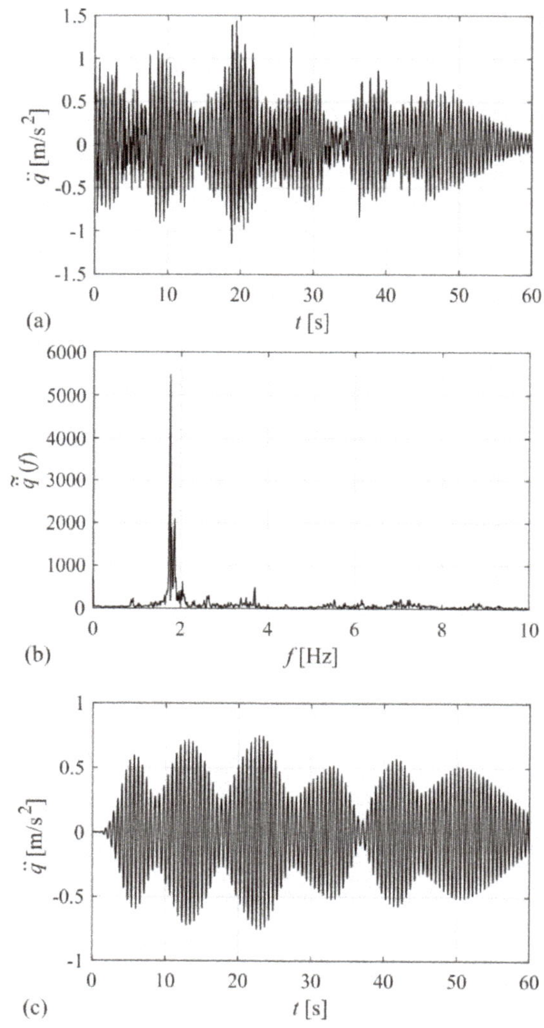

Figure 9. The measured (**a**) and filtered (**c**) vertical acceleration and Fourier spectrum (**b**) by accelerometer 2 (V2) due to a single pedestrian crossing with a step frequency of 1.75 Hz.

Table 4 reports the maximum acceleration responses obtained from these single-pedestrian walking load tests. The maximum transverse and vertical acceleration responses of 0.156 and 0.830 m/s^2, respectively, are associated with a step frequency of 2.05 Hz.

Table 4. Maximum acceleration responses due to different walking load tests [m/s^2].

Walking Tests	Step Frequency		
Accelerometer	1.5 Hz	1.75 Hz	2.05 Hz
V1	0.407	0.655	0.703
V2	0.504	0.755	0.794
V3	0.264	0.673	0.589
V4	0.298	0.647	0.546
V5	0.352	0.552	0.737
V6	0.360	0.472	0.730
V7	0.671	0.478	0.573
V8	0.692	0.468	0.800
V9	0.770	0.312	0.830
V10	0.808	0.184	0.770
L1	0.038	0.035	0.061
L3	0.032	0.033	0.156
L7	0.028	0.032	0.054
L9	0.045	0.030	0.060

The vibration serviceability of the footbridge under pedestrian walking loads can be evaluated based on the SÉTRA guidelines [28]. The guideline classifies footbridges according to four comfort levels based on maximum acceleration responses in both horizontal and vertical directions (Table 5). The vertical and lateral peak accelerations measured on the footbridge both fall within the mean comfort level. This result, obtained for a single pedestrian crossing, raises concerns about the comfort level due to the crossing of multiple pedestrians, which will be investigated in Section 5.

Table 5. Comfort levels and corresponding acceleration limits defined by SÉTRA [28].

Comfort Level	Vertical Acceleration Limit [m/s^2]	Horizontal Acceleration Limit [m/s^2]
Maximum	<0.5	<0.15
Mean	0.5–1	0.15–0.3
Minimum	1–2.5	0.3–0.8
Unacceptable	>2.5	>0.8

4. Finite Element Model of the Footbridge

The FE model of the footbridge was built with ANSYS software [32] to investigate numerically the dynamic behavior of the structure. The model was built based on the field surveys and then updated according to ambient vibration test results.

The main cables and the hangers were modeled using the 3D spar element "LINK180". The effective steel area, density, modulus of elasticity, and Poisson's ratio of each spiral strand of the main cables were taken as 945 mm^2, 7850 kg/m^3, 160 GPa, and 0.3, respectively. Furthermore, the longitudinal and transversal floor beams were modeled using the 3D elastic beam element "BEAM188" considering the modulus of elasticity and density as 210 GPa and 7850 kg/m^3, respectively. The same material properties were assigned to the hangers. The timber boards and handrails were assumed as nonstructural elements and modeled with the concentrated mass element "MASS21". The amount of concentrated mass applied at the deck nodes is estimated by assuming the values of density for timber boards and handrails as 700 and 7850 kg/m^3, respectively. Table 6 summarizes the properties of all elements employed to build the numerical model.

Table 6. Element properties.

Structural Element	ANSYS Element	A [m²]	I_y [m⁴]	I_z [m⁴]	Mass [kg]
Cable	LINK180	0.284×10^{-2}	-	-	-
Hanger	LINK180	0.729×10^{-3}			-
Transversal beam	BEAM188	1.434×10^{-3}	2.18×10^{-6}	1.58×10^{-7}	-
Longitudinal beam	BEAM188	1.192×10^{-3}	1.08×10^{-6}	1.02×10^{-7}	-
Handrail	MASS21	-	-	-	15
Timber board	MASS21	-	-	-	60

Pylons were not modeled due to the high stiffness assumed at the pylon saddles: hence, the cables were restrained at the pylon position by means of rigid constraints that allow sliding in the longitudinal direction. The ends of the cables were restrained to the ground by pinned supports. Moreover, it was assumed that the hangers were hinged to the main cables and floor beams. Longitudinal and transversal beams were connected with fixed joints. Finally, both ends of the longitudinal beams were restrained to translations and rotations. The general view of the FE model of the footbridge is shown in Figure 10.

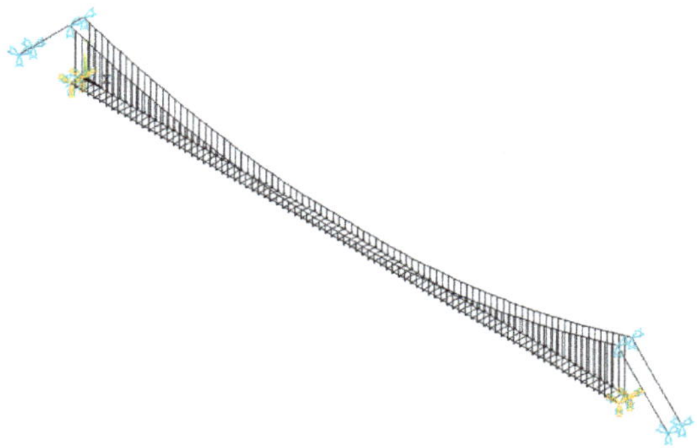

Figure 10. General view of the FE model of the footbridge.

4.1. Preliminary Static Analysis

Suspended footbridges are characterized by non-linear behavior, due to the well-known geometric effects of the main cables [33]. Therefore, in their structural analysis, it is first necessary to determine the geometric configuration resulting from dead load and cable prestress. In the case of the Ramello footbridge, it is worth recalling that the surveyed geometry refers to the deformed configuration of the footbridge under dead load and prestress.

Since tension force in the cables has not been measured, it has been determined through a parametric study. Non-linear static analyses have been carried out for different values of the cable pretension T in the range 0–360 kN and vertical deflections q have been measured in different sections of the footbridge. Figure 11 plots the vertical deflections at the abscissas $x = L/2$ and $x = L/3$ along the deck for different values of the cable prestress. The value of prestress corresponding to the minimum deflection at both monitored positions has been retained for successive dynamic analyses. Specifically, a tension force in the main cables T = 180 kN was selected.

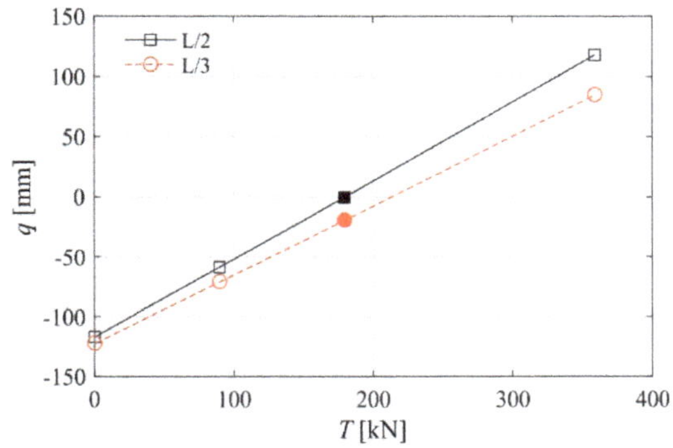

Figure 11. Vertical deflections at $x = L/2$ and $x = L/3$ for different values of cable prestress (filled markers identify the selected value of prestress).

4.2. Modal Analysis

A modal analysis was performed on the footbridge model to extract its dynamic properties. The modal analysis was conducted after a nonlinear static analysis of the footbridge, subjected to dead loads and the pretension of the cables, in order to determine the geometric tangent stiffness matrix [2,34]. The natural frequencies of the first thirty global modes are summarized in Table 7.

Table 7. Modal parameters of the footbridge obtained through FE modeling.

Mode	Mode Shape	Frequency [Hz]	Mode	Mode Shape	Frequency [Hz]
1	LS	0.304	19	VS	1.469
2	VA	0.435	21	TA	1.499
3	TA	0.504	23	TS	1.692
4	LA	0.568	24	VS	1.699
5	VS	0.607	25	LA	1.723
6	TS	0.706	26	VA	1.766
7	LS	0.848	29	TA	1.961
8	VA	0.858	30	LS	2.010
11	TA	0.989	31	TA	2.015
12	VS	1.046	32	VS	2.030
15	LA	1.145	33	VA	2.256
16	TS	1.207	35	LA	2.301
17	VA	1.302	36	TS	2.303
18	LS	1.432	38	VS	2.532

From a direct inspection of Tables 2 and 7, it is evident that ambient vibration tests did not allow the identification of the complete set of the lateral and torsional modes, but only of a very limited number of them. This is due to the evanescent excitation of such modes, provided only by the wind, which was very weak during the ambient vibration tests. For this reason, the validation of the numerical model, based on a comparison of modal parameters (i.e., natural frequencies and mode shapes) with the experimental estimates obtained from the SSI method, is limited to the vertical bending modes. The correlation of numerical and experimental mode shapes is investigated through the Modal Assurance Criterion (MAC), which is expressed as [35]:

$$\mathrm{MAC}\left(\phi_{\mathrm{E},i}, \phi_{\mathrm{N},j}\right) = \frac{\left(\phi_{\mathrm{E},i}^{\mathrm{T}} \phi_{\mathrm{N},j}\right)^2}{\left(\phi_{\mathrm{E},i}^{\mathrm{T}} \phi_{\mathrm{E},i}\right)\left(\phi_{\mathrm{N},j}^{\mathrm{T}} \phi_{\mathrm{N},j}\right)}, \quad (1)$$

where $\phi_{E,i}$ is a vector that represents the *i*-th mode shape extracted experimentally from field vibration tests, $\phi_{N,j}$ is a vector that represents the *j*-th numerical mode shape, and T stands for transpose. Generally, MAC values greater than 0.8 mean a very good match between the two mode shapes. Furthermore, the correlation of the two modes in terms of natural frequency can be investigated by calculating the percentage frequency error based on the following expression:

$$\Delta f = \frac{f_N - f_E}{f_E} \times 100[\%], \tag{2}$$

where f_E and f_N are the experimental and numerical natural frequencies, respectively. According to Table 8, there is generally a good agreement between the numerically and experimentally identified modal parameters, with MAC values higher than 0.9 for almost all the considered modes, and frequency errors generally lower than 5%. This outcome demonstrates that the number of sensors and setups was sufficient to correctly detect the vertical bending modes, including higher-order ones. Moreover, the very good match between experimental and numerical mode shapes allows us to hypothesize that the footbridge is not interested in localized damage, despite the diffused surface deterioration of the structural elements. Figure 12 plots the mode shapes of the four vertical bending modes, whose frequencies fall within the range of walking excitation, while Figure 13 compares the experimental and numerical mode shapes reported in Table 8.

Table 8. Comparison of numerical and experimental modal parameters for bending vibration modes.

Numerical Results		Experimental Results		Δf [%]	MAC
Mode	**Frequency (Hz)**	**Mode**	**Frequency (Hz)**		
2	0.435	2	0.560	−22.32	0.985
5	0.607	3	0.612	−2.09	0.996
8	0.858	4	0.894	−4.07	0.994
12	1.046	7	1.089	−3.99	0.893
17	1.302	8	1.344	−1.38	0.975
19	1.469	9	1.531	−3.39	0.913
26	1.766	10	1.803	−0.80	0.990
32	2.030	11	1.980	2.53	0.941
33	2.256	12	2.310	−1.92	0.994
38	2.532	13	2.520	−1.09	0.736

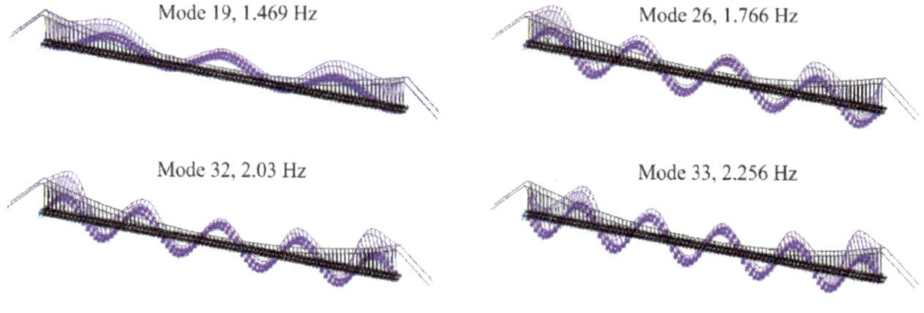

Figure 12. Mode shape of modes sensitive to walking excitation (obtained through FE modeling).

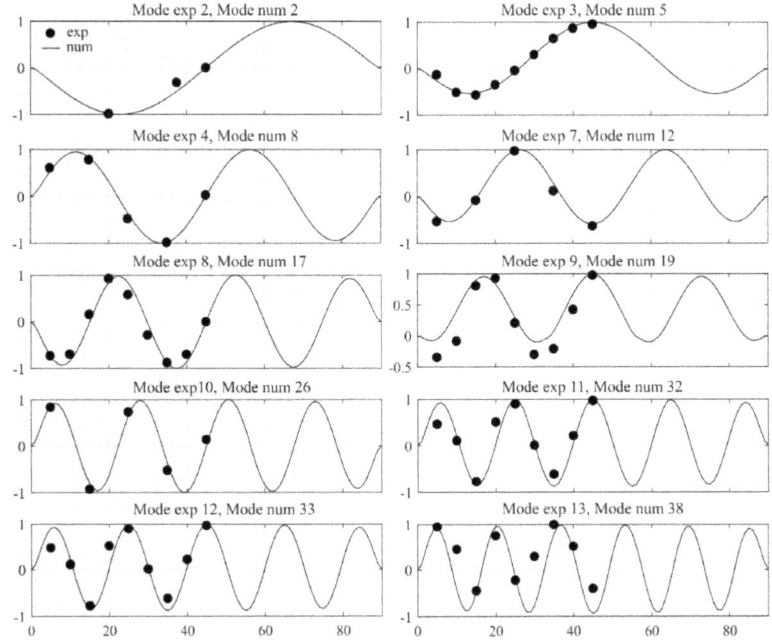

Figure 13. Comparison between experimental and numerical mode shapes.

5. Vibration Serviceability Assessment

The dynamic response of the footbridge under single and multiple pedestrian loads is calculated in order to assess its vibration serviceability. It is assumed that the footbridge system is a linear mono-dimensional system, whose dynamics are described by the equation of motion:

$$m(x)\frac{\partial^2 q(x,t)}{\partial t^2} + \mathcal{C}\left[\frac{\partial q(x,t)}{\partial t}\right] + \mathcal{L}[q(x,t)] = f_p(x,t), \tag{3}$$

where $q(x,t)$ is the displacement of the footbridge, x is the abscissa along the bridge deck and t is the time, $m(x)$ is the structural mass per unit length, $\mathcal{C}$ is the damping operator, $\mathcal{L}$ is the stiffness operator, $f_p(x,t)$ is the external force per unit length. Under the hypothesis of classical damping, Equation (3) is usually solved by applying the principal transformation and assuming that the dynamic response is dominated by one mode of vibration:

$$q(x,t) = \varphi_j(x)p_j(t), \tag{4}$$

where φ_j is the j-th mode shape of the footbridge and p_j is the corresponding principal coordinate. The equation of motion of the j-th principal coordinate p_j is expressed as:

$$\begin{gathered} \ddot{p}_j(t) + 2\xi_j\omega_j\dot{p}_j(t) + \omega_j^2 p_j(t) = \tfrac{F_j(t)}{M_j}, \\ F_j(t) = \int_0^L f_p(x,t)\varphi_j(x)dx, \end{gathered} \tag{5}$$

where M_j, ω_j, ξ_j, and $F_j(t)$ are the modal mass, circular natural frequency, modal damping ratio, and modal force of the j-th mode, respectively, and L is the span length.

5.1. Single Pedestrian Excitation

The mathematical model usually adopted to describe the vertical dynamic load due to a single pedestrian walking is defined by the following expression ([36,37]):

$$f_p(x,t) = F_p(t)\cdot\delta(x - ct), \tag{6}$$

where c is the pedestrian speed, δ is Dirac delta function, and $F_p(t)$ is the time-varying vertical force induced by a single pedestrian, which is usually defined as a sum of Fourier harmonic components:

$$F_p(t) = G + \sum_{h=1}^{H} G\alpha_h sin(2\pi h f_s t + \phi_h), \tag{7}$$

where G is the static weight of the pedestrian, h is the order number of the harmonic, H is the total number of contributing harmonics, α_h is the dynamic load factor (DLF) of the h-th harmonic, f_s the step frequency (Hz), and ϕ_h is the phase angle of the h-th harmonic. It is generally accepted and confirmed by experimental tests that the dynamic response of footbridges is mainly affected by the first walking harmonic. For normal walking speeds, the DLF of the first vertical harmonic is $\alpha_1 \cong 0.4$ and pedestrian velocity can be calculated as $c \cong 0.9 f_s$ (e.g., [30]).

The dynamic response of the footbridge due to a single pedestrian crossing can be obtained by numerically solving the equation of motion (5), with $f_p(x,t)$ expressed as in Equations (6) and (7).

As an alternative, the dynamic response of the footbridge can be approximately calculated based on the analytical solution proposed by Piccardo and Tubino [38], which predicts the dynamic response of the footbridge due to a single pedestrian crossing in resonant conditions. Specifically, the j-th principal coordinate p_j is calculated in the nondimensional form, as follows:

$$\begin{gathered} p_j(\tilde{t}) = A_j(\tilde{t}) cos(\tilde{t}) \\ A_j(\tilde{t}) = -\frac{\alpha_h G}{2M_j\omega_j^2}\left[\int_0^{\tilde{t}} \varphi_j\left(\tilde{\Omega}_c\tilde{\tau}\right)\exp(\xi_j\tilde{\tau})d\tilde{\tau}\right]\exp(-\xi_j\tilde{t}), \end{gathered} \tag{8}$$

where the nondimensional parameters $\tilde{t}$ and $\tilde{\Omega}_c$ are defined as:

$$\tilde{t} = \omega_j t; \quad \tilde{\Omega}_c = \frac{c}{\omega_j L}. \tag{9}$$

In order to predict the dynamic response of the footbridge due to a single pedestrian crossing in the experimental tests described in Section 3.3, the dynamic response of the footbridge to a single pedestrian walking at 1.5, 1.75, and 2.05 Hz is estimated numerically (Equation (5)) and analytically (Equation (8)) considering as the mode of interest j the one whose frequency is the nearest to the excitation frequency, i.e., experimental modes 9, 10, and 11, respectively. In both cases, the adopted damping ratio is the average damping identified from the ambient vibration tests reported in Table 2. The obtained peak accelerations are reported in Table 9, compared with experimental results. Both numerical and analytical predictions are not very accurate, especially for step frequencies of 1.5 and 2.05 Hz, with a maximum error of about 35%. In order to obtain a more accurate prediction of the peak acceleration response, the damping ratios were modified within the range of values identified in the different setups (Table 3) in order to obtain a quite perfect match between numerical and experimental maximum accelerations. The modified damping ratios for the considered three bending modes are 2.7%, 0.78%, and 1.8%, respectively. Figure 14 plots an example of the vertical acceleration responses due to a single pedestrian crossing with a step frequency of 1.75 Hz obtained through numerical and analytical approaches with average and modified damping. From a comparison between Figure 14a,b, it can be deduced that assuming a modified damping ratio (0.78%) lower than the average value

(1.732%, see Table 3), both the analytical and the numerical vertical responses increase. However, the increase in the numerical response (f_s = 1.75 Hz) is larger than the increase in the analytical one (f_s = 1.803 Hz). This circumstance is due to the fact that the modal force associated with the moving harmonic load (Equations (5)–(7)) can be decomposed into two harmonic components, one of which is closer to the resonance condition when f_s = 1.75 Hz. Furthermore, comparing Figures 14b and 9, it can be deduced that the experimentally measured time histories and the ones obtained numerically and analytically are not in perfect agreement. The difference can be due to many factors, such as the slight variations in the walking speed and step frequency in experimental tests, that are not taken into account by the analytical force model in Equations (6) and (7). Table 9 reports the peak accelerations obtained with the numerical and analytical approach by adopting the modified damping ratios. The results show that the modified damping generally allows us to obtain a more accurate analytical prediction of the peak response with a maximum error of about 16%. The obtained results confirm that, for the present footbridge, the dynamic response to a single pedestrian is mainly dominated by a single mode and that the analytical model in Equation (8) is able to predict the peak response with sufficient accuracy, despite the fact that resonant conditions were not perfectly achieved during the experimental tests.

Table 9. Comparison of peak acceleration responses [m/s^2] obtained through experimental, numerical, and analytical approaches for a single pedestrian crossing at different step frequencies.

		$\xi_{Average}$		$\xi_{Modified}$	
f_s [Hz]	Exp	Numerical	Analytical	Numerical	Analytical
1.5	0.81	0.906	0.964	0.798	0.885
	Error [%]	13.12	20.39	−0.37	10.45
1.75	0.76	0.4917	0.485	0.729	0.629
	Error [%]	−34.87	−35.76	−3.41	−16.69
2.05	0.83	0.895	0.786	0.835	0.744
	Error [%]	7.78	−5.33	0.57	−10.34

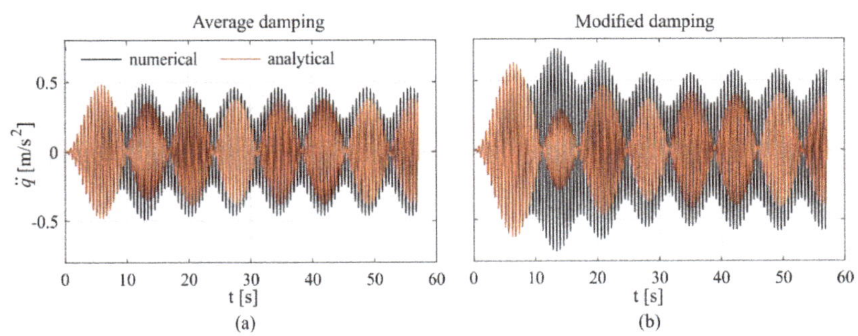

Figure 14. Time history of the acceleration responses by the numerical and analytical solutions predicted with average (**a**) and modified damping (**b**) during a single pedestrian crossing with a step frequency of 1.75 Hz.

5.2. Multiple Pedestrian Excitation

The serviceability of the footbridge is assessed based on the approach proposed by the SÉTRA guidelines [28]. According to SÉTRA, footbridges are classified into four classes, from urban footbridges with heavy traffic (Class I) to seldom-used footbridges (Class IV). Despite the Ramello footbridge belonging to Class IV, for which dynamic calculations are not required, the guideline suggests considering at least Class III for extremely lively footbridges to ensure a minimum amount of risk control. For Class III footbridges, serviceability assessment should be performed under the action of a sparse crowd, characterized by a pedestrian density ρ equal to 0.5 ped/m^2. A resonant uniformly distributed harmonic load $F_v(t)$ [N/m^2] is defined as follows:

$$F_v(t) = \alpha G \cos(2\pi f_j t) N_{eq} \psi, \quad N_{eq} = 10.8\sqrt{\xi_j \rho/(BL)}, \quad \psi = \begin{cases} (f_j - 1.25)/0.45 & 1.25 < f_j < 1.7 \\ 1 & 1.7 < f_j < 2.1 \\ 1 - (f_j - 2.1)/0.2 & 2.1 < f_j < 2.3 \end{cases}, \tag{10}$$

where αG = 280 N, N_{eq} is the equivalent number of perfectly synchronized pedestrians per square meter generating the 95th percentile of the peak acceleration response induced by random pedestrians, f_j is the natural frequency of the j-th mode, B is the deck width, and ψ is a reduction factor to consider that the risk of resonance reduces if the footbridge frequency is outside the interval of 1.7–2.1 Hz for vertical vibrations. This load should be applied for each vertical mode at risk with the same sign as the one of the considered mode shape to obtain the most unfavorable effect. Moreover, the modal mass should be estimated while also taking into account the mass of pedestrians. The peak acceleration of the footbridge can be predicted with the following expression:

$$\ddot{q}_{max,v} = \frac{\alpha G N_{eq} B \int_0^L \varphi_j(x)\mathrm{d}x}{2\xi_j M_{j,tot}}, \tag{11}$$

where $M_{j,tot}$ is the total modal mass of the footbridge and pedestrians.

The peak accelerations calculated for the three modes at resonance risk are reported in Table 10. According to Table 5, all the obtained values fall in the range of unacceptable comfort. It is worth pointing out that the simplified procedure proposed by the SÉTRA guidelines often leads to an overestimation of the structural response in the vertical direction since human–structure interaction is not taken into account. However, the very high values of peak accelerations suggest the need to further investigate the footbridge dynamic behavior under more realistic loading scenarios and to evaluate the possibility of installing suitable countermeasures.

Table 10. Peak accelerations due to sparse traffic according to SÉTRA guidelines.

	f_j = 1.53 Hz	f_j = 1.80 Hz	f_j = 1.98 Hz
$\ddot{q}_{max,v}$ [m/s^2]	5.57	8.20	7.56

6. Conclusions

This paper investigated the dynamic behavior and vibration serviceability of a historic suspension footbridge based on non-invasive low-cost modal testing and a numerical model.

The comparison among different frequency- and time-domain techniques for modal parameters extraction has shown that when using low-cost sensors with low vibration levels, the time-domain SSI method allows for the extraction of a larger number of modes. In particular, the operational modal analysis allowed the identification of 11 vibration modes, including one lateral, two torsional, and eight bending mode shapes within the frequency range of 0–2.6 Hz. The comparison of the modal characteristics of the numerical model and the ones estimated experimentally showed that ambient vibration tests carried out with low-cost sensors allowed a reliable identification of the bending modes, but not of the torsional and lateral ones, which were very weakly excited by ambient actions. The accordance between the experimentally identified natural frequencies and the ones obtained from the numerical model demonstrates that the stiffness of the elements corresponds to the one evaluated assuming standard values of the elastic modulus of steel, excluding a significant degradation of material properties. Furthermore, the good accordance between the experimental and numerical mode shapes confirms that the global structural behavior of the footbridge is well-captured by the numerical model and excludes significant local damages to the structural elements that would reflect on the identified mode shapes.

The footbridge is characterized by four lowly damped bending vibration modes in the range of step frequency typical of normal walking, and thus it is very sensitive to human-induced vibrations. The comparison between experimental accelerations and the ones estimated numerically based on a moving harmonic load model of a single pedestrian confirmed the reliability of such an approximated loading model. Finally, the serviceability assessment according to the SÉTRA guidelines showed that the footbridge would have an unacceptable comfort level under the crossing of a sparse crowd. The very high values of peak accelerations suggest the need to further investigate the footbridge dynamic behavior under more realistic loading scenarios and to evaluate the possibility of installing suitable vibration countermeasures (e.g., [39–41]).

In summary, the presented results have evidenced the following advantages (+) and drawbacks (−) of the non-invasive low-cost technology adopted:

+ The reliable identification of natural frequencies and mode shapes, in conjunction with the SSI modal identification technique;

+ The potential to draw considerations on the structural damage on the basis of the comparison between experimental and numerical modal properties;

+ The accurate estimation of the acceleration level for vibration serviceability assessment;

− Modal identification is limited to vertical bending modes;

− Modal damping ratios identified from ambient vibrations are very dispersed.

The last two issues are worthy of further investigation and the possibility to perform forced vibration tests or adopt higher-level equipment should be taken into consideration in order to also identify lateral and torsional modes and better estimate damping ratios.

Author Contributions: Conceptualization, F.T.; methodology, F.T.; software, E.B., A.M. and F.V.; validation, F.V.; formal analysis, F.T. and F.V.; resources, F.T.; data curation, E.B. and A.M.; writing—original draft preparation, E.B.; writing—review and editing, F.T. and F.V.; visualization, F.V.; supervision, F.T.; project administration, F.T. All authors have read and agreed to the published version of the manuscript.

Funding: This research received no external funding.

Institutional Review Board Statement: Not applicable.

Informed Consent Statement: Not applicable.

Data Availability Statement: Data sharing is not applicable to this article.

Acknowledgments: The Authors greatly acknowledge the technical support by Giancarlo Cassini, Giuseppe Riotto and Davide Burlando during the experimental campaign.

Conflicts of Interest: The authors declare no conflict of interest.

References

1. Re, L. *Cable Suspended: The Suspension Bridges of the 19th Century in Valsesia*; Lindau: Turin, Italy, 1993. (In Italian)
2. Bruno, L.; Venuti, F.; Nascè, V. Pedestrian-induced torsional vibrations of suspended footbridges: Proposal and evaluation of vibration countermeasures. *Eng. Struct.* **2012**, *36*, 228–238. [CrossRef]
3. Gentile, C.; Gallino, N. Ambient vibration testing and structural evaluation of an historic suspension footbridge. *Adv. Eng. Softw.* **2008**, *39*, 356–366. [CrossRef]
4. Brincker, R.; Ventura, C. *Introduction to Operational Modal Analysis*; John Wiley & Sons, Ltd.: Chichester, UK, 2015.
5. Salawu, O.S.; Williams, C. Review of full-scale dynamic testing of bridge structures. *Eng. Struct.* **1995**, *17*, 113–121. [CrossRef]
6. Scislo, L.; Guinchard, M. Non-Invasive Measurements of Ultra-Lightweight Composite Materials Using Laser Doppler Vibrometry System. In Proceedings of the 26th International Congress on Sound and Vibration, ICSV 2019, Montreal, QC, Canada, 7–11 July 2019.
7. Brownjohn, J.M.W. Vibration characteristics of a suspension footbridge. *J. Sound Vib.* **1997**, *202*, 29–46. [CrossRef]
8. Živanović, S.; Pavic, A.; Reynolds, P. Modal testing and FE model tuning of a lively footbridge structure. *Eng. Struct.* **2006**, *28*, 857–868. [CrossRef]

9. Ren, W.-X.; Harik, I.E.; Blandford, G.E.; Lenett, M.; Baseheart, T.M. Roebling suspension bridge. II: Ambient testing and live-load response. *J. Bridge Eng.* **2004**, *9*, 119–126. [CrossRef]
10. Caetano, E.; Cunha, A.; Magalhaes, F.; Moutinho, C. Studies for controlling human-induced vibration of the Pedro e Ines footbridge Portugal. Part 1 Assessment of Dynamic Behaviour. *Eng. Struct.* **2010**, *32*, 1069–1081.
11. Hong, A.L.; Ubertini, F.; Betti, R. Wind analysis of a suspension bridge: Identification and finite-element model simulation. *J. Struct. Eng.* **2011**, *137*, 133–142. [CrossRef]
12. Van Nimmen, K.; Lombaert, G.; De Roeck, G.; Van den Broeck, P. Vibration serviceability of footbridges: Evaluation of the current codes of practice. *Eng. Struct.* **2014**, *59*, 448–461. [CrossRef]
13. Tubino, F.; Carassale, L.; Piccardo, G. Human-induced vibrations on two lively footbridges in Milan. *J. Bridge Eng.* **2016**, *21*, C4015002. [CrossRef]
14. Lai, E.; Gentile, C.; Mulas, M.G. Experimental and numerical serviceability assessment of a steel suspension footbridge. *J. Constr. Steel Res.* **2017**, *132*, 16–28. [CrossRef]
15. Lackny, L.; Scislo, L.; Guinchard, M. Application of Probabilistic Power Spectral Density Technique to Monitoring the Long-Term Vibrational Behaviour of CERN Seismic Network Stations. *Vib. Phys. Syst.* **2020**, *31*, 2020311.
16. Brincker, R.; Zhang, L.M.; Andersen, P. Modal identification from output-only systems using frequency domain decomposition. *Smart Mater. Struct.* **2001**, *10*, 441–445. [CrossRef]
17. Kordestani, H.; Xiang, Y.-Q.; Ye, X.-W.; Yun, C.-B.; Shadabfar, M. Localization of damaged cable in a tied-arch bridge using Arias intensity of seismic acceleration response. *Struct Control Health Monit.* **2020**, *27*, e2491. [CrossRef]
18. Van Overschee, P.; De Moor, B. *Subspace Identification for Linear Systems: Theory, Implementation, Applications*; Springer: New York, NY, USA, 1996.
19. Cao, M.S.; Sha, G.G.; Gao, Y.F.; Ostachowicz, W. Structural damage identification using damping: A compendium of uses and features. *Smart Mater. Struct.* **2017**, *26*, 043001. [CrossRef]
20. Casas, J.R.; Moughty, J.J. Bridge damage detection based on vibration data: Past and new developments. *Front. Built Environ.* **2017**, *3*, 1–12. [CrossRef]
21. Racic, V.; Pavic, A.; Brownjohn, J.M.W. Experimental identification and analytical modelling of human walking forces: Literature review. *J. Sound Vib.* **2009**, *326*, 1–49. [CrossRef]
22. Živanović, S.; Pavic, A.; Reynolds, P. Vibration serviceability of footbridges under human-induced excitation: A literature review. *J. Sound Vib.* **2005**, *279*, 1–74. [CrossRef]
23. Brownjohn, J.M.W.; Pavic, A.; Omenzetter, P. A spectral density approach for modelling continuous vertical forces on pedestrian structures due to walking. *Can. J. Civ. Eng.* **2004**, *31*, 65–77. [CrossRef]
24. Piccardo, G.; Tubino, F. Simplified procedures for vibration serviceability analysis of footbridges subjected to realistic walking loads. *Comput. Struct.* **2009**, *87*, 890–903. [CrossRef]
25. Caprani, C.C.; Keogh, J.; Archbold, P.; Fanning, P. Enhancement for the vertical response of footbridges subjected to stochastic crowd loading. *Comput. Struct.* **2012**, *102–103*, 87–96. [CrossRef]
26. Chen, J.; Wang, J.; Brownjohn, J.M.W. Power spectral density model for pedestrian walking load. *J. Struct. Eng.* **2019**, *145*, 04018239. [CrossRef]
27. Wang, J.; Chen, J.; Yokoyama, Y.; Xiong, J. Spectral model for crowd walking load. *J. Struct. Eng.* **2020**, *146*, 04019220. [CrossRef]
28. SETRA. *Footbridges–Assessment of Vibrational Behaviour of Footbridges under Pedestrian Loading*; Technical Department for Transport, Roads and Bridges Engineering and Road Safety, Ministry of Transport and Infrastructure: Paris, France, 2006.
29. De Roeck, G.; Peeters, B. *MACEC2. 0—Modal Analysis on Civil Engineering Constructions*; Department of Civil Engineering, Catholic University of Leuven: Leuven, Belgium, 1999.
30. FIB. *Guidelines for the Design of Footbridges*; Bulletin 32, International Federation for Structural Concrete: Lausanne, Switzerland, 2005.
31. Alampalli, S. Effects of testing, analysis, damage, and environment on modal parameters. *Mech. Syst. Signal Process.* **2000**, *14*, 63–74. [CrossRef]
32. ANSYS. *Academic Teaching Mechanical User Manual*; Release 2022 R1; ANSYS: Canonsburg, PA, USA, 2022.
33. Brownjohn, J.M.W. Observations on non-linear dynamic characteristics of suspension bridges. *Earthq. Eng. Struct. Dyn.* **1994**, *23*, 1351–1367. [CrossRef]
34. Bruno, L.; Venuti, F.; Scotti, A. Limit of hanger linearity in suspension footbridge dynamics: A new section model. *J. Sound Vib.* **2011**, *330*, 6387–6406. [CrossRef]
35. Maia, N.M.M.; Silva, J.M.M. *Theoretical and Experimental Modal Analysis*; Research Studies Press: Taunton, UK, 1997.
36. Allen, D.E.; Murray, T.M. Design criterion for vibrations due to walking. *Eng. J.* **1993**, *30*, 117–129.
37. Bachmann, H.; Ammann, W. *Vibrations in Structures Induced by Man and Machines*; Iabse: Zurich, Switzerland, 1987; Volume 3.
38. Piccardo, G.; Tubino, F. Dynamic response of Euler-Bernoulli beams to resonant harmonic moving loads. *Struct. Eng. Mech.* **2012**, *44*, 681–704. [CrossRef]
39. Banas, A.; Jankowski, R. Experimental and numerical study on dynamics of two footbridges with different shapes of girders. *Appl. Sci.* **2020**, *10*, 4505. [CrossRef]

40. Bayat, E.; Bayat, M.; Hafezzadeh, R. Numerical performance assessment of Tuned Mass Dampers to mitigate traffic-induced vibrations of a steel box-girder bridge. *Struct. Eng. Mech.* **2021**, *78*, 125–134.
41. Van Nimmen, K.; Verbeke, P.; Lombaert, G.; De Roeck, G.; Van Den Broeck, P. Numerical and experimental evaluation of the dynamic performance of a footbridge with tuned mass dampers. *J. Bridge Eng.* **2016**, *21*, C4016001. [CrossRef]

buildings

Article

Experimental Study and Numerical Simulation Analysis on Vertical Vibration Performances of 12 m Span Wood Truss Joist Floors

Zhanyi Zhang [1], Shuangyong Wang [2], Hao Deng [2] and Haibin Zhou [1,2,*]

[1] Research Institute of Wood Industry, Chinese Academy of Forestry, Beijing 100091, China
[2] National Engineering Research Center of Wood Industry, Beijing 102300, China
* Correspondence: zhouhb@caf.ac.cn

Abstract: Walking-induced vibration control in wood floors is a critical issue attracting the attention of many researchers and engineers. This paper presents an experimental study applying static deflection tests, modal tests, and pedestrian load tests to a series of full-scale 12 m span tooth plates connected to wood truss joist floors with strongbacks and partition walls. A comparison of the calculation error of vibration parameters between the theoretical formula and a numerical model was also conducted. The results show that strongbacks and partition walls effectively reduce both the vertical displacement and the root means acceleration at the center of the floor under pedestrian load but increases the natural frequency. The partition wall can achieve a better vibration-reduction effect than strongbacks. The error of the finite element model is higher than that of the theoretical formula. Using the theoretical formula in engineering wood floor design is recommended.

Keywords: wood floor; floor vibration; vibration serviceability; numerical methods

Citation: Zhang, Z.; Wang, S.; Deng, H.; Zhou, H. Experimental Study and Numerical Simulation Analysis on Vertical Vibration Performances of 12 m Span Wood Truss Joist Floors. *Buildings* **2022**, *12*, 1455. https://doi.org/10.3390/buildings12091455

Academic Editors: Jun Chen and Haoqi Wang

Received: 25 July 2022
Accepted: 2 September 2022
Published: 15 September 2022

1. Introduction

A raised wood floor system is generally designed in residential low-rise construction to elevate the living space off the ground or downstairs with the benefit of a high degree of industrial prefabrication. The vertical vibration performance of wooden floors is essential to residential building quality. However, raised wood floors are sensitive to residents' daily activity or other dynamic loads, and annoying vibrations arise from their low mass compared with steel or concrete floors. Research shows that if the natural vibration frequency of the floor is between 4 and 8 Hz, residents will feel discomfort and anxiety due to the similar resonances of the wood floor with organs of people, which affects the comfort and livability of wooden buildings [1]. Currently, the serviceability design of wood-framed floors is usually based on limiting the relevant parameters such as deflection, acceleration, natural frequency, or their combinations [2]. This approach is practical for vibration control of small and medium span floors; whether it suits large span floors remains unclear. Therefore, it is necessary to analyze the vertical vibration performances of large-span wooden floors through relevant experimental studies and summarize a general design method for the serviceability of wooden floors.

It is necessary to evaluate wood floor theoretical models with sufficient field test data to study the dynamic performance of wood floors. Previous research studies related to full-scale tests are as follows. Khokhar et al. [3] conducted experimental tests on 4.2 m laminated veneer lumber (LVL) joist wood floors and compared different types of between-joists bracing on the effects of vibrational serviceability. Jarnerö et al. [4] assessed the dynamic performances of 5.1 m wood floors experimentally in the laboratory with different boundary conditions and in field tests at different stages of construction. Weckendorf et al. [5] presented an experimental study of low amplitude dynamic responses on 5.5 m cross-laminated-timber (CLT) floors. Ding et al. [6] conducted vibration tests on 6 m spruce-pine-fir (SPF) timber joist floors. Wang [7] investigated the structural behavior of 6.1 m two-way wood truss floors. Zhou et al. [8,9] analyzed the vibration performances of 4.7 m solid lumber joist floors and 8.26 m engineered wood truss joist floors. Xue [10] presented an experimental study on 6 m wooden truss joist floors. Rajendra Rijal et al. [11] compared the modal behaviors of 6 m and 8 m timber floors. Studying other factors related to the vibration performance of wood floors is also necessary. Onysko et al. [12] conducted massive vibrational serviceability tests on floors with a span of less than 10 m. Foy Cdric et al. [13] conducted modal tests on two 4 m wood floors in free boundary conditions and built a numerical model to carry out the parametric study. Fuentes et al. [14] presented an experimental study of a 7.2 m wood floor. Xue et al. [15] studied the effects of joist spacing and bracing elements on 6 m wood truss joist floors. Persson et al. [16] analyzed the influence of uncertain parameters on the modal properties of 7.2 m plywood truss floors. Yujian Dong and Lilin Cao. [17] proposed a model to determine the human-induced response of a 9 m steel-wood composite floor. Zhang and Yang [18] compared loading methods on floor vibration due to individual walking styles. Sepideh Ashtari [19] analyzed the difference between the rigid and flexible connections of a 10.8 m CLT floor.

The classical measurement approach using a modal hammer in the case of experimental modal analysis is time-consuming and laborious. Some modern and contactless methods have been developed. LukaszScislo [20] applied a 3D scanning vibrometry system, a non-contact measurement method, to obtain natural frequencies and modal shapes of ultra-light structures. The results show that this system is helpful for modal analysis of high fragility and low weight structures without contact by using the excitation of the loudspeaker. Emilio Di Lorenzo et al. [21] investigated the use of digital image correlation (DIC) for modal analysis. DIC is a non-contact full-field image analysis technique that uses high-speed and high-resolution cameras to measure structures' strains and displacements to derive the structure's modal characteristics.

Vibration serviceability research of wood floors is usually concerned with trusses ranging from 3 m to 12 m. However, the static and dynamic performances related to the vibration serviceability of wood truss joist floors longer than 9 m have not been clarified to date. This paper analyzes the vertical vibration performances of 12 m wood truss floors by field tests. Tests at this scale are rare in related research. It also discusses the effects of strong-backs and partition walls on vibration responses. Laboratory studies based on numerical simulations are used to improve our understanding of the complexities of the vibration response of large span raised wood floor systems. A 12 m finite element model of a wood floor is built to predict modal behaviors and unit point load deflection. The simulation results are compared with theoretically predicted results to evaluate the finite model of wood floors. This paper intends to contribute to understanding the vibration performances of large-span wood floors for future vibration serviceability research and the engineering application of large-span wood truss joist floors.

2. Overview of the Wood Floor and Test Methods

2.1. Floor Configurations

The floor was designed according to the standard of the Canadian National Building Code with a deflection of not more than L/360 under a uniform load of 1.9 kPa. Based on the edge of the surrounding wall, the design length of the floor was 12.11 m, the design width was 6.09 m, and it was built on a wall with a height of 1.85 m, as shown in Figure 1. The wall frame material was SPF material, covered with OSB board, and the walls were assembled with 50 mm nails at intervals of 300 mm. The wooden truss joists consisted of J-level SPF material, the section size was 38 mm × 89 mm, and the top and bottom chords were connected by SPF finger joints and glued in the thickness direction. Tooth plates connected the nodes of the truss. The dimensions of the truss and the tooth plates are shown in Figure 2. The 21 trusses were arranged on the wall with a spacing of 300 mm. The ends of each truss were nailed obliquely with the top plate of the wall by two 125 mm drill-tail screws. The rim boards were made of LVL, 38 mm thick and 500 mm high. The rim board and the truss were vertically nailed with three 70 mm drill-tail screws. The rim boards and the outermost wooden truss joists were placed on the wall, which was obliquely nailed symmetrically on both sides at intervals of 600 mm with the top plate of the wall by 90 mm self-tapping screws. The details of nail connections are shown in Figure 3. The sheathing material was 15 mm OSB board, 2440 mm long and 1220 mm wide. The major direction of the OSB sheathing was placed perpendicular to the joist. The OSB subfloor was arranged half-staggered from each other, as shown in Figure 4. The OSB subfloor was connected to the joist using 50 mm screws of 150 mm around the perimeter and 300 mm in the field. The distance between the screws and the OSB subfloor edge was greater than 10 mm.

Figure 1. Construction of test floor.

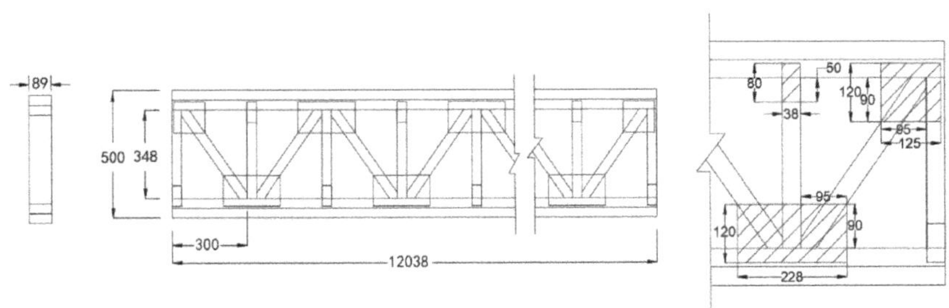

Figure 2. Dimensions of timber truss (unit: mm).

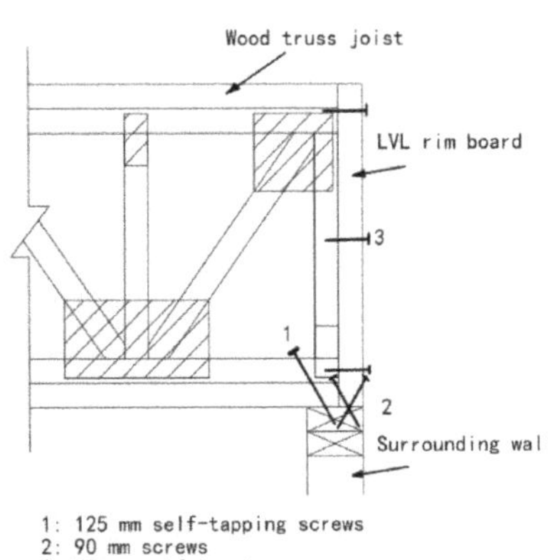

Figure 3. Connection between the floor and the wall.

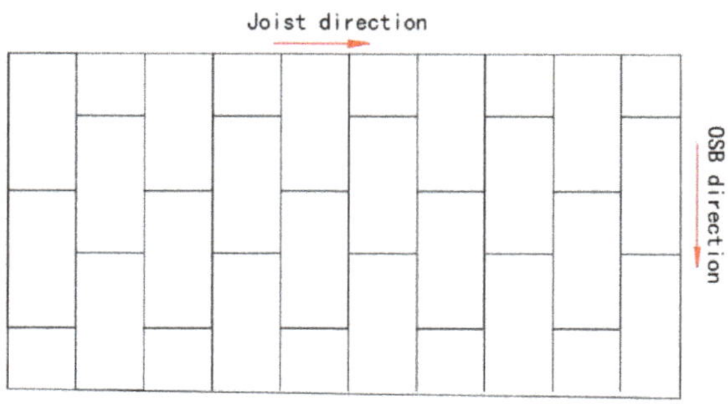

Figure 4. OSB layout.

Different test floors were designed by installing strongbacks or partition walls into the initial floor system. The information for each group of test floors is shown in Table 1, and the schematic diagram is shown in Figure 5. The strongbacks were constructed from 40 mm × 140 mm cross-sectional size, J-level SPF specification material selected by visual inspection. It is installed in the vertical direction of the joists and connected to the truss web rod with three 80 mm wood screws from the top, middle, and bottom in the height direction of the strongbacks, as shown in Figure 6. The door on the surrounding wall could transport the partition wall. It is connected with the interface at the corresponding position on the surrounding walls, as shown in Figure 7.

Table 1. Details of test floor structures.

Floor	Details
T1	No strongbacks and partition wall
T2	Double strongback rows at mid-span
T3	Double strongback rows at mid-span and one strongback row each at one-sixth span and one-third span
T4	One partition wall each at one-fourth span
T5	One partition wall at mid-span

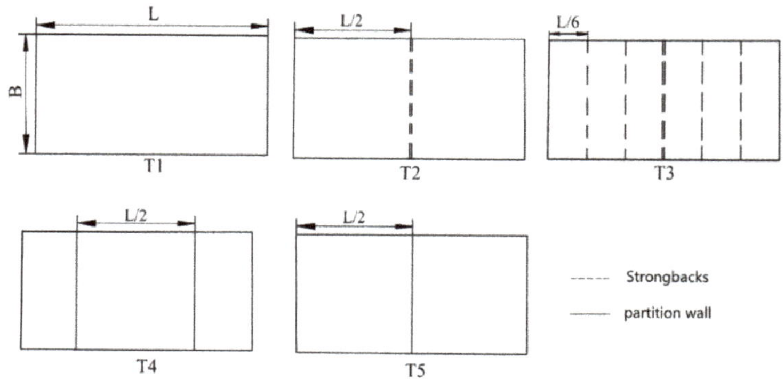

Figure 5. Schematic representation of test floors.

The studs of the partition wall and the surrounding wall were fastened with two 90 mm drill screws at intervals of 600 mm, and the top and bottom plates of the partition wall and the surrounding wall were fastened with two 50 mm drill screws.

Figure 6. Double strongbacks rows at mid-span.

Figure 7. Partition wall.

2.2. *Test and Analysis Methods*

2.2.1. Test Methods

1. Static load deflection test

The static load deflection of 1 kN is a common parameter in the vibration design of wooden floors. The test apparatus included a Mitutoyo brand ID-C150XB dial indicator with an accuracy of 0.001 mm and a range of 0 to 50.8 mm, seven 180 cm steel hangers, and 1 kN weights consisting of five 20 kg iron discs. The layout of the test measuring points is shown in Figure 8. Except for T5, which was measured at the quarter-span marked as line B, the rest of the floors were measured at the mid-span marked as line A. The measuring point was the intersection of the measuring line and the joist, and the loading point coincides with the measuring point on the J10 joist.

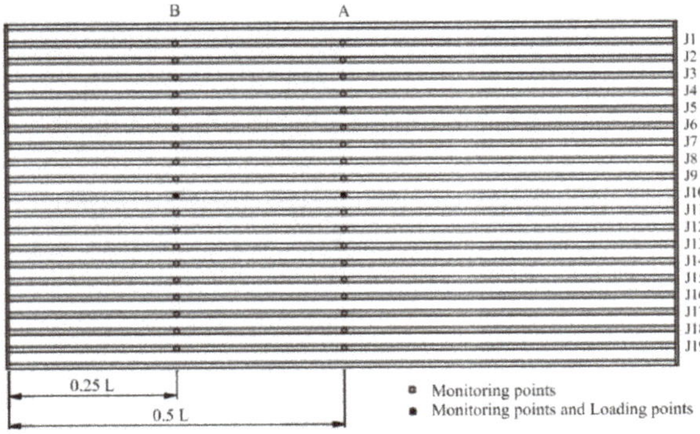

Figure 8. Distribution of the monitoring points and loading points.

There were 21 joists on the test floor, except for the two joists nailed to the wall on both sides. There was a total of 19 monitoring points A maximum of 7 points could be measured simultaneously, so the measurements were performed three times. The measurement process is shown in Figure 9. First, seven steel hangers were placed under the monitoring point, and dial indicators were fixed on each steel hanger, ensuring that the dial indicator probe was in contact with the monitoring point marked on the lower surface of the joist. The zero-adjustment operation was then performed. After that, the test personnel moved the heavy object to the loading point and left the floor. The display data was recorded when it was stable and summarized as the static load deflection curve of the 1 kN concentrated load at each joist measuring point of the floor.

Figure 9. 1 kN static load deflection test.

2. Modal test

The modal test apparatus included the INV9314 test hammer (COINV Inc., Beijing, China) with a reference sensitivity of 50.5 uV/N, the INV3020C signal acquisition and analyzer with the DASP-V10 software platform (COINV Inc., Beijing, China) that can collect and analyze vibration signals, and four INV9828 piezoelectric acceleration sensors (COINV Inc., Beijing, China) with a sensitivity of 50 mV·s^2/m, and a range of 100 m/s^2.

A total of 20 × 6 = 120 monitoring points were marked on the floor, as shown in Figure 10. The bottom of the four sensors was then coated with beeswax to ensure they could connect closely with the floor surface and placed on measuring points 1–4. An excitation point was at the center of the building, and if the excitation effect was not clear, it was selected at the quarter span. A rubber head hammer was used to excite the floor three times to obtain the acceleration response under the excitation action. The sensor was then moved to measuring points 5–8, the floor was excited three times, and the process was repeated until all the points were measured. The test personnel pounded the floor on the prefabricated wooden beams to ensure that there was no additional mass on the floor, as shown in Figure 11. The collected acceleration responses were processed by fast Fourier transform (FFT) to obtain the frequency response function (FRF), and then the first three order natural frequencies, damping, and mode shapes of the floor were extracted from the FRF.

3. Pedestrian load test

The pedestrian load test apparatus was the same as the modal test. The walking paths for each group of floors are shown in Figure 12. The testers weighed 85 kg, and a metronome was used to adjust the walking frequency to around 2 Hz. By walking along three paths, horizontal (H), vertical (S), and oblique (X), walking excitation was applied to the floor. In order to prevent the test personnel from accidentally touching the sensor while walking, the walking path was set at 60 cm wide. The sensor was placed at point A on the central joist of the floor. When the tester started to walk, the data was recorded by the signal acquisition analyzer, and the recording stopped when the tester reached the end of the path, as shown in Figure 13. From the pedestrian load test, the time-history curve of the acceleration response of the floor under walking load was obtained, and the root-mean-square (RMS) acceleration at the center of the floor was determined.

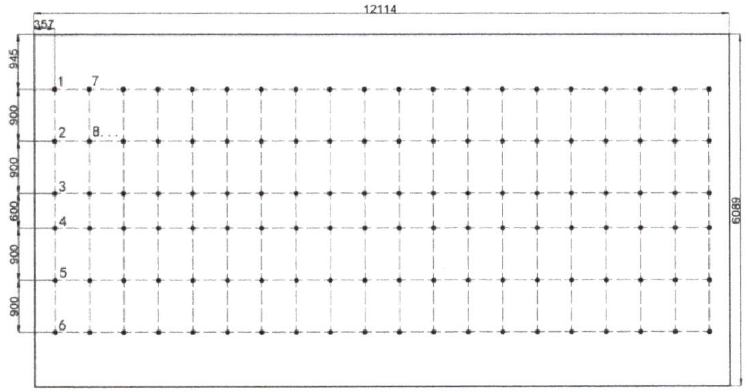

Figure 10. Distribution of the monitoring points (unit: mm).

Figure 11. The modal test.

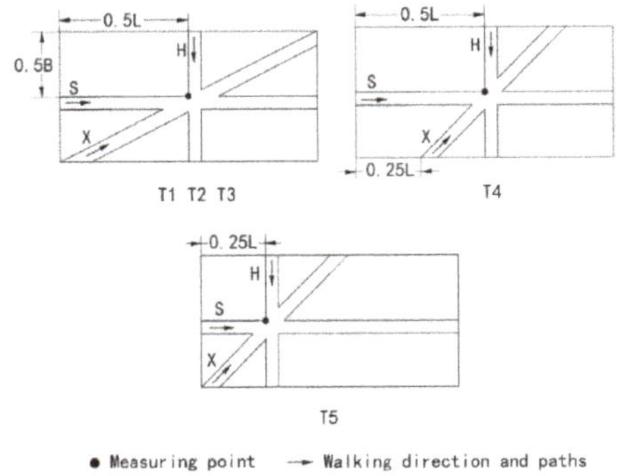

Figure 12. Layout plan of pedestrian load test.

Figure 13. Pedestrian load test.

2.2.2. Analysis Methods

1. Finite element method

The finite element model of the 12 m wood truss floor consists of the truss and the sheathing. Each truss member adopts the Beam 188 element, and the sheathing adopts the Shell 181 element. The material parameters of the SPF and OSB sheathing are shown in Table 2. All the nodes are set as hinges. The common boundary conditions of the floor include simply supported on four sides, fixed on four sides, and simply supported on both sides and fixed on both sides. The nail connection between the ends of the truss and the wall can be seen as a half-stiffness node. So, it is used in the numerical model that the fixed support on the edge joists and the simply supported connection between the other joists. 1 kN concentrated load is applied at the center of the floor.

Table 2. Parameters of the engineered wood materials.

Properties	SPF	OSB
	38 mm × 89 mm	15 mm Thickness
ρ (kg/m^3)	497	650
ν	0.49	0.45
E_L (MPa)	8700	4280
E_R (MPa)	900	2080
E_T (MPa)	700	20.8
G_{LR} (MPa)	500	1000
G_{LT} (MPa)	500	50
G_{RT} (MPa)	30	50

2. Theoretical method

The test floor is simplified as in Figure 14. Based on the Timoshenko ribbed plate theory, the first-order vibration frequency of the wooden floor and the deflection of the wooden floor under the concentrated load at the center of the slab can be calculated. This paper adopts the first-order natural frequency calculation formula from [22] and the deflection calculation formula from [23].

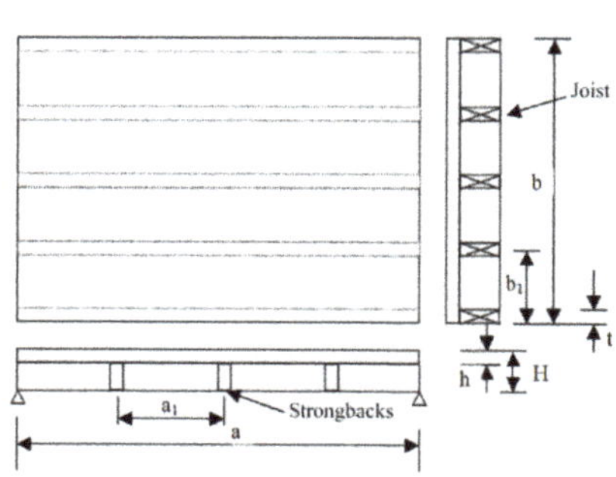

Figure 14. Timber floor ribbed slab model.

3. Results and Discussion

3.1. Influence of Strongbacks and Partition Wall on the Vertical Deflection of the Floor

The 1 kN concentrated load deflection of each group of floors is shown in Figure 15. Among them, the 1 kN static load deflection of T1 was the largest (1.376 mm), which is less than L/250, and meets the bending deflection limit for floor beams according to GB 50005-2017 "Standards for Design of Timber Structures" [24]. The deflection of the T2 floor was 39.3% lower than that of the T1 floor. It shows that the strongbacks can significantly increase the stiffness of the floor. GB 50005-2017 [24] also stipulates that the spacing of the strongbacks in the span direction should not be greater than 2.1 m. It can be seen from Figure 15 that the static load deflection of T3, which meets this condition, is further reduced compared with T2 after installing two rows of strongbacks. Compared with T1, the deflection of T3 decreased by 47.5%. For the long-span wooden floor, when the distance between the strongbacks was not more than 2.1 m, the rigidity of the floor was significantly improved compared to when the strongbacks were not installed. Compared with T1, the deflection of T4 decreased by 61.3%, indicating that installing a partition wall can significantly increase the stiffness of the floor. Compared with T1, the deflection of T5 decreased by 64.4%, which was similar to T4. It may be related to the space with a span of 6 m divided by the partition wall.

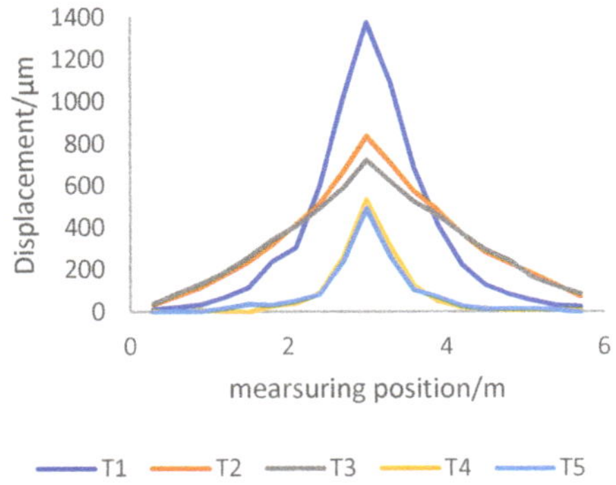

Figure 15. The measured deflection of test floors.

3.2. *Influence of Strongbacks and Partition Walls on Modal Performances of Floors*

As shown in Figure 16, the first three vibration modes of the five groups of floors were similar, up and down in the form of sine waves along the width direction. The first three natural frequencies and damping of the five groups of floors are shown in Table 3. The first-order vibration frequency of T1 is 6.8 Hz, less than 8 Hz, which is the limit to vibration comfort. Compared with T1, the first-order frequencies of T2 and T3 increased by 3.0% and 6.9%, respectively, the second-order frequencies increased by 21.6% and 36.0%, respectively, and the third-order frequencies increased by 55.2% and 80.3%, respectively. These results indicate that the installation of strongbacks improves the first three frequencies of the floor, and the effect on the second and third frequencies is noticeable. The test results are consistent with the installation of strongbacks for 6 m-span wood truss joist floors [10]. It was demonstrated that the first-order frequency of the floor is mainly controlled by the stiffness of the floor parallel to the joist direction. After installing the partition wall, the first three natural frequencies of the floor were significantly improved, and the spacing of adjacent order frequencies was also improved. It indicates that partition wall increases the stiffness of the floor and improves vibration comfort. Compared with T1, the first-order frequency of T5 was increased by 77.6%, which indicates that the installation position of the partition wall has a different effect on the stiffness of the floor, and the partition wall in the middle has the best effect on the natural vibration frequency. Compared with T4, the first-order frequency of T5 was reduced by 2.5%. The maximum span of T4 and T5 divided by the partition wall was 6 m. At the same time, the first-order frequency of T4 and T5 was close, indicating that the maximum distance between the walls perpendicular to the joist direction affects the first-order frequency of the floor. Compared with T4, the second and third order frequencies of T5 increased by more than 10%, indicating that the increase of longitudinal stiffness by installing a partition wall at a 6 m span is better than at 3 m and 9 m spans.

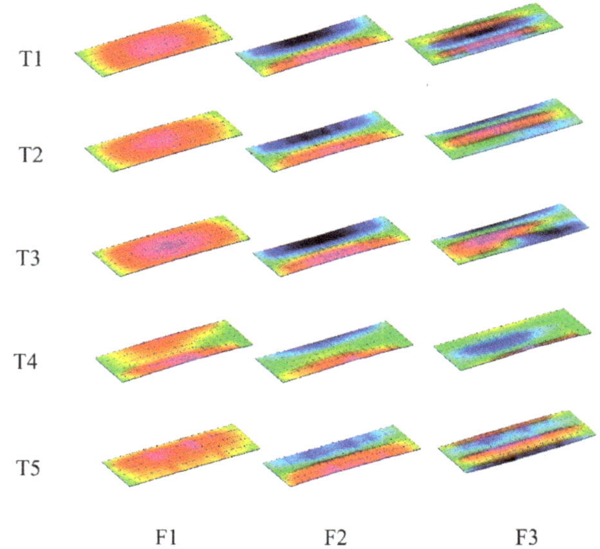

Figure 16. First three mode shapes obtained from the experiment.

Table 3. The first three-order natural vibration frequencies and damping of each test floor.

Floor	f_1 (Hz)	ζ_1 (%)	f_2 (Hz)	ζ_2 (%)	f_3 (Hz)	ζ_3 (%)
T1	6.8	1.3	8.1	0.8	9.1	1.3
T2	7.1	1.4	9.9	1.4	14.1	1.9
T3	7.3	0.9	11.1	2.0	16.4	1.4
T4	12.5	3.5	15.1	2.4	16.8	1.9
T5	12.2	3.4	16.8	3.5	19.4	1.7

3.3. Influence of Strongbacks and Partition Wall on Pedestrian Load Response of Floor

The pedestrian load consists of a series of single footfall loads. The time history curve of floor vibration response is obtained by recording the vibration response caused by pulse excitation under a continuous walking load. The time history curve of vibration acceleration response at the center of the floor is shown in Figures 17–21 for each floor along three walk paths. When the natural frequency of the floor was greater than 8–10 Hz, the floor produced transient vibration and decayed rapidly. The amplitude of transient vibration is related to the stiffness and quality of the floor. When the natural frequency of the floor is less than 8–10 Hz, the floor may produce resonance, and the amplitude is related to damping. From the time history curve, it can be found that T1, T2, and T3 floors with natural frequencies less than 10 Hz had obvious resonance under walking load. The acceleration response raised by the pedestrian load remained at a high level even at the beginning of walking, then slowly reduced after the person stopped walking. The T4 and T5 floors with natural frequencies greater than 10 Hz generated transient vibration under walking load and decayed rapidly. The acceleration response was high only when the foot fell, and the peak acceleration increased fast when walking past the sensor.

The RMS acceleration of central vibration of all floors along three walking paths is shown in Table 4. It shows that the RMS vibration acceleration at the center of each floor was greater than that in the H direction when walking in the S direction. When walking in the three paths, the RMS vibration acceleration at the center of the T1, T2 and, T3 floors was about twice that of T4 and T5 floors. The main reason is that the natural vibration frequencies of these three groups of floors were low, and the floors were resonant under walking excitation. Compared with the pedestrian load test results of the same type of

6 m-span timber floor, the RMS acceleration of a 12 m-span floors in all directions was reduced by more than 60% [10]. However, the level of RMS acceleration of all wood floors is too large to satisfy the vibration serviceability Standard in ISO 10137 [25].

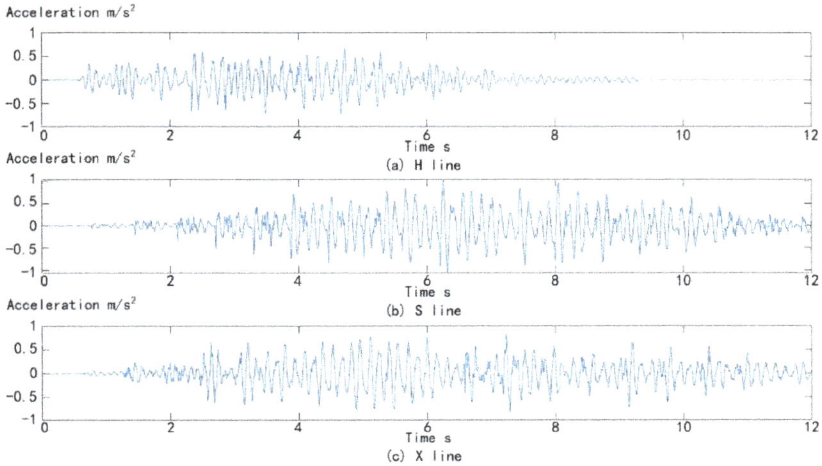

Figure 17. Acceleration response of T1 floor.

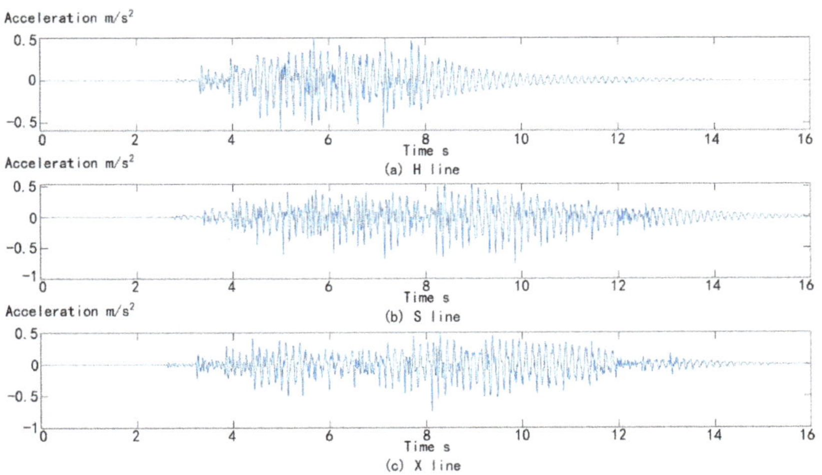

Figure 18. Acceleration response of T2 floor.

Table 4. RMS acceleration of the floor under a single person walking load (unit: m/s^2).

Floor	H Line	S Line	X Line
T1	0.176	0.255	0.197
T2	0.111	0.131	0.123
T3	0.127	0.170	0.157
T4	0.051	0.070	0.050
T5	0.045	0.064	0.051

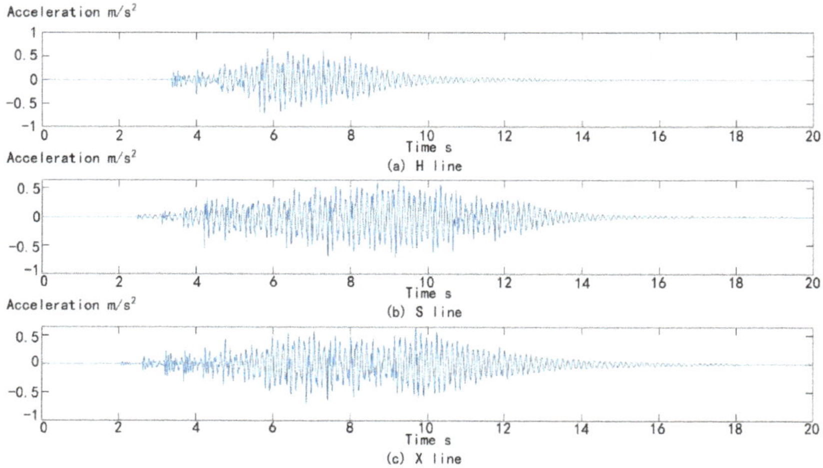

Figure 19. Acceleration response of T3 floor.

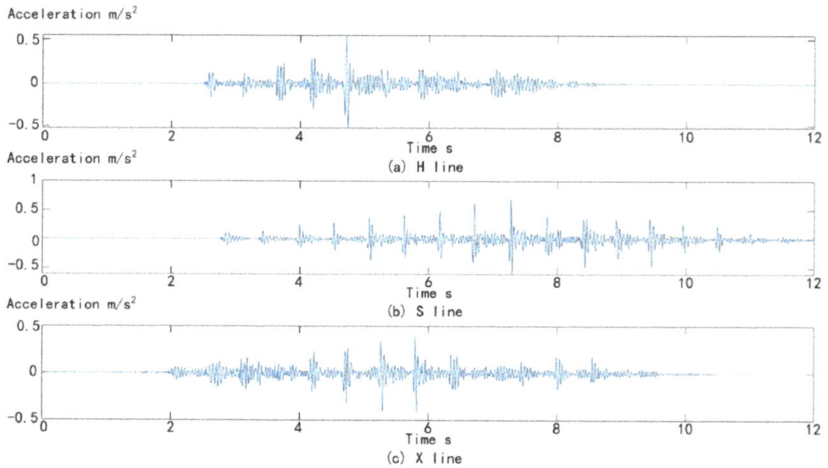

Figure 20. Acceleration response of T4 floor.

After two rows of strongbacks were installed in the center of the floor span, the RMS vibration acceleration at the center of the floor decreased when walking along the three paths. It was related to the increase in floor stiffness after the installation of strongbacks. When the strongbacks were installed two meters from the center of the floor span, the RMS vibration acceleration at the center of the floor did not decrease but increased. In summary, strongbacks in the center of the floor span reduce the RMS vibration acceleration at the center of the floor. It decreases the transient vibration response of the floor caused by the walking load. For floors with lower natural frequencies, the priority is to increase the natural frequency and then find other ways to reduce the vibration response.

After the installation of the partition wall, the RMS acceleration of vertical vibration of the floor under walking load was significantly reduced. Compared with T1, the RMS vibration acceleration at the center of the T4 floor decreased by 70.8% when walking in the H direction, 72.4% when walking in the S direction, and 74.8% when walking in the X direction. Compared with T1, the RMS vibration acceleration at the center of the T5 floor decreased by 74.4% when walking in the H direction, 75% when walking in the S direction, and 74.3% when walking in the X direction. The different positions of the partition wall had different effects on the improvement of the fundamental natural frequency of the floor, so the reduction of the RMS acceleration was also different. By comparison, installing the partition wall in the center of the floor span was more effective at reducing the RMS vibration acceleration in the center of the floor. The natural frequencies of T4 and T5 were similar, and there was little difference in the RMS vibration acceleration in the center of the floor when walking in various directions. It shows that the RMS vibration acceleration at the center of the floor may be related to the maximum distance between the walls perpendicular to the joist direction. The larger the distance, the smaller the RMS vibration acceleration at the center of the floor.

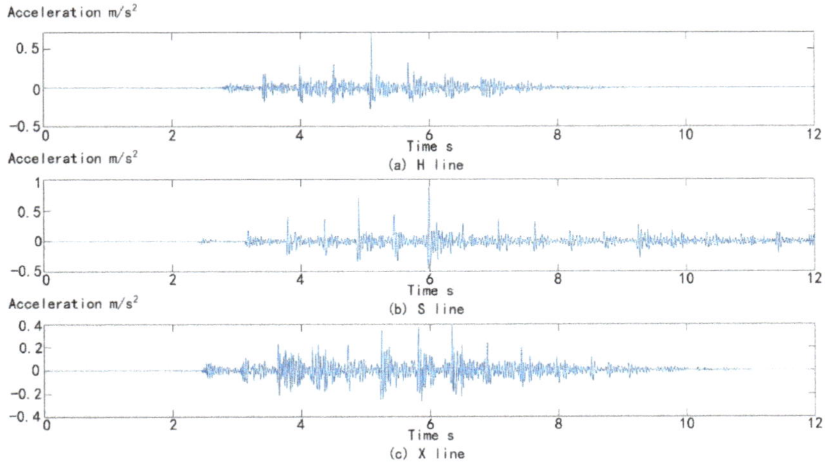

Figure 21. Acceleration response of T5 floor.

3.4. *A Comparison of the Numerical Simulation and Theoretical Results*

The predicted value of T1 deflection can be obtained by the finite element software static analysis and theoretical formula solution. The comparison between the two predicted values and the measured value is shown in Table 5. It shows that the theoretical prediction value of 1.59 mm is 15% higher than the measured value of 1.376 mm, which may be attributed to the calculation formula being different from the actual floor. The boundary conditions of the theoretical model assume simply supported on four sides, whereas the outermost joist, in actuality, is nailed to the surrounding wall. Therefore, the boundary conditions of the existing wood floor system are complex. The tooth plate joints of the actual wood truss joist are semi-rigid connections, so the overall stiffness of the floor is higher than theoretically predicted, which leads to the theoretically predicted deflection being large.

Table 5. Comparison between the measured and predicted deflection values under 1kN static load at the center of the T1 floor.

Predicted Values (mm)		Experimental Value (mm)
Theoretical Predicted Value	**Simulation Value**	
1.59	1.79	1.376

3.4.1. Simulation of 1 kN Static Load

The simulation results of the deflection of the T1 floor under the concentrated load of 1 kN acting on the floor center are shown in Figure 22. The deflection at the center of the floor is the largest, and the calculated result is 1.79 mm, which is 30% higher than the measured value. The error is larger than the experimental value, which is similar to the floor simulation results of Shen [26]. The deflection distribution is oval. Parallel to the span direction, the deflection decays slowly faster when vertical to the span direction. The simulation predicted value of the floor is greater than the theoretical predicted value, and there is a substantial difference from the measured value. Therefore, the model could be further optimized by increasing the rim board model and refining the truss joints from hinged joints to semi-rigid connections.

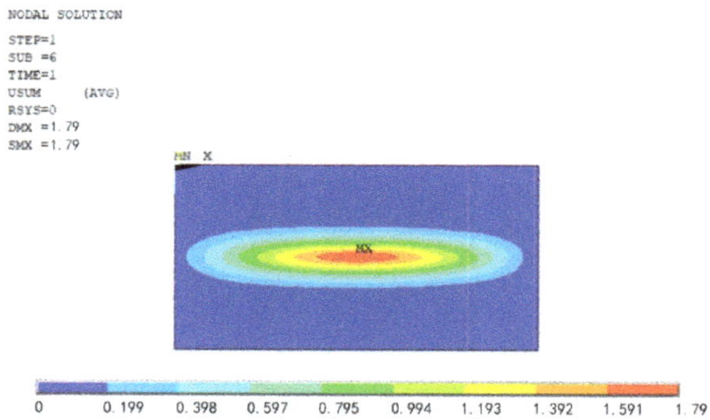

Figure 22. Vertical displacement distribution of timber floor model (unit: mm).

3.4.2. Modal Test Simulation

The first three vibration modes of the finite element floor are shown in Figure 23, and the first three natural frequencies are 5.88 Hz, 6.03 Hz and 6.37 Hz. The theoretical predicted value of the first-order natural frequency of the floor can be calculated. A comparison of the predicted and measured values of the first-order natural frequency of the floor is shown in Table 3. It can be seen from Table 6 that the predicted values obtained by the two methods are lower than the measured values. There is little difference between the theoretical predicted values and the measured values, indicating that it is feasible to estimate the first-order vibration frequency of the floor by using the theoretical formula. The first three modes of the simulated floor coincide with the measured floor. The difference between the predicted value and the measured value of the first-order frequency is 13.5%, which is close to the error of the finite element model of the 6 m wooden truss joist floor established by Shen [26]. However, the error of the second-order and third-order natural frequency values is more than 20%. Therefore, when the boundary conditions are the fixed support on the edge joists and the simply supported connection between the other joists, the modal performance of the floor can be predicted. However, the accuracy of the first three natural frequencies is still unacceptable.

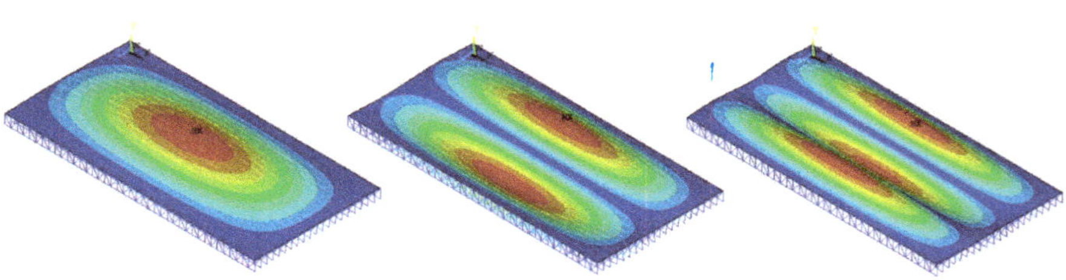

Figure 23. The first three modes of the numerical model of the floor.

Table 6. Comparison between the measured and predicted values of the first-order natural frequency of wood floors.

Predictive Value (Hz)		Experimental Value (Hz)
Theoretical Predictive Value	Simulation Value	
6.47	5.88	6.8

4. Conclusions

In this paper, laboratory tests including static load deflection tests, modal tests, and pedestrian load tests were conducted to reveal the vibration performances of 12 m wood truss joist floors. The selective action law of strongbacks and partition walls on the vibration performance of large span wood floors is discussed. Combined with numerical simulation, the prediction accuracies of the finite element model and theoretical formula on the static load deflection and modal performance of the floor are compared.

The strongbacks increase the stiffness of the floor perpendicular to the joist direction, which can significantly increase the high-order natural frequency of the floor, but has little effect on the fundamental natural frequency. Installation of two rows of strongbacks in the midspan can significantly reduce the RMS vibration acceleration at the center of the floor under walking excitation. However, the RMS acceleration increases with increasing numbers of strongbacks. Installation of a partition wall perpendicular to the joist under the floor can improve the overall stiffness of the floor, and the effect on the fundamental natural vibration frequency and RMS acceleration of the floor is better than that of the strongbacks.

The 1 kN static load deflection of the 12 m span wood floor is 1. 376 mm. The prediction error of the theoretical calculation formula based on the ribbed plate model of the wooden floor is 15.6%, and the error of the finite element model is 30.1%. The fundamental natural vibration frequency of the 12 m span floor is 6.8 Hz, the error of theoretical calculation is 4.9%, and the error of the finite element model is 13.5%. In engineering applications, The vibration performances can be estimated by the theoretical formula and qualitatively analyzed by a finite element model.

This study presents a method to build a large-scale test platform of wood floors. This assessment of the vibration performances of wood floors will benefit the engineering application of large-span wood truss joist floors. Further research will be conducted on other methods to increase the stiffness of wood floors and the effect of different loads, such as multi-person walking loads, on the vibration performance of large span wood truss joist floors. A more accurate model also will be developed to predict the vibration responses of wood floors.

Author Contributions: Z.Z.: conceptualization, field measurements, numerical modelling, original draft preparation; S.W., H.D.: floor construction and field measurements; H.Z.: conceptualization, supervision, funding acquisition, review, and editing. All authors have read and agreed to the published version of the manuscript.

Funding: This study was funded by Inner Mongolia S&T Project (2022YFDZ0002) and National Natural Science Foundation (31770603).

Institutional Review Board Statement: Not applicable.

Informed Consent Statement: Not applicable.

Data Availability Statement: Not applicable.

Acknowledgments: The testing facility and material in this research were supported by the National Engineering Research Center of Wood Industry. These contributions are gratefully acknowledged. For the experimental study, the help of the following individuals is acknowledged: Du Yonghe, Wang Zhiyong, Han Xu, Wang Jiahui, Wang Duoyu.

Conflicts of Interest: The authors declare no conflict of interest.

References

1. Zhang, X.D.; Gao, J.; Yan, W.M. Effects of Environmental Vibration on Human Health. *J. Environ. Health* **2008**, *25*, 74–76.
2. Hu, L.J. Design guide for wood-framed floor systems. *Can. For. Serv.* **2007**, *32*, 1–60.
3. Khokhar, A.; Chui, Y.; Smith, I. Influence of stiffness of between-joists bracing on vibrational serviceability of wood floors. In Proceedings of the World Conference on Timber Engineering, Auckland, New Zealand, 15–19 July 2012.
4. Jarnerö, K.; Brandt, A.; Olsson, A. Vibration properties of a timber floor assessed in laboratory and during construction. *Eng. Struct.* **2015**, *82*, 44–54. [CrossRef]
5. Weckendorf, J.; Ussher, E.; Smith, I. Dynamic response of CLT plate systems in the context of timber and hybrid construction. *Compos. Struct.* **2016**, *157*, 412–423. [CrossRef]
6. Ding, Y.W.; Fu, H.Y.; Wang, Z.; Wang, Y.L.; Xie, W.B.; Zhao, T.C.; Si, S.G. Vibration test and comfort analysis of wooden building floor structure under environmental and impact excitations. *Build. Struct.* **2020**, *50*, 100–106.
7. Wang, B. Structural Behavior of Two-Way Wood Truss Floors. Master's Thesis, Harbin Institute of Technology, Harbin, China, July 2012.
8. Zhou, H.B.; Jiang, Z.H.; Fei, B.H.; Ren, H.Q. Design methods to control vibration of solid lumber joist floors. *J. Chongqing Jianzhu Univ.* **2008**, *30*, 49–52.
9. Zhou, H.B.; Jiang, Z.H.; Fei, B.H.; Ren, H.Q.; Hu, L.J. Study on vibration performance of engineered wood truss joist floors. *J. Chongqing Jianzhu Univ.* **2008**, *30*, 109–113.
10. Xue, S. Study on the Vibration Performance of the Composite Floor Combined with Wooden Truss Joist. Master's Thesis, Chinese Academy of Forestry, Beijing, China, June 2019.
11. Rijal, R.; Samali, B.; Shrestha, R.; Crews, K. Experimental and analytical study on dynamic performance of timber floor modules (timber beams). *Constr. Build. Mater.* **2016**, *122*, 391–399. [CrossRef]
12. Onysko, D.; Hu, L.; Jones, E.; Lenardo, B.D. Serviceability design of residential wood-frame floors in Canada. *Wood Des. Focus* **2000**, *11*, 3–8.
13. Foy, C.; Monnier, J.; Auchet, S.; Dauchez, N. Vibro-acoustic characterization of wood floors and influence of the constructive principle. In Proceedings of the INTER-NOISE and NOISE-CON Congress and Conference Proceedings, Madrid, Spain, 16–19 June 2019; Volume 259, pp. 3064–3073.
14. Fuentes, S.; Fournely, E.; Bouchaïr, A. Experimental study of the in-plan stiffness of timber floor diaphragms. *Eur. J. Environ. Civ. Eng.* **2014**, *18*, 1106–1117. [CrossRef]
15. Xue, S.; Zhang, Z.Z.; Zhang, Z.Z.; Zhou, H.B.; Shen, Y.L. Effects of strongbacks and strappings on vibrations of timber truss joist floors. *Shock Vib.* **2021**, *2021*, 6630719. [CrossRef]
16. Persson, P.; Frier, C.; Pedersen, L.; Andersen, L.; Manuel, L. Influence of uncertain parameters on modal properties of wood floors. In *Advances in Engineering Materials, Structures and Systems: Innovations, Mechanics and Applications*; CRC Press: Boca Raton, FL, USA, 2019; pp. 127–132. [CrossRef]
17. Yujian, D.; Lilin, C. Acceleration response analysis of a steel-wood composite floor system under human-induced vibration. *J. Sci. Res. Rep.* **2021**, *27*, 60–71. [CrossRef]
18. Zhang, S.G.; Yang, N. Comparison of loading methods of floor vibration due to individual walking. *J. Beijing Jiaotong Univ.* **2013**, *37*, 152–156+161.
19. Sepideh, A. *In-Plane Stiffness of Cross-Laminated Timber Floors*; The University of British Columbia: Vancouver, BC, Canada, 2012; pp. 43–68.
20. LukaszScislo. Non-invasive measurements of ultra-lightweight composite materials using laser doppler vibrometry system. In Proceedings of the 26th International Congress on Sound and Vibration, Montreal, QC, Canada, 7–11 July 2019.
21. Di Lorenzo, E.; Mastrodicasa, D.; Wittevrongel, L.; Lava, P.; Peeters, B. Full-Field Modal Analysis by Using Digital Image Correlation Technique. In *Rotating Machinery, Optical Methods & Scanning LDV Methods*; Springer: Berlin/Heidelberg, Germany, 2020; Volume 6, pp. 119–130.
22. *ISO 18324*; Timber Structures-Test Methods-Floor Vibration Performance. ISO: Geneva, Switzerland, 2016.
23. Xiong, H.B.; Kang, J.H.; He, M.J. *Light Wood Frame Construction*, 1st ed.; Tongji Press: Shanghai, China, 2018; pp. 260–261.
24. *GB 50005-2017*; Standards for Design of Timber Structures. China Architecture & Building Press: Beijing, China, 2017.
25. *ISO 10137*; Bases for Design of Structures—Serviceability of Buildings and Walkways against Vibrations. ISO: Geneva, Switzerland, 2007.
26. Shen, Y.L.; Zhou, H.B.; Xue, S.; Zhang, J.C. A comparison on numerical simulation models for vibrational performances of the wood truss joist floor system. *Shock Vib.* **2021**, *2021*, 5560554. [CrossRef]

buildings

MDPI

Article

Finite Element Model for Vibration Serviceability Evaluation of a Suspended Floor with and without Tuned Mass Dampers

Zhihao Wang [1,*], Luyao Song [1], Zhipeng Cheng [1,2,*], Hui Yang [3], Jinlong Wen [1] and Meng Qi [1]

[1] School of Civil Engineering and Communication, North China University of Water Resources and Electric Power, Zhengzhou 450045, China
[2] Key Laboratory of Urban Security and Disaster Engineering of Ministry of Education, Beijing University of Technology, Beijing 100124, China
[3] China Construction Third Engineering Bureau Group Co., Ltd., Wuhan 430064, China
* Correspondence: wangzhihao@ncwu.edu.cn (Z.W.); chengzhipeng1107@163.com (Z.C.)

Abstract: This study aims to provide an accurate finite element (FE) modeling method for structural vibration serviceability evaluation of the suspended floor under human-induced excitation. The fundamental dynamic characteristics and human-induced vibration responses of a typical suspended floor were first measured via a series of field tests. Subsequently, the overall and local equivalent FE models of the suspended floor were respectively established, and their applicability was then verified by comparing the predicted dynamic characteristics and responses of the suspended floor with the corresponding field test results. Finally, passive tuned mass dampers (TMDs) were designed for vibration control of the suspended floor using the local equivalent FE model, and the applicability of the local FE model in assessing the vibration serviceability of the suspended floor with TMDs was further confirmed via pedestrian-induced vibration tests. Results demonstrate that the simplified local equivalent FE model proposed in this study can well replace the complicated overall FE model to evaluate the vibration serviceability of the suspended floor with and without TMDs.

Keywords: suspended floor; vibration serviceability evaluation; finite element (FE) model; field test; tuned mass damper (TMD)

Citation: Wang, Z.; Song, L.; Cheng, Z.; Yang, H.; Wen, J.; Qi, M. Finite Element Model for Vibration Serviceability Evaluation of a Suspended Floor with and without Tuned Mass Dampers. *Buildings* **2023**, *13*, 309. https://doi.org/10.3390/buildings13020309

Academic Editor: Agathoklis Giaralis

Received: 20 December 2022
Revised: 9 January 2023
Accepted: 13 January 2023
Published: 20 January 2023

1. Introduction

Currently, lightweight building materials and diversified building functions present a promising prospect in the development of modern architecture [1–3]. To meet people's demand for the flexibility and usability of the architectural space, an increasing number of large-span and lightweight structures have been constructed. However, these flexible structures with low natural frequencies are often prone to pedestrian-induced excessive vibration, thus facing the issue of structural vibration serviceability [4–7]. Consequently, it is necessary to accurately evaluate the vibration serviceability of flexible structures during their design phase.

The accuracy of numerical methodologies for evaluating the vibration serviceability of large-span structures is dependent on efficient finite element (FE) modeling methods, computational precision of the modal parameters, and evaluation criteria of the structural vibration serviceability [8–10]. The influences of non-structural components with weak constraint effect of glass curtain walls on the vibration characteristics of the cantilevered floor have been investigated by Zhu et al. [11,12], demonstrating that the finite element (FE) modeling method neglecting the weak constraint effect of glass curtain walls may lead to conservative design results for structural vibration serviceability evaluation. It was also confirmed that appropriate modal parameters are the prerequisite for numerical analysis on structural dynamic response, which significantly dominates the accuracy of structural vibration serviceability evaluation [13–17]. Additionally, available guidelines including ISO 10137:2007 [18,19], AISC Design Guide #11 [20], and JGJ/T441-2019 [21] stipulated limits for the fundamental frequency of the structure and the perceptibility of humans. Nevertheless, the criteria of each guideline for assessing structural vibration serviceability are not unified. For example, evaluation index for structural vibration serviceability in AISC Design Guide #11 and JGJ/T441-2019 is peak acceleration, while that in ISO 10137: 2007 is the root mean square (RMS) acceleration.

For structures that cannot satisfy the specified limits of vibration serviceability evaluation criteria, tuned mass dampers (TMDs) have been widely used to suppress their excessive vibration [22–24]. To improve the control performance of TMDs, lots of novel TMDs, such as semi-active TMD (SA-TMD) [25], semi-active independent variable mass TMD (SAVM-TMD) [26], adaptive variable mass passive TMD (APVM-TMD) [27], and rotational inertial double-TMD (RID-TMD), have been well developed [28]. Moreover, the effectiveness of TMDs in mitigating human-induced vibrations of the floor was well verified numerically and experimentally [29,30].

It is necessary to establish a reliable finite element (FE) model for vibration serviceability evaluation [31,32] and optimum TMD design [33,34] of the suspended floor. Compared with the traditional column-supported or wall-supported floor system, a suspended floor system is lifted by hanging pillars or cables, which makes its vertical stiffness weaker than the traditional floor, and it is more sensitive to human-induced dynamic loads [35,36]. Lv et al. [37] conducted field tests on a suspended floor to investigate dynamic characteristics and human-induced vibration serviceability, which has provided a fundamental test basis to fill the lack of vibration serviceability research on the suspended floor. However, there are some non-negligible factors in achieving precise finite element modeling and numerical analysis for the dynamic characteristics and responses of the suspended floor, especially the structural inter-story coupling effect. Costa-Neves et al. [38] found that there was coupled phenomenon between different floors in the overall FE model for multi-story structures. Wang et al. [39] compared the vertical dynamic characteristics of a large-span suspended steel space frame-glass composite floor (SSSF-GCF) between the overall FE model and local equivalent FE model, indicating that the inaccurately simplified global FE model significantly would underestimate vertical human-induced acceleration response of the floor. Consequently, it is urgent to develop a precise FE modeling method for evaluating the vibration serviceability of the suspended floor.

This study presents experimental measurements and numerical analyses on the vibration serviceability of a 36 m long suspended floor in Nanjing Global Trading Plaza, China. First, the dynamic characteristics of the suspended floor were experimentally identified. Subsequently, the overall and corresponding local equivalent FE models of the suspended floor were developed. Finally, the applicability of the local FE model for vibration serviceability evaluation of the suspended floor with passive tuned mass dampers (TMDs) was demonstrated.

2. Field Test

In this section, the fundamental dynamic characteristics and human-induced vibration responses of a typical suspended floor were measured through ambient excitation and human-induced vibration tests, respectively. The natural frequency and damping ratio of the suspended floor in the first mode were identified.

2.1. Basic Overview of the Structure

As shown in Figure 1, the 50 m height suspended corridor of Nanjing Global Trading Plaza spans 36 m between two ultra-high-rise towers, including zones A, B, and C. The suspended part consists of a steel frame-support structure, where the load is transferred by hanging pillars into the core tubes or giant frames. The steel structural components (such as steel beams, frame columns, and hanging pillars) are made of Q345 steel, whose elastic modulus, Poisson's ratio, and density are 2×10^{11} N/mm, 0.3, and 7850 kg/m^3, respectively. The ribbed floor is made of C30 concrete, whose elastic modulus, Poisson's ratio, and density are 3×10^{10} N/mm, 0.2, and 2400 kg/m^3, respectively. The component parameters of the second-story suspended floor in zone B are listed in Table 1.

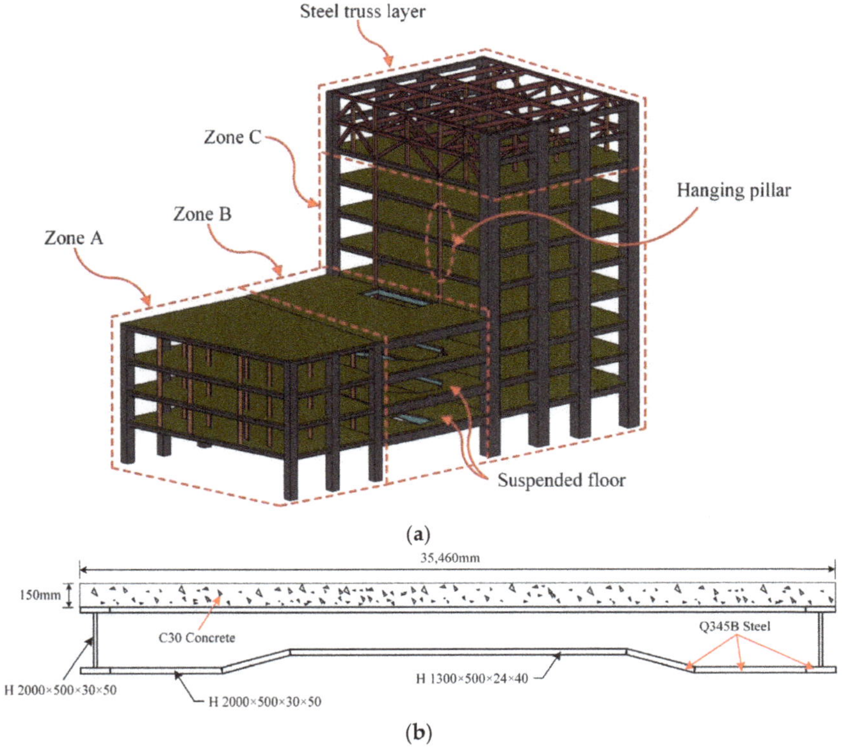

Figure 1. The suspended structure: (**a**) structural configuration; (**b**) details of the second-story suspended floor in zone B.

Table 1. Component parameters of the second-story suspended floor in zone B.

Structural Components	Sectional Specification	Section Type	Remarks
GKL-1	H1000 × 500 × 30 × 40	Welding H-shape	Frame beam
GKL-2	H1300 × 500 × 24 × 40	Welding H-shape	Frame beam
GKL-3	H1450 × 500 × 30 × 50	Welding H-shape	Frame beam
GCL-2	H1500 × 400 × 20 × 30	Welding H-shape	Secondary beam
GCL-3	H700 × 300 × 13 × 24	Welding H-shape	Secondary beam
GCL-4	H600 × 350 × 8 × 14	Welding H-shape	Secondary beam
DZ-1	H1000 × 500 × 20 × 34	Welding H-shape	Hanging pillars
DZ-2	B500 × 500 × 35 × 35	Welding Box-shape	Hanging pillars

2.2. Dynamic Characteristic Test

The schematic diagram of the test setup for measuring the fundamental dynamic characteristics of the suspended floor is illustrated at four measuring points in Figure 2. As illustrated in Figure 3a, a 16-channel portable COINV-DASP-V10 data acquisition instrument is employed to record the vibration data with a 204.8 Hz sampling frequency, which contains performance indexes including a resolution of 24-bit, a dynamic range of 0 dB~120 dB, and the highest sampling frequency of 51.2 kHz. As shown in Figure 3b, an ultra-low frequency version type of the DH610V magnetoelectric vertical acceleration sensor was used to measure the vertical acceleration response of the suspended floor, which contains performance indexes including nominal sensitivity of 0.3 V/m·s^{-2}, a maximum measurement range of 20 m·s^{-2}, a frequency range of 0.25 Hz~100 Hz, and a test resolution of 3×10^{-3} mm/s^2 [40].

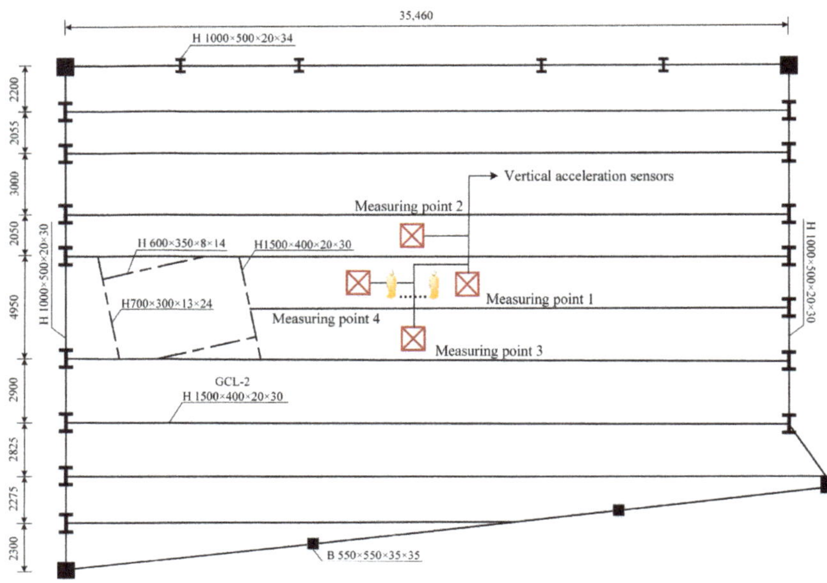

Figure 2. The schematic diagram of the test setup for the dynamic characteristic test.

(**a**) (**b**)

Figure 3. The schematic diagram of the test setup: (**a**) data acquisition instrument; (**b**) magnetoelectric vertical acceleration sensor.

2.2.1. Eigenfrequency Analysis

Figure 4 shows the acceleration response and corresponding fast Fourier transformation (FFT) spectra of the suspended floor at measurement point 1 under ambient excitation. It is noteworthy that taking a lowpass filter to remove the mixed noise of the frequency above 10 Hz [41–43]. In Figure 4b, the peak frequency is about 3.54 Hz, which is the first modal fundamental frequency of the suspended floor.

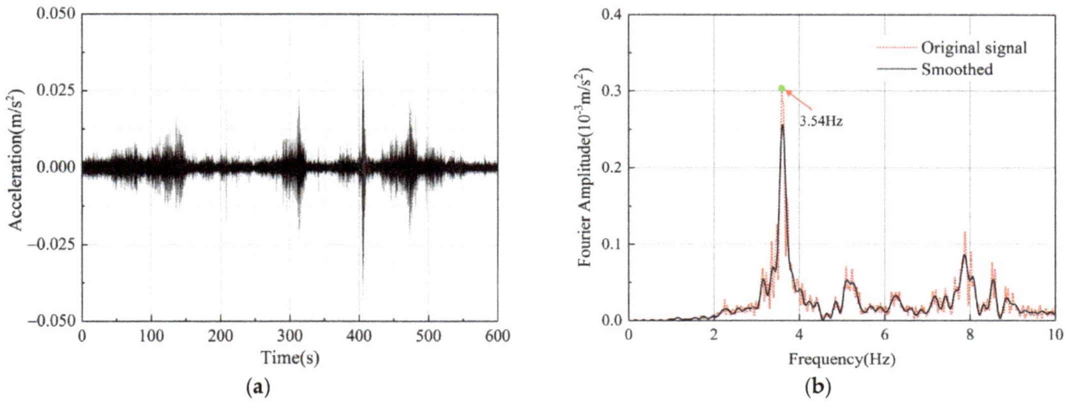

Figure 4. Acceleration response and corresponding FFT spectra of the suspended floor at the measurement point 1 under ambient excitation: (**a**) acceleration time history; (**b**) FFT spectra.

2.2.2. Damping Ratio

Free vibration tests were conducted by applying heel impact on the structure [44–46]. A typical free-decay acceleration response of the suspended floor at test point 1 is presented in Figure 5. The modal damping ratio of the suspended floor was identified using the exponential function to fit the envelope curve of the free-decay acceleration response. Accordingly, the first-order vertical modal damping ratio of the suspended floor was determined as 2.10%.

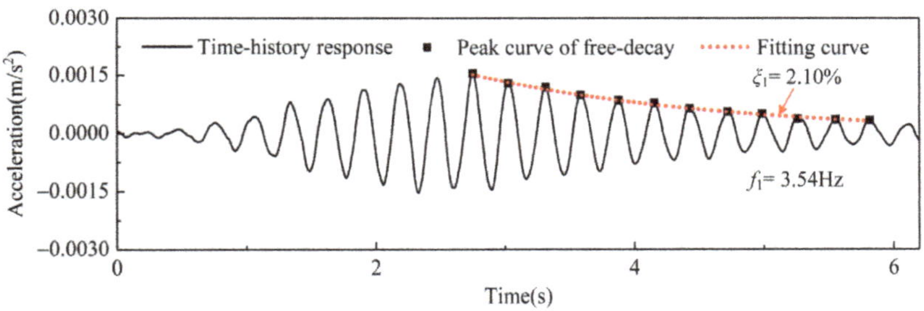

Figure 5. Free-decay acceleration response of the suspended floor at test point 1.

2.3. Human-Induced Vibration Test

2.3.1. Test Scenarios

According to the American Road Traffic Capacity Manual HCM2000 [47], the densely populated state of 1 pedestrian/m^2 corresponds to 4 people walking synchronously, which is equivalent to 20 people moving freely [48]. For the case of queue congestion, the crowd density is determined as 2 pedestrians/m^2, and the maximum number of pedestrians is 40, equivalent to 200 people moving freely. As illustrated in Figure 6b, the pedestrian-induced load is applied in the scope of 10 m × 10 m as a vibration-sensitive location.

2.3.2. Test Results

Since root mean square (RMS) acceleration is commonly applied to quantify the amplitude of the structural acceleration response, it is introduced to characterize the dynamic response of the suspended floor [18,19], which can be calculated by:

$$a_{RMS} = \sqrt{\frac{1}{T}\int_0^T a_w^2(t)dt} \quad (1)$$

where $a_w(t)$ denotes the frequency weighted acceleration; T denotes the duration of the vibration caused by excitation.

Figure 7 presents the dynamic responses of the suspended floor when the pedestrians walk in place (WP) and run in place (RP). Cases WP-1~WP-4 represent one to four pedestrians on the floor. For the cases of pedestrians walking in place, the RMS accelerations of cases WP-2, WP-3, and WP-4 are 1.21, 1.52, and 1.77 times that of case WP-1, respectively. For the cases of pedestrians running in place, the RMS accelerations of cases RP-2, RP-3, and RP-4 are 1.56, 2.14, and 2.37 times that of case RP-1, respectively. It can be found that the per-person RMS acceleration decreases as the number of pedestrians increases, which is attributed to the fact that it is difficult to synchronize the step frequency for multi-person scenarios.

(a) (b)

Figure 6. Field tests of human-induced vibration: (**a**) single pedestrian walking in place; (**b**) 40 pedestrians walking in place.

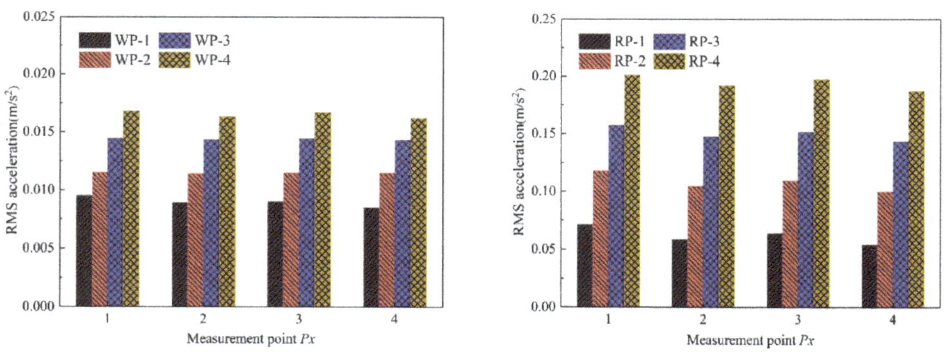

(a) (b)

Figure 7. Comparisons of RMS accelerations at different measuring points under human-induced dynamic loads: (**a**) walking in place (1.77 Hz); (**b**) running in place (3.54 Hz).

3. Finite Element Modeling of the Suspended Floor

In this section, the overall and local equivalent FE models of the suspended floor were first established, respectively. Subsequently, the dynamic characteristics and responses of the suspended floor predicted with the two models were compared with the corresponding field test results, which verified the applicability of the two models in assessing the vibration serviceability of the suspended floor.

3.1. Overall FE Model

3.1.1. Model Parameters

Using the usual mesh refinement techniques implemented in ANSYS software, the overall FE model of the suspended floor was established in this study. The steel structural components such as beams and columns were modeled by the three-dimensional beam elements (BEAM188) with tension, compression, torsion, and bending capabilities, while the reinforced concrete slab was modeled by the shell finite element (SHELL63) with both bending and membrane capabilities. Moreover, considering the effect of non-structural components on structural vibration characteristics, glass, external walls, and partition walls were regarded as the additional stiffness applied to the floor [16]. Due to the similarity of material physical parameters between the concrete slab and decorative surface, the effect of the decorative layer on the natural frequency was represented by the principle of equivalent stiffness [8], and thus the thickness of the floor was determined as 200 mm. Additionally, taking into account the rationality of the structural dynamic calculation, the additional live load on the structure was regarded as the vibration participating mass. According to the equivalent mass principle, the density of C30 concrete for modal analysis was adjusted to 2931 kg/m^3, and its elastic modulus was magnified 1.35 times.

3.1.2. Boundary Condition

Suspended structures have no vertical grounding components at the location of hanging pillars, resulting in insufficient supporting stiffness and a significant inter-story coupling effect. Hence, to ensure the accuracy of the numerical analysis for structural dynamic responses, the precise definition of structural boundary conditions is crucial.

Figure 8 shows the overall FE model of the suspended floor. Considering the complete interaction between the concrete slab and rib-beams, the suspended floor's FE model coupled all nodes between the slab and beams to prevent the occurrence of any slip. Near the right side of the high-rise tower, the deformation joint was represented by releasing slight displacement along the X-direction. Additionally, the structural characteristic of the beam-to-hanging pillar connections in the suspended floor was simulated by the hinges.

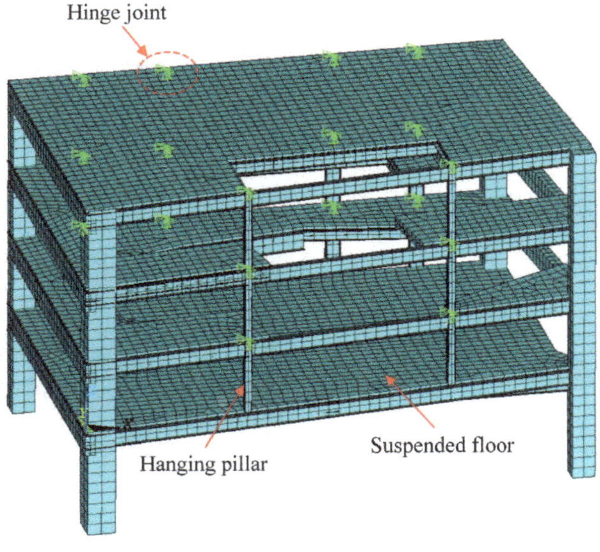

Figure 8. The overall FE model of the suspended floor in zone B.

3.2. Local Equivalent FE Model

Compared with the overall FE model, the local equivalent FE model has the advantages of model simplicity, computation efficiency, and practicality [39]. It is found that the impact of inter-story coupling effects on evaluating the structural vibration serviceability can be attributed to the superposition of vertical acceleration between different stories. Accordingly, a local equivalent FE model is further proposed for the suspended floor, which is illustrated in Figure 9. The modeling process is divided into the following three steps. First, the hanging pillar was simplified as the equivalent mass distributed along its vertical position. Second, the COMBIN14 spring element with three degrees of freedom per node was used to simulate the elastic support with adjustable vertical stiffness at the lifting point, which helps to improve the synergy between beams and columns. Third, the axial stiffness of the hanging pillars was used to estimate the vertical elastic support stiffness.

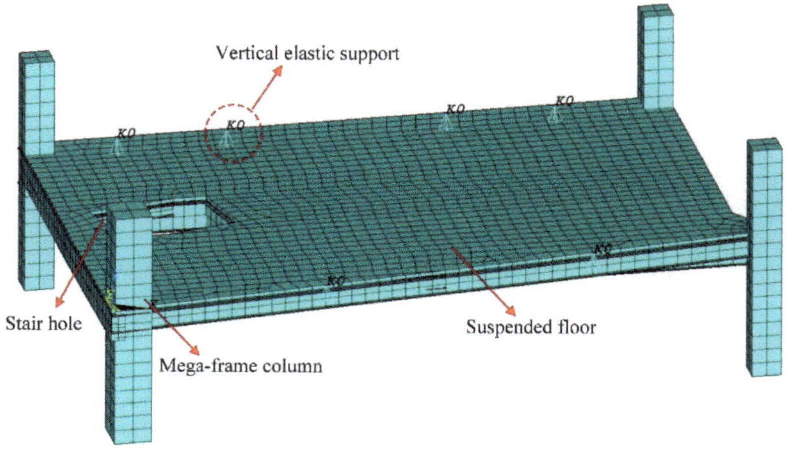

Figure 9. The local equivalent FE model of the suspended floor.

3.3. Dynamic Characteristics Comparisons

3.3.1. Fundamental Frequency

Table 2 compares the fundamental frequencies of the second-story suspended floor obtained by numerical analysis and field test. As listed in Table 2, the error of the suspended floor's fundamental frequency between the overall FE model and the corresponding field test result is only 0.56%, while the error of the suspended floor's fundamental frequency between the local equivalent FE model and the corresponding field test result is only 0.28%.

Table 2. Comparison of the fundamental frequencies of the second-story suspended floor obtained by numerical analysis and field test.

FE Model Type	Numerical Analysis (Hz)	Field Test (Hz)	Error (%)
Overall FE model	3.56	3.54	0.56
Local equivalent FE model	3.53		0.28

3.3.2. Mode Shape

Figure 10 demonstrates the first-order mode shape of the suspended floor predicted with the two FE models. The inter-story coupling effect in the overall FE model mainly behaves in the vibration of the top floor slab, which increases the modal participation mass. Notice that the first-order modal masses of the overall FE model and the local equivalent FE model are calculated to be 125,207.67 kg and 121,009.81 kg, respectively.

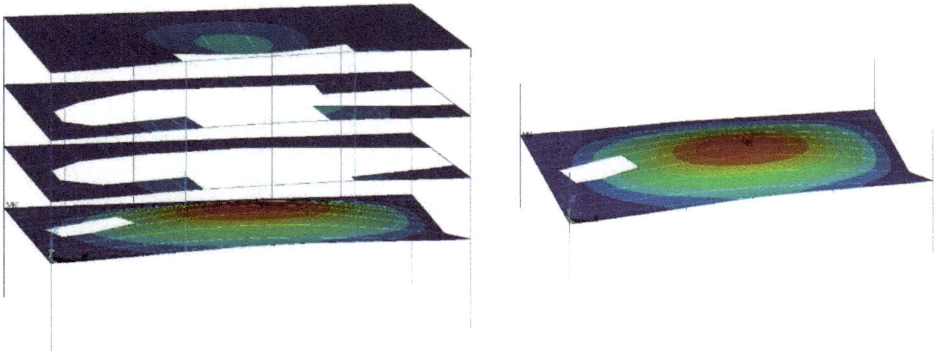

(a) (b)

Figure 10. Mode shapes of the suspended floor predicted with the two FE models: (**a**) overall FE model; (**b**) local equivalent FE model.

As an important index to indicate the relevance of the different vibration mode shapes, the modal assurance criterion (MAC) is normally used to evaluate the modal similarity of two dynamic systems [49]. The formula for calculating MAC value can be expressed as:

$$\mathrm{MAC}\left(\boldsymbol{\Phi}_i, \boldsymbol{\Phi}_j\right) = \frac{\left|\boldsymbol{\Phi}_i^{\mathrm{T}}, \boldsymbol{\Phi}_j\right|^2}{\left(\boldsymbol{\Phi}_i^{\mathrm{T}}, \boldsymbol{\Phi}_i\right)\left(\boldsymbol{\Phi}_j^{\mathrm{T}}, \boldsymbol{\Phi}_j\right)} \tag{2}$$

where $\boldsymbol{\Phi}_i$ and $\boldsymbol{\Phi}_j$ are the structural mode shape of the overall FE model and the local equivalent FE model, respectively.

The MAC values between the overall and the local equivalent FE model come to 99.77%, indicating that the mode shapes of the suspended floor predicted with two FE models are approximately consistent.

3.4. Comparisons of Numerical and Measured Dynamic Responses of the Suspended Floor

To verify the applicability of the two models in assessing the vibration serviceability of the suspended floor, the pedestrian-induced loads were applied in the FE model of the suspended floor to obtain its numerical dynamic responses, and the numerical results were further compared with the corresponding field test results.

Figures 11–13 compare the numerical and measured vertical acceleration time-history responses and vertical RMS acceleration of the suspended floor subjected to pedestrian-induced loads under cases WP-1, WP-2, and WP-4. With the increasing number of pedestrians, numerical acceleration responses are significantly higher than the corresponding field test results. It can be concluded that there are minor differences in pedestrian walking frequency, which enhances the randomness of the crowd and reduces the actual vibration responses of the suspended floor. Additionally, it has been found that the predicting value considering the crowd-structure interaction (CSI) effect is much closer to the measured acceleration than that without considering the CSI effect [50]. It is noteworthy that the RMS acceleration responses predicted with the overall FE model are in good agreement with those of the local equivalent FE model, and thus the simplified local equivalent FE model can replace the complicated overall FE model to predict the practical dynamic responses of the floor subjected to pedestrian excitations.

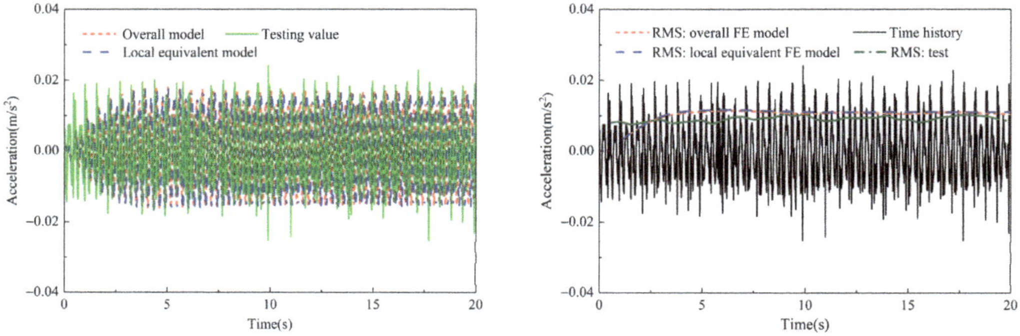

(**a**) (**b**)

Figure 11. Comparison of numerical and measured acceleration time-history responses and 20 s RMS acceleration of the suspended floor under case WP-1: (**a**) acceleration time-history responses; (**b**) RMS acceleration.

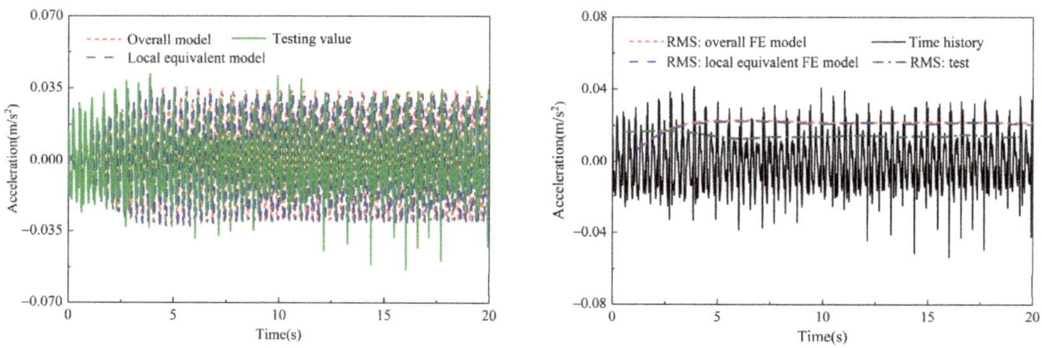

(**a**) (**b**)

Figure 12. Comparison of numerical and measured acceleration time-history responses and 20 s RMS acceleration of the suspended floor under case WP-2: (**a**) acceleration time-history responses; (**b**) RMS acceleration.

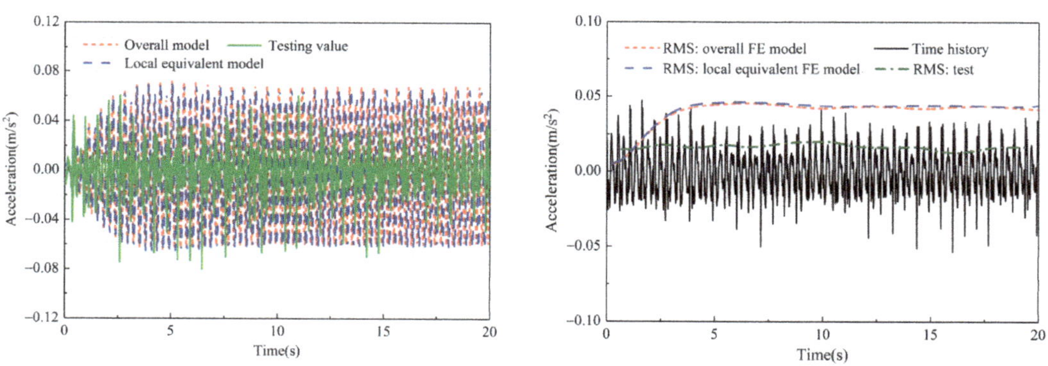

(**a**) (**b**)

Figure 13. Comparison of numerical and measured acceleration time-history responses and 20 s RMS acceleration of the suspended floor under case WP-4: (**a**) acceleration time-history responses; (**b**) RMS acceleration.

4. Vibration Serviceability Evaluation of the Suspended Floor with TMDs

In this section, vibration serviceability evaluation of the suspended floor with TMDs was conducted. Using the local equivalent FE model, passive tuned mass dampers (TMDs) were designed for vibration control of the suspended floor. In order to confirm the applicability of the local FE model in assessing the vibration serviceability of the suspended floor with TMDs, the numerical and measured human-induced dynamic responses of the suspended floor with TMDs were further compared.

4.1. Vibration Serviceability Evaluation

Figure 14 compares the measured RMS accelerations in all test cases and the reference limits under different environments specified in ISO 10137. It can be observed that the RMS accelerations of the cases of WP-1, WP-2, WP-3, and WP-4 with a 1.77 Hz step frequency meet the vibration serviceability limits for residences, office buildings, and workshops, while the RMS accelerations of the cases of RP-1, RP-2, RP-3, and RP-4 with a 3.54 step frequency exceed the vibration serviceability limits for the workshop. If the structures cannot meet the vibration serviceability criteria shown in Figure 14, efficient control devices need to be adopted to control their vibrations. As one of the most common and effective control devices, passive TMD is widely used to control structural vibrations [29,30]. Therefore, passive TMDs were designed for vibration control of the suspended floor to improve its vibration serviceability in the following sections.

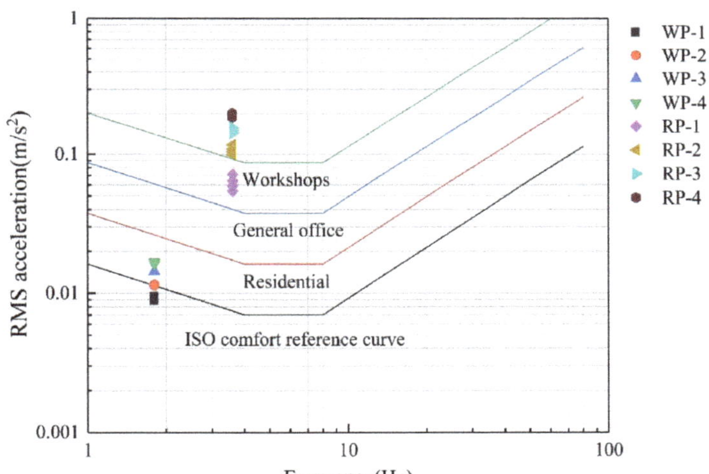

Figure 14. Vibration serviceability evaluation of the measured RMS accelerations according to ISO 10137: 2007.

4.2. Numerical Analysis for TMD Parameter Design

The stiffness and damping parameters of TMDs for vibration control of the suspended floor were calculated according to the optimal frequency ratio f_{opt} and optimal damping ratio ξ_{opt}, which can be expressed as [51]:

$$f_{opt} = \frac{1}{1+\mu} \tag{3}$$

$$\xi_{opt} = \sqrt{\frac{3\mu}{8(1+\mu)}} \tag{4}$$

where μ denotes the mass ratio of TMDs installed on the structure.

Considering the engineering purpose of cost saving and easy installation, the mass ratio of TMDs installed on the structure is determined to be $\mu = 3\%$, and its specific design parameters are listed in Table 3. The number of TMDs was determined to be five, and the installation locations of TMDs on the suspended floor are shown in Figure 15.

Table 3. The design parameters of TMDs for mitigating vibration of the suspended floor.

Mass (kg)	Stiffness (N/m)	Damping (N·s/m)	Frequency (Hz)	Number
720	349,956.78	1495.17	3.54	5

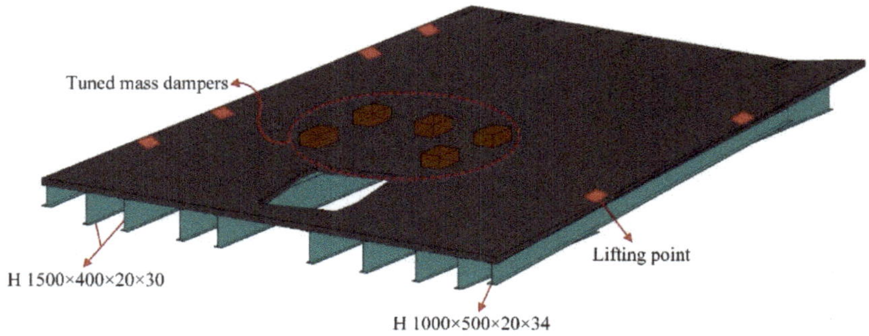

Figure 15. Installation locations of TMDs on the suspended floor.

Based on the local equivalent FE model of the suspended floor, the dynamic characteristics of the suspended floor with TMDs were investigated. Five MASS21 element nodes representing the quality of TMDs were established below 30 cm of the TMDs installation position. The COMBIN14 element was also adopted to simulate the spring-damper components of TMDs, whose spring stiffness and damping parameter were defined via the real constant. Meanwhile, pedestrian-induced loads were reloaded to the local equivalent FE model of the suspended floor.

Figure 16 compares the acceleration time-history response of the suspended floor with and without TMD under case WP-1. As illustrated in Figure 16, the acceleration response of the suspended floor with TMDs is significantly reduced in comparison with the suspended floor without TMDs.

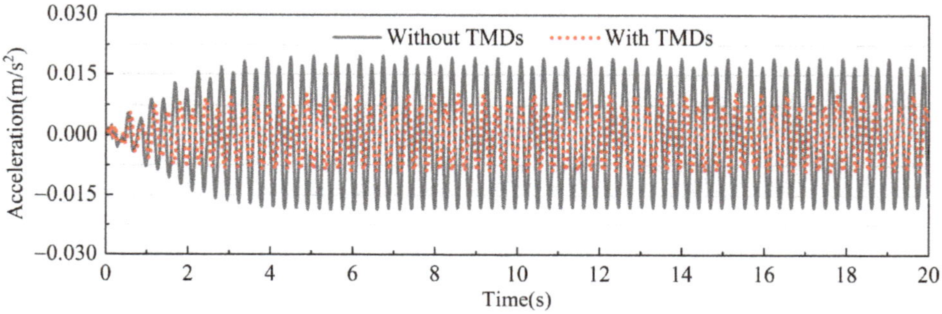

Figure 16. Acceleration time-history responses of the suspended floor with and without TMD under case WP-1.

4.3. Dynamic Testing for the Suspended Floor with TMDs

To further verify the effectiveness of TMDs in improving the vibration serviceability of the suspended floor, dynamic testing for the suspended floor with TMDs under pedestrian loads was conducted. The test setup, measuring point arrangement, and loading mode adopted in this section are the same as that adopted in Section 2. On-site complete installations of TMDs are illustrated in Figure 17.

Figure 17. TMDs installed on-site.

For the case of walking in place, it can be observed from Figure 18 that the peak acceleration decreased from 0.0095 m/s^2 to 0.0061 m/s^2. The maximum RMS value induced by a single pedestrian is reduced by approximately 35.79%, indicating that TMDs can effectively mitigate the structural vibration response.

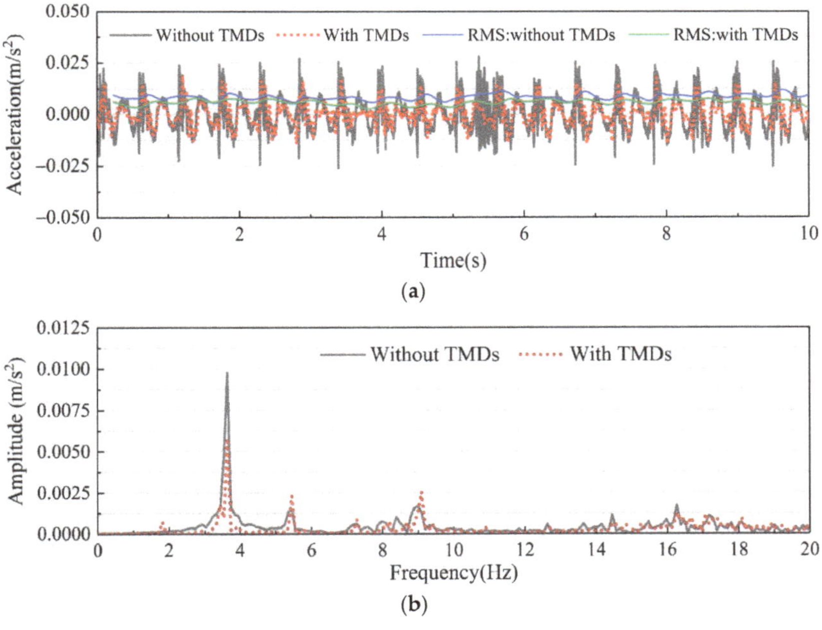

Figure 18. Comparison of the measured acceleration time time-history responses and corresponding FFT spectrum of the suspended floor with and without TMDs under case WP-1: (**a**) acceleration time-history response; (**b**) FFT spectrum.

4.4. Comparisons of Numerical and Measured Dynamic Responses of the Suspended Floor with TMDs

Table 4 compares the numerical and measured RMS acceleration of the suspended floor with and without TMDs under pedestrian-induced loads. Compared with the suspended floor without TMDs, the reduction rates of RMS accelerations of the suspended floor with TMDs are 43.93%, 43.66%, and 43.29% for numerical simulation under the cases of WP-1, WP-2, and WP-4, respectively, while the reduction rates of RMS accelerations of the suspended floor with TMDs are 35.79%, 34.78%, and 35.71%% for field tests under the cases of WP-1, WP-2, and WP-4, respectively. The difference between numerical and measured acceleration responses is mainly attributed to the high sensitivity of TMD control performance to its frequency and damping.

Table 4. Comparison of numerical and measured RMS acceleration of the suspended floor with and without TMDs under pedestrian loads.

Test Conditions	TMDs Installation	RMS Acceleration (m/s^2)			
		Testing	Reduction Rate (%)	Simulation	Reduction Rate (%)
WP-1	W/O	0.0095	35.79	0.0107	43.93
	W/	0.0061		0.0060	
WP-2	W/O	0.0115	34.78	0.0213	43.66
	W/	0.0075		0.0120	
WP-4	W/O	0.0168	35.71	0.0425	43.29
	W/	0.0108		0.0241	

Note: W/O and W/ represent the suspended floor without and with TMDs, respectively.

The RMS accelerations of the suspended floor with and without TMDs under case WP-1 are further compared in Figure 19. For the suspended floor without TMDs, the suspended floor's RMS acceleration responses predicted by its overall FE model and local equivalent FE model under case WP-1 reach 0.0104 m/s^2 and 0.0107 m/s^2, respectively, which agree well with the corresponding field test result of 0.0095 m/s^2. For the suspended floor with TMDs, the suspended floor's RMS acceleration predicted with its overall FE model and local equivalent FE model under case WP-1 reach 0.0054 m/s^2 and 0.0060 m/s^2, respectively, which are in good agreement with the corresponding field test result of 0.0061 m/s^2. Therefore, it can be concluded that the simplified local equivalent FE model can replace the complicated overall FE model to predict the human-induced dynamic responses of the suspended floor.

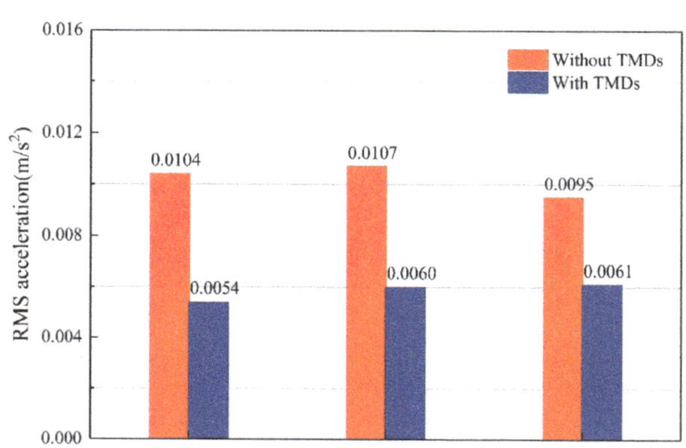

Figure 19. Comparison of RMS acceleration predicted with two FE models and corresponding field test results of the suspended floor with and without TMDs under case WP-1.

5. Conclusions

Recently, the vibration serviceability of the suspended floor has received more and more attentions from engineers and researchers. In fact, unlike traditional floor systems, there are some non-negligible factors in achieving precise finite element modeling and numerical analysis for the dynamic characteristics and responses of the suspended floor, especially the structural inter-story coupling effect. Thus, it is urgent to develop a precise FE modeling method for evaluating the vibration serviceability of the suspended floor. Under this circumstance, taking a 36 m long suspended floor system as a case study, this study proposes accurate finite element (FE) modeling methods for human-induced vibration serviceability evaluation of the suspended floor. The main findings are summarized as follows:

(1) The overall FE model can be established and updated by addressing the effect of decorative layer and structural additional mass, and the connection relationship of beam-to-hanging pillars in the suspended floor is suggested to be hinged. The fundamental frequency of the suspended floor and the vertical RMS acceleration responses of single pedestrian walking predicted with the overall FE model are found to be consistent with the field test results.

(2) The local equivalent FE model of the suspended floor can be proposed by simplifying the hanging pillar as the elastic support with adjustable vertical stiffness. The fundamental frequency and mode shape of the suspended floor predicted with the local equivalent FE model are found to be consistent with the overall FE model. Moreover, the RMS acceleration responses predicted with the local equivalent FE model agree well with the corresponding experimental results for the case of single pedestrian walking.

(3) The simplified local equivalent FE model can replace the complicated overall FE model to evaluate the vibration serviceability of the suspended floor. For the suspended floor with and without TMDs in this study, the error of RMS acceleration responses between two FE models and corresponding field test results ranges from 1.64% to 11.48% when a single pedestrian is walking in place.

Author Contributions: Conceptualization, Z.W.; methodology, Z.W.; software, L.S. and J.W.; validation, Z.W., Z.C. and J.W.; formal analysis, L.S.; investigation, L.S.; resources, Z.W.; data curation, Z.W., L.S. and Z.C.; writing—original draft preparation, Z.W., L.S., Z.C., H.Y., J.W. and M.Q.; writing—review and editing, Z.W., L.S. and Z.C.; supervision, Z.W.; project administration, Z.W. and H.Y.; funding acquisition, Z.W. and H.Y. All authors have read and agreed to the published version of the manuscript.

Funding: The authors greatly acknowledge the financial support from the National Natural Science Foundation of China (Grant No. 51878274), the Technology R & D Project of China Construction Co., Ltd. (Grant No. CSCEC-2021-Q-63), and the Academic Degrees & Graduate Education Reform Project of Henan Province (2021SJGLX015Y).

Data Availability Statement: Data are contained within the article. Additional supporting data presented in this study are available upon request from the corresponding authors.

Conflicts of Interest: The authors declare no conflict of interest.

References

1. Hassanieh, A.; Chiniforush, A.A.; Valipour, H.R.; Bradford, M.A. Vibration behaviour of steel-timber composite floors, part (2): Evaluation of human-induced vibrations. *J. Constr. Steel Res.* **2019**, *158*, 156–170. [CrossRef]
2. Ferrante, C.; Andrade, S.; Lima, L.; Vellasco, P. Analytical study and experimental tests on innovative steel-concrete composite floorings. *J. Constr. Steel Res.* **2019**, *168*, 105868. [CrossRef]
3. Ahmed, I.M.; Tsavdaridis, K.D. The evolution of composite flooring systems: Applications, testing, modelling and eurocode design approaches. *J. Constr. Steel Res.* **2019**, *155*, 286–300. [CrossRef]
4. Xie, Z.; Hu, X.; Du, H.; Zhang, X. Vibration behavior of timber-concrete composite floors under human-induced excitation. *J. Build. Eng.* **2020**, *32*, 101744. [CrossRef]

5. Abdeljaber, O.; Hussein, M.; Avci, O.; Davis, B.; Reynolds, P. A novel video-vibration monitoring system for walking pattern identification on floors. *Adv. Eng. Softw.* **2020**, *139*, 102710. [CrossRef]
6. Van Engelen, N.C.; Graham, J. Comparison of prediction and measurement techniques for pedestrian-induced vibrations of a low-frequency floor. *Struct. Control Health Monit.* **2019**, *26*, e2294. [CrossRef]
7. Cai, Y.; Gong, G.; Xia, J.; He, J.; Hao, J. Simulations of human-induced floor vibrations considering walking overlap. *Appl. Sci.* **2020**, *2*, 19. [CrossRef]
8. Devin, A.; Fanning, P.J.; Pavic, A. Modelling effect of non-structural partitions on floor modal properties. *Eng. Struct.* **2014**, *91*, 58–69. [CrossRef]
9. Jarnerö, K.; Brandt, A.; Olsson, A. Vibration properties of a timber floor assessed in laboratory and during construction. *Eng. Struct.* **2015**, *82*, 44–54. [CrossRef]
10. Jiang, L.; Cheng, R.; Zhang, H.; Ma, K. Human-induced-vibration response analysis and comfort evaluation method of large-span steel vierendeel sandwich plate. *Buildings* **2022**, *12*, 1228. [CrossRef]
11. Zhu, Q.; Liu, K.; Liu, L.; Du, Y.; Zivanovic, S. Experimental and numerical analysis on serviceability of cantilevered floor based on human-structure interaction. *J. Constr. Steel Res.* **2020**, *173*, 106184. [CrossRef]
12. Liu, K.; Liu, L.; Zhu, Q.; Liu, Y.; Zhou, F. Dynamic testing and numerical simulation of human-induced vibration of cantilevered floor with tuned mass dampers. *Structures.* **2021**, *34*, 1475–1488. [CrossRef]
13. Wei, X.; Wan, H.P.; Russell, J.; Živanović, S.; He, X. Influence of mechanical uncertainties on dynamic responses of a full-scale all-FRP footbridge. *Compos. Struct.* **2019**, *223*, 110964. [CrossRef]
14. Silva, J.G.; Andrade, S.A.; Lopes, E.D. Parametric modelling of the dynamic behaviour of a steel–concrete composite floor. *Eng. Struct.* **2014**, *75*, 327–339. [CrossRef]
15. Wei, X.; Russell, J.; Živanović, S.; Mottram, J.T. Measured dynamic properties for FRP footbridges and their critical comparison against structures made of conventional construction materials. *Compos. Struct.* **2019**, *223*, 110956. [CrossRef]
16. Miskovic, Z.; Pavic, A.; Reynolds, P. Effects of full-height nonstructural partitions on modal properties of two nominally identical building floors. *Can. J. Civ. Eng.* **2009**, *36*, 1121–1132. [CrossRef]
17. Huang, H.; Gao, Y.; Chang, W.S. Human-induced vibration of cross-laminated timber (CLT) floor under different boundary conditions. *Eng. Struct.* **2020**, *204*, 110016. [CrossRef]
18. *ISO 2631-1*; Mechanical Vibration and Shock-Evaluation of Human Exposure to Whole-Body Vibration. International Organization for Standardisation: Genève, Switzerland, 1997.
19. *ISO 10137*; Bases for Design of Structures-Serviceability of Buildings and Walkways against Vibration. International Organization for Standardisation: Genève, Switzerland, 2007.
20. *AISC811-97*; Steel Design Guide Series 11: Floor Vibrations Due to Human Activity. American Institute of Steel Construction: Chicago, IL, USA, 1997.
21. *JGJ/T441-2019*; Technical Standard for Human Comfort of the Floor Vibration. China Construction Industry Press: Beijing, China, 2019.
22. Mokrani, B.; Tian, Z.; Alaluf, D.; Meng, F.; Preumont, A. Passive damping of suspension bridges using multi-degree of freedom tuned mass dampers. *Eng. Struct.* **2017**, *153*, 749–756. [CrossRef]
23. Li, Q.; Fan, J.; Nie, J.; Li, Q.; Chen, Y. Crowd-induced random vibration of footbridge and vibration control using multiple tuned mass dampers. *J. Sound Vib.* **2010**, *329*, 4068–4092. [CrossRef]
24. Chen, J.; Han, Z.; Xu, R. Effects of human-induced load models on tuned mass damper in reducing floor vibration. *Adv. Struct. Eng.* **2019**, *22*, 2449–2463. [CrossRef]
25. Wang, L.; Nagarajaiah, S.; Shi, W.; Zhou, Y. Semi-active control of walking-induced vibrations in bridges using adaptive tuned mass damper considering human-structure-interaction. *Eng. Struct.* **2021**, *244*, 112743. [CrossRef]
26. Maślanka, M. Optimised semi-active tuned mass damper with acceleration and relative motion feedbacks. *Mech Syst. Signal Process.* **2019**, *130*, 707–731. [CrossRef]
27. Wang, L.; Shi, W.; Zhang, Q.; Zhou, Y. Study on adaptive-passive multiple tuned mass damper with variable mass for a large-span floor structure. *Eng. Struct.* **2020**, *209*, 110010. [CrossRef]
28. Wang, P.; Chen, J.; Han, Z. Rotational inertial double tuned mass damper for human-induced floor vibration control. *Struct. Eng. Mech.* **2022**, *82*, 283–294.
29. Van Nimmen, K.; Verbeke, P.; Lombaert, G.; Roeck, G.; Van den Broeck, P. Numerical and experimental evaluation of the dynamic performance of a footbridge with tuned mass dampers. *J. Bridge Eng.* **2016**, *21*, C4016001. [CrossRef]
30. Zhang, Z.; Guo, T.; Yang, K. Simulation and measurement of human-induced vibrations of the Beijing Olympic watchtower with tuned mass dampers. *J. Perform. Constr. Facil.* **2017**, *31*, 04017095. [CrossRef]
31. Robertson, M.; Holloway, D.; Taoum, A. Vibration of suspended solid-timber slabs without intermediate support: Assessment for human comfort. *Aust. J. Struct. Eng.* **2018**, *19*, 266–278. [CrossRef]
32. Caballero-Garatachea, O.; Juárez-Luna, G.; Sandoval-Hernández, M.E.R. Methods for the vibration analysis of reinforced concrete precast one-way joist slab floor systems under human walking. *J. Build. Eng.* **2021**, *43*, 103217. [CrossRef]
33. Jimenez-Alonso, J.F.; Saez, A. Motion-based design of TMD for vibrating footbridges under uncertainty conditions. *Smart Struct. Syst.* **2018**, *21*, 727–740.

34. Lievens, K.; Lombaert, G.; De Roeck, G.; Van den Broeck, P. Robust design of a TMD for the vibration serviceability of a footbridge. *Eng. Struct.* **2016**, *123*, 408–418. [CrossRef]
35. Chulahwat, A.; Mahmoud, H. A combinatorial optimization approach for multi-hazard design of building systems with suspended floor slabs under wind and seismic hazards. *Eng. Struct.* **2017**, *137*, 268–284. [CrossRef]
36. Shahabi, A.B.; Ahari, G.Z.; Barghian, M. Suspended columns for seismic isolation in structures (SCSI): Experimental and numerical studies. *Earthq. Struct.* **2020**, *19*, 17–28.
37. Lv, Q.; Lu, Y.; Liu, Y. Vibration serviceability of suspended floor: Full-scale experimental study and assessment. *Structures* **2021**, *34*, 1651–1664. [CrossRef]
38. Costa-Neves, L.F.; Silva, J.S.; Lima, L.R.O.; Jordão, S. Multi-storey, multi-bay buildings with composite steel-deck floors under human-induced loads: The human comfort issue. *Comput. Struct.* **2014**, *136*, 34–46. [CrossRef]
39. Wang, Z.; Qi, X.; Huang, Y.; Fan, B.; Li, X. Dynamic behavior of a suspended steel space frame-glass composite floor. *Adv. Civ. Eng.* **2021**, 8382585. [CrossRef]
40. Komarizadehasl, S.; Komary, M.; Alahmad, A.; Lozano-Galant, J.A.; Ramos, G.; Turmo, J. A novel wireless low-cost inclinometer made from combining the measurements of multiple MEMS gyroscopes and accelerometers. *Sensors* **2022**, *22*, 5605. [CrossRef]
41. Komarizadehasl, S.; Huguenet, P.; Lozano, F.; Lozano-Galant, J.A.; Turmo, J. Operational and analytical modal analysis of a bridge using low-cost wireless Arduino-based accelerometers. *Sensors* **2022**, *22*, 9808. [CrossRef]
42. Zhang, X.; Li, Q.; Wang, Y.; Wang, Q. Vibration of a U-shaped steel-concrete composite hollow waffle floor under human-induced excitations. *Adv. Struct. Eng.* **2020**, *23*, 2996–3008. [CrossRef]
43. Wang, D.; Wu, C.; Zhang, Y.; Li, S. Study on vertical vibration control of long-span steel footbridge with tuned mass dampers under pedestrian excitation. *J. Constr. Steel Res.* **2019**, *154*, 84–98. [CrossRef]
44. An, Q.; Ren, Q.; Liu, H.; Yan, X.; Chen, Z. Dynamic performance characteristics of an innovative cable supported beam structure–concrete slab composite floor system under human-induced loads. *Eng. Struct.* **2016**, *117*, 40–57. [CrossRef]
45. Abeysinghe, C.M.; Thambiratnam, D.P.; Perera, N.J. Dynamic performance characteristics of an innovative hybrid composite floor plate system under human-induced loads. *Compos. Struct.* **2013**, *96*, 590–600. [CrossRef]
46. Varela, W.D.; Battista, R.C. Control of vibrations induced by people walking on large span composite floor decks. *Eng. Struct.* **2011**, *33*, 2485–2494. [CrossRef]
47. Transportation Research Board. *Highway Capacity Manual*; Transportation Research Board: Washington, DC, USA, 2000.
48. Fujino, Y.; Pacheco, B.M.; Nakamura, S.I.; Warnitchai, P. Synchronization of human walking observed during lateral vibration of a congested pedestrian bridge. *Earthq Eng. Struct Dyn.* **1993**, *22*, 741–758. [CrossRef]
49. Jaishi, B.; Ren, W.X. Structural finite element model updating using ambient vibration test results. *J. Struct. Eng.* **2005**, *131*, 617–628. [CrossRef]
50. Zeng, D.; Wang, H.; Chen, J. Dynamic reliability analysis of large-span structures under crowd bouncing excitation. *Buildings* **2022**, *12*, 332. [CrossRef]
51. Den Hartog, J.P. *Mechanical Vibrations*; Courier Corporation: New York, NY, USA, 1985.

buildings

MDPI

Article

Human-Induced-Vibration Response Analysis and Comfort Evaluation Method of Large-Span Steel Vierendeel Sandwich Plate

Lan Jiang [1,2], Ruoheng Cheng [1,2], Huagang Zhang [3,*] and Kejian Ma [3]

[1] Hubei Provincial Engineering Technology Research Center for Power Transmission Line, China Three Gorges University, Yichang 443002, China
[2] College of Electrical Engineering & New Energy, China Three Gorges University, Yichang 443002, China
[3] Space Structures Research Center, Guizhou University, Guiyang 550025, China
* Correspondence: hgzhang@gzu.edu.cn

Abstract: A steel Vierendeel sandwich plate used as a large-span lightweight floor structure for vibration comfort during crowd gatherings was considered. Taking the steel Vierendeel sandwich plate in Guizhou Museum as an example, through finite element transient analysis, the effects of the structural damping, pedestrian self-weight, floor span, surface concrete slab thickness, and structural parameters on the floor's acceleration response distribution were deeply studied. According to the distribution characteristics of the acceleration response, a distribution model function was constructed, and a distribution Gauss model of the relationship between the peak acceleration response and the position of the steel Vierendeel sandwich plate was established. A field test of the sandwich plate under human-induced fixed-point excitation was carried out, and the model fitting results were compared with the actual test results. The results showed that the Gaussian model could effectively estimate the peak acceleration response at different positions on the floor. In addition, according to the distribution model, a comfort evaluation method based on the comfort assurance rate was proposed that could greatly reduce the representative value of the acceleration evaluation. The research results provide a reference for the comfort evaluation and corresponding vibration-reduction design of long-span steel Vierendeel sandwich plates.

Keywords: Vierendeel sandwich plate; comfort evaluation; human-induced load; acceleration response distribution

Citation: Jiang, L.; Cheng, R.; Zhang, H.; Ma, K. Human-Induced-Vibration Response Analysis and Comfort Evaluation Method of Large-Span Steel Vierendeel Sandwich Plate. *Buildings* **2022**, *12*, 1228. https://doi.org/10.3390/buildings12081228

Academic Editors: Jun Chen and Haoqi Wang

Received: 20 June 2022
Accepted: 9 August 2022
Published: 12 August 2022

1. Introduction

With the progress of building technology and the application of high-strength materials, "Large span, low self weight and low damping" is the development direction of building structures, and various types of long-span floor forms have been invented [1,2]. The Vierendeel sandwich plate is a new type of structure that is widely used in long-span industrial and public buildings. Crowd aggregation is inevitable during normal use of floors. A floor is characterized by light weight and small vertical stiffness. A large vibration response is easily produced under a pedestrian load. At the least, it will cause people's discomfort, and at the worst, it will lead to fatigue damage of the floor and reduce the service life of the structure [3,4]. Historically, the Millennium Bridge in London, Techno Mart building in Korea, and other projects had to be stopped due to excessive human-induced-vibration response. It can be seen that the structural vibration caused by a human-induced load has become a problem that must be considered in the design of long-span structures.

A hollow sandwich plate is a bidirectional stress hollow structure that is composed of a surface concrete slab, top and bottom chords, and shear connectors. Compared with the general frame structure, bidirectional stress, good integrity, and the use of less steel are its characteristics. The structure of a Vierendeel sandwich plate is shown in Figure 1.

It has a wide application prospect in large-span multistory buildings [5]. As a large-span, lightweight floor structure, a steel Vierendeel sandwich plate may be sensitive to vibration due to its own structural characteristics. It is necessary to deeply study the influence of human-induced loads on Vierendeel sandwich plates, as well as the comfort evaluation method.

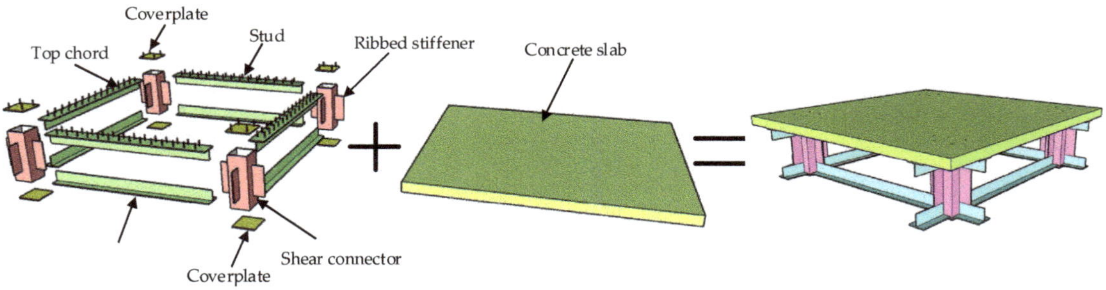

Figure 1. Physical model diagram of Vierendeel sandwich plate.

At present, the human-induced-load model and comfort evaluation standards are the main research directions of domestic and foreign scholars on the human-induced vibration of long-span floors. For the study of loads, the load model is mainly established on the basis of the single-step drop test. Based on a large number of tests, a variety of periodic walking load models have been proposed by researchers [6–8]. On this basis, Chen Jun et al. used the probability density evolution method to analyze the impact of load randomness on the vibration response of a floor. It was considered that the randomness of the pedestrian load had a significant influence on the vibration response of the floor, and the randomness of the pedestrian load should be considered in the comfort evaluation of the floor [9]. In terms of comfort assessment, the current project was mainly implemented with reference to some national or industrial standards. Standards set by the American Institute of Steel Construction [10] and Precast/Prestressed Concrete Institute [11], as well as in the UK Concrete Society's *Technical Report No. 43* [12] and the Concrete Centre's *CCP-016* [13] are widely used. Human-induced loads are divided into general walking loads and rhythmic loads under AISC and PCI standards, and different acceleration limits and frequency limits are given. *CCP-016* states that when the first natural vibration frequency of a floor is lower than 4.2 times the fundamental frequency of the pedestrian load, resonance will occur; and when the first natural frequency of the floor is greater than 4.2 times the fundamental frequency of the pedestrian load, the floor vibration is mainly caused by effective impact. Two methods that can be used to evaluate the structural comfort are proposed in the UK standards. One is the evaluation method based on the response factor, which assumes that the floor vibration is continuous and of the same amplitude. Another evaluation method is based on the vibration dose value, which considers a possible pause in the vibration process and the impact of different vibration amplitudes on human comfort, and can be used for long-term evaluation of the comfort degree. In a comfort evaluation, the analytical or numerical calculation results are usually directly compared with the standard values used in the engineering community to determine the comfort level. However, the span of a long-span floor can reach tens of meters, and pedestrian comfort is affected by the spatial and temporal distribution characteristics of the stimulated points and the feeling points. Therefore, it is of great significance to study the human-induced-vibration response distribution of long-span floors for reasonable evaluation of comfort performance.

Many scholars have studied the vibration comfort of structures. The most basic research on comfort is the human motivation model. The human-induced motivation mode is the most basic research of comfort. In 1961, Harper completed the earliest walking load test with a force-measuring plate, and stated that the walking load curve was M-shaped [6]. Subsequently, several researchers tested the load of human walking using the direct or indirect method [7,14] and analyzed the influence of walking speed, shoe type, ground characteristics, and other factors on the load model. Chen et al. adopted optical motion-capture technology in which the reflective marks of key parts of the human body were captured by high-speed infrared cameras [15]. This was used to identify the human motion space trajectory to obtain human walking parameters. At present, the time-domain model of load is mainly used in the comfort analysis of structures. The Fourier series model was the most commonly used in existing pedestrian load model research. The dynamic load factor is included in the Fourier series model, and the load mode is directly affected by the value of the dynamic load factor. Therefore, the dynamic load factor has been studied by many researchers [8,16]. In conclusion, it is generally believed that the vertical first-order dynamic load factor is around 0.3~0.5. In terms of the comfort analysis method, the time-domain analysis method based on finite elements is an effective method. Zhu et al. considered the interaction between pedestrians and structures, and used an ANSYS finite element software simulation to study the structural vibration comfort of a two-story cantilevered steel truss floor deck in the Gansu Science and Technology Museum as the engineering background [17]. Cao et al. conducted an experimental study of the human-induced vibration of a large-span composite floor based on a single-person foot-load model in order to meet the comfort requirements and control the floor [18]. Wang et al. carried out an human-induced-vibration test and an ANSYS finite element analysis of a large-scale glulam arch bridge model in order to study the human-induced-vibration characteristics of a wooden-structure pedestrian bridge [19]. Peak acceleration is used for comfort performance evaluation. Based on the walking route method, the vibration comfort performance of a steel Vierendeel sandwich plate was analyzed by Jiang et al. under different walking routes [20]. Based on the existing load model, a finite element numerical calculation method was adopted in the above research to analyze the maximum response of different types of structures under human-induced excitation and to evaluate the comfort. However, for long-span floors, the vibration responses of different positions on the floor will differ greatly under human-induced excitation. If the maximum response is simply used to evaluate the comfort of the entire floor, the evaluation will inevitably be too conservative. Therefore, for large-span floor structures such as a Vierendeel sandwich plate, it is necessary to analyze the vibration-response-distribution characteristics of the floor under human-induced excitation in order to provide a basis for the reasonable evaluation of comfort performance.

For the evaluation of comfort, there are mainly two methods: one is to limit the natural vertical vibration frequency of the floor above a certain value, which is called the frequency threshold method. Another method requires that the dynamic response (such as acceleration and speed) of the floor under a given human-induced load does not exceed a certain limit, which is called the dynamic response threshold method. The dynamic response threshold method is widely used because it considers multiple factors of floor vibration and can better evaluate comfort. In the evaluation, the vibration acceleration response is often used as the index. Under the British BS 5400 standard [21], the peak acceleration of the structure is used as the pedestrian comfort limit index, and a function of the vertical first-order frequency of the floor is given as the acceleration limit. Similar acceleration limits are also given in the European EN 1990 standard [22]. The International Organization for Standardization stipulates in its ISO 10137 that when the acceleration response is less than the vibration comfort limit, the comfort is considered to meet the requirements; otherwise, the comfort is considered to not meet the requirements [23]. The German EN03 standard states that pedestrian comfort cannot be simply divided into comfort and discomfort. The vibration comfort level should be divided in detail according to the natural vibration frequency and structural acceleration response [24]. The Chinese specifications GB50010-2010 [25] and JGJ3-2010 [26] refer to ISO standards, and the peak acceleration limits for floors with different natural frequencies are given. Among the above standards, the evaluation of peak acceleration is adopted in BS 5400, EN 1990, EN03, and Chinese standards, and the evaluation of peak acceleration and RMS acceleration is adopted in ISO 10137. In the application of standards, Fiore et al. proposed a practical probabilistic method for evaluating bridge reliability based on a histogram. Useful estimates of the probability of exceeding the predefined human sensitivity limit were provided by histograms [27]. The above published standards provides references for the evaluation of the human-induced-vibration comfort of large-span Vierendeel sandwich plates.

In this paper, the steel Vierendeel sandwich plate at the Guizhou Museum was taken as the research object. According to the functional characteristics of steel Vierendeel sandwich plates, the response characteristics of the floor with time and space were analyzed using a time-domain method, and the effects of different factors on the acceleration-response distribution were studied. A corresponding distribution mathematical model was constructed, and a comfort evaluation method based on the floor area comfort assurance rate was proposed.

2. Distribution Rule of Vibration Response of Vierendeel Sandwich Plates

2.1. Finite Element Model

Orthogonally placed Vierendeel sandwich plates are widely used in engineering, and are the most representative. A steel Vierendeel sandwich plate was used as the analysis object in this study. A steel Vierendeel sandwich plate is a two-way stress structure, and its plane aspect ratio is generally close to one. The size parameters of the structure are shown in Table 1, and schematic diagram showing the size of each component of the structure is presented in Figure 2. In order to analyze the vibration response characteristics of a steel Vierendeel sandwich plate under human-induced excitation, a typical size floor slab was designed, and the basic size of the structure is shown in Table 2. The surface concrete slab structure adopted C30 grade concrete, and the steel structures such as the top and bottom chords and the shear connectors adopted Q345B steel. The material parameters are shown in Table 3. Jiang et al. showed that the dynamic characteristics of a solid–shell model of a steel Vierendeel sandwich plate were closest to the actual structure [28]. Therefore, a solid–shell finite element model of the steel Vierendeel sandwich plate was established in ANSYS software, as shown in Figure 3. Since the top (bottom) chords, shear connectors, and ribbed stiffener of the T-section steel were all thin-walled structures, four-node spatial elastic shell elements were used. The thickness of surface concrete slab was larger than the span of the slab, so eight-node 3D solid elements were used. In practical engineering, a surface concrete slab is reliably connected to the top chords through studs. In the finite element model, the top chords and concrete slab adopted a shell element and a solid element, respectively, and the numbers of node degrees of freedom of these two element types were different. Therefore, the degrees of freedom of the top chords' flange and the concrete slab's nodes at the corresponding positions were coupled to realize the deformation coordination between different elements. Steel Vierendeel sandwich plates are generally rigidly connected to steel columns at the surrounding nodes. Ref. [28] discussed the influence of three different boundary conditions on the dynamic characteristics of the floor. The results showed that the dynamic characteristics of the floor were closer to the measured values when fixed constraints were applied to the intersection of the grid. In summary, in the numerical calculation, the translational and rotational degrees of freedom of the grid intersections around the floor were constrained (Figure 2a).

Table 1. Structural size parameter symbols.

Parameter Name	Symbol	Parameter Name	Symbol
Floor span	L	Grid size	a
Grid number	n	Total height (excluding concrete slab)	h
Chord height	h_1	Chord width	b_1
Chord flange thickness	t_f	Chord web thickness	t_w
Shear connector thickness	t_p	Ribbed stiffener width	b_t
Concrete slab thickness	δ	Shear connector width	b_2
Ribbed stiffener thickness	t_t	-	-

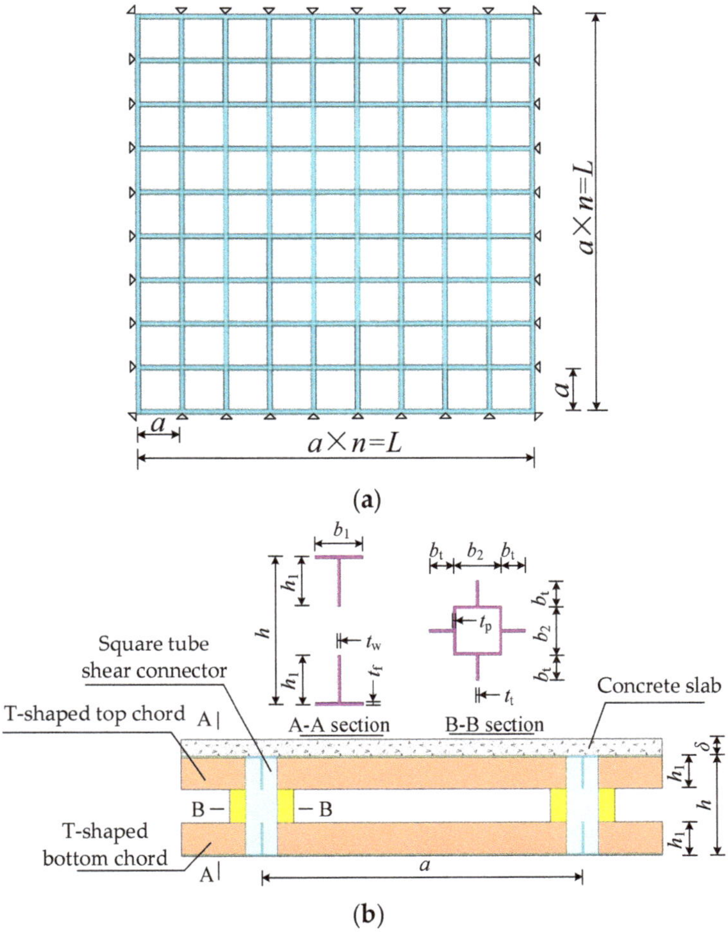

Figure 2. Structural size diagram: (**a**) plane layout; (**b**) structural details.

Table 2. Basic parameters of model structure.

Span/L (m)	**Grid Size/a (m)**	**Overall Height/h (mm)**	**Chord Height/h_1 (mm)**	**Chord Width/b_1 (mm)**
18	2	600	200	200
Chord flange thickness/t_f(mm)	Ribbed stiffener width/b_t (mm)	Shear connector thickness/t_p (mm)	Ribbed stiffener thickness/t_t (mm)	Concrete slab thickness/δ (mm)
8	100	6	8	100

Table 3. Material physical parameters.

Material	**Modulus of Elasticity (MPa)**	**Poisson's Ratio**	**Density (kg/m^3)**
Q345 steel	2.06×10^5	0.3	7850
C30 concrete	3.00×10^4	0.2	2500

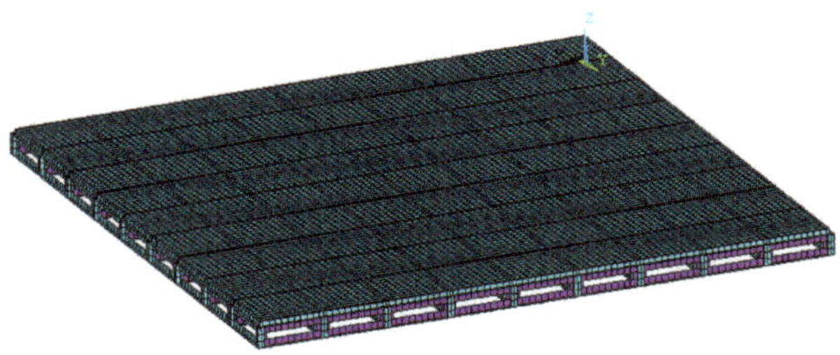

Figure 3. Finite element model of Vierendeel sandwich plate.

The human-induced excitation load was applied in the finite element model of the floor, and a finite element transient dynamic response analysis was carried out to acquire the distribution rule of the acceleration response. The step-by-step integration method was used in the analysis, and the Rayleigh damping model was used; the damping ratio reference value was 0.02. While considering that a multi-person excitation condition is generally expressed as the product of the calculation results of the single-person excitation condition and the effect coefficient, we analyzed the corresponding characteristics of the floor under fixed-point excitation; that is, under the condition of marching on the spot by a single person. The load model adopted the walking excitation parameter model recommended by the International Association for Bridge and Structural Engineering (IABSE), as shown in Equation (1):

$$F_P(t) = G[1 + \sum_{i=1}^{3} \alpha_i \sin(2i\pi f_s t - \Phi_i)] \quad (1)$$

where F_P is the exciting force, G is the weight of human beings, i is the ith order, α_i is the dynamic load factor of the ith order load frequency, f_s is the walk frequency, and Φ_i is the phase angle of the ith order load frequency. The dynamic load factor has been studied by many researchers; Blanchard proposed a first-order sine harmonic model with a dynamic load factor of 0.257 [29]. Based on three-dimensional motion capture technology and a large amount of data, Chen Jun gave the value of the dynamic load factor: $\alpha_1 = 0.235f_s - 0.2010$, $\alpha_2 = 0.0949$, $\alpha_3 = 0.0523$, which was in line with the body characteristics of Chinese people [8], so this value was used as the calculation condition of this paper.

Matsumoto analyzed the probability distribution characteristics of walking frequency through a random sampling test and found that human walking frequency obeyed the normal distribution, with a mean value of 2.0 Hz and a standard deviation of 0.173 Hz [30]. Han X believed that human self-weight obeyed a normal distribution, with a mean of 700 N and a standard deviation of 145 N [31]. The value range of load parameters is shown in Table 4. Therefore, in the analysis of human-induced-vibration response characteristics, the walking frequency was 2 Hz and the weight of the human was 700 N.

Table 4. Value range of load parameters.

Parameter	Mean Value	Variation Range
Walking frequency/Hz	2.0	1.6/1.8/2.0/2.2/2.4
Human weight/N	700	555/600/650/700/750/800/845

2.2. Response Characteristics of Human-Induced Vibration

To perform a transient analysis of the acceleration response of the floor, the position of both the pedestrian load and the structural vibration response receiver should be determined first. There are two main principles for selecting loading points: one is whether the response generated by the excitation at this point is the most unfavorable, and the other is whether the excitation at this point occurs easily under actual working conditions. The location of the vibration receiver also follows two principles: whether the position of the receiver is the most unfavorable, and where it is prone to occur. In view of this, two issues were mainly analyzed in this paper: the effect of different point of excitations on the peak response and the acceleration response distribution at different positions under the same point of excitation.

2.2.1. Effect of Substructure on Floor Acceleration

As a special form of floor structure, a Vierendeel sandwich plate has great differences in the section size at each point of the plane. Figure 4 shows a grid diagram of the Vierendeel sandwich plate. In the figure, the points of shear connectors on the slab are shown by D, F, G and I; the points of the top chords on the Vierendeel beam are shown by B, C, E and H; and A is the point at the center of the plane of the concrete slab.

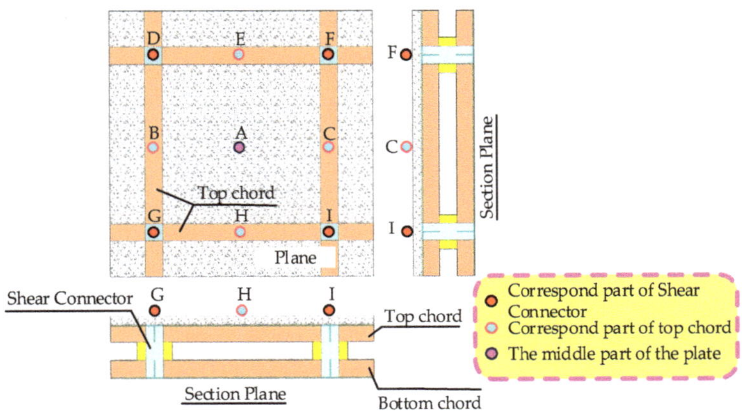

Figure 4. Schematic diagram of typical parts of floor.

In order to analyze whether the acceleration response at different positions on the floor was affected by the characteristics of the substructure, the peak acceleration responses at the different positions shown in Figure 4 were compared, as shown in Figure 5. It can be seen in Figure 5 that under the conditions of different concrete slab thicknesses, an approximately linear distribution was displayed by the DEF and BAC values. This showed that the peak acceleration response on the Vierendeel sandwich plate was hardly affected by the different positions, and the effect of this factor could be ignored when selecting the sensing point.

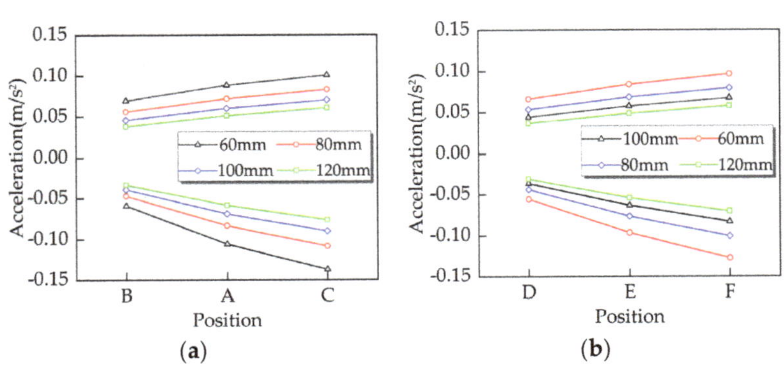

Figure 5. The effect of different positions on the plane on the peak acceleration: (**a**) point DEF; (**b**) point BAC.

2.2.2. Effect of Point of Excitation Position on Floor Acceleration

To study the effect of the point-of-excitation position on the peak dynamic response of the floor, nine typical points of excitation (points A~I) were determined on the floor; these positions are shown in Figure 6.

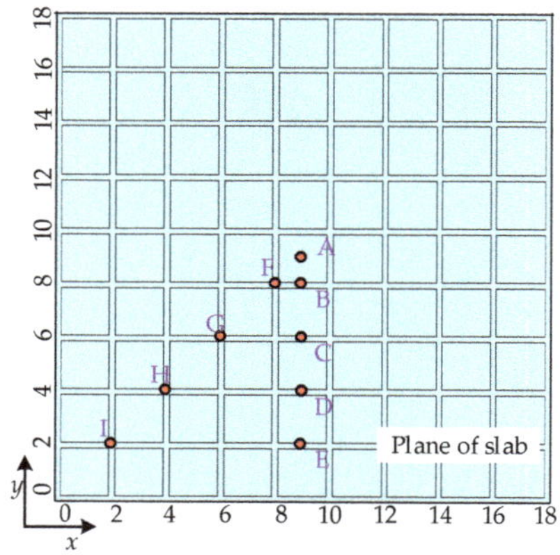

Figure 6. Distribution diagram of points of excitation.

The peak acceleration distribution of each part of the floor under fixed-point excitation when the point of excitation was located at the geometric center of the floor (point A) is shown in Figure 7. It can be seen in the figure that the acceleration response was funnel-shaped on the floor: the closer to the point of excitation, the greater the absolute value of the peak acceleration.

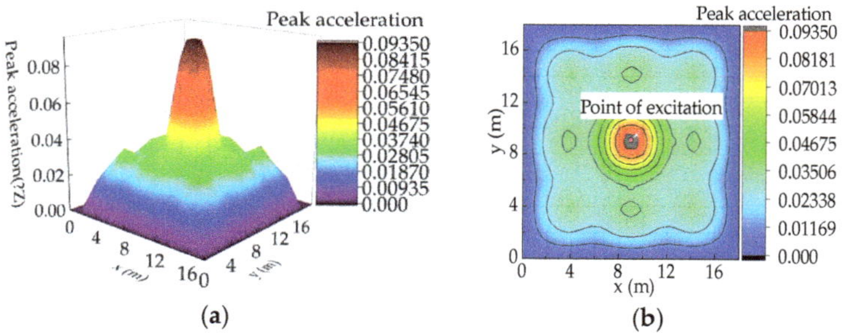

Figure 7. Acceleration distribution of point A: (**a**) peak acceleration diagram; (**b**) peak acceleration contour map.

When the point of excitation was located on the nongeometric center of the floor (points B~I), the dynamic response of the floor was calculated and the acceleration peaks at different positions of the floor were extracted; the contour map was drawn as shown in Figure 8. As can be seen in Figure 8, in general, the closer to the point of excitation, the greater the floor's response peak, and its maximum value is located at the point of excitation. In addition, the closer to the point of excitation, the denser the contour line. The acceleration peak on the floor decreased exponentially with the increase in the distance between the sensing point and the point of excitation. It should be noted that when the point of excitation was close to the constrained edge of the floor (points E and I), the acceleration peak of floor was no longer at the point of excitation, but was near the point of excitation.

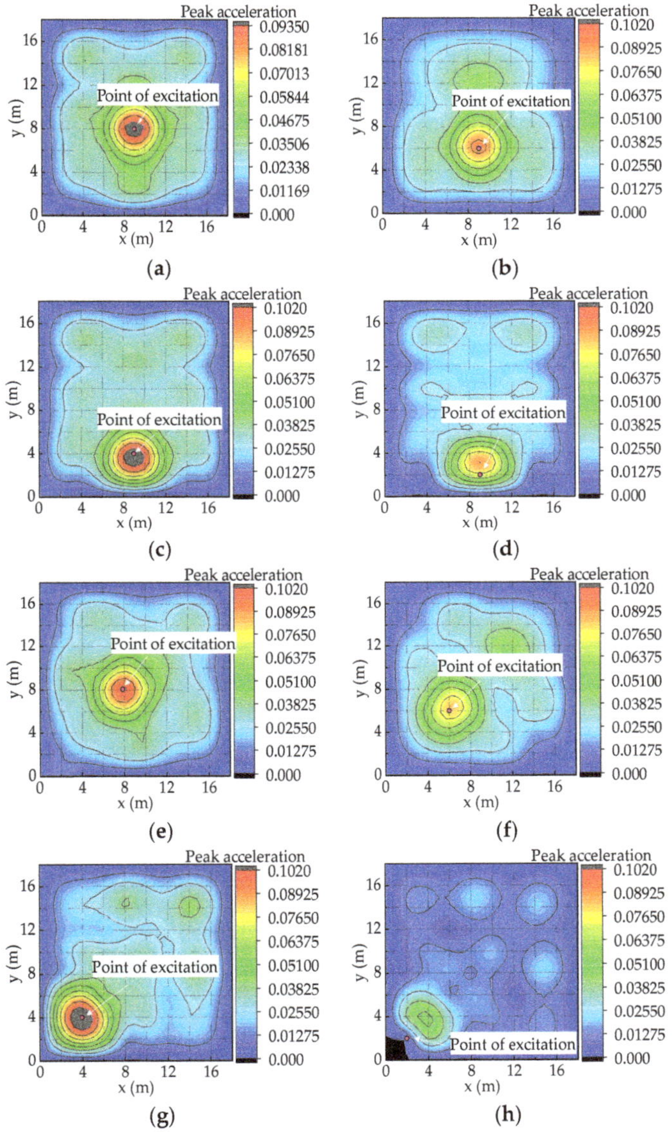

Figure 8. Contour map of peak acceleration: (**a**) point B; (**b**) point C; (**c**) point D; (**d**) point E; (**e**) point F; (**f**) point G; (**g**) point H; (**h**) point I.

2.2.3. Effect of Damping on Floor Acceleration

The amount of mechanical energy loss in a floor system is described as damping, and is usually expressed as the ratio of actual damping to critical damping; that is, the damping ratio. During the calculation, the first two natural frequencies and corresponding damping ratios were directly defined, then the damping in the model was calibrated using mass and stiffness matrix modifiers, which can be obtained using Equations (2) and (3):

$$\alpha = \frac{2\omega_1\omega_2(\omega_1\xi_2 - \omega_2\xi_1)}{{\omega_1}^2 - {\omega_2}^2} \quad (2)$$

$$\beta = \frac{2(\omega_1 \xi_1 - \omega_2 \xi_2)}{\omega_1{}^2 - \omega_2{}^2} \tag{3}$$

where α and β are the mass and stiffness matrix modifiers, respectively; ω_i is the natural frequency of the ith order; and ξ_i is the damping ratio of the ith mode (i = 1, 2).

The damping matrix [C] is a linear combination of the mass matrix $[M]$ and the stiffness matrix $[K]$. This damping, which is called Rayleigh damping, can be calculated using Equation (4):

$$[C] = \alpha[M] + \beta[K] \tag{4}$$

Floor damping includes floor damping and nonstructural damping. Structural materials, the floor system, the building structure, and other factors will affect the damping; as affected by various factors, floor damping is generally between 2% and 10%. The effect of damping change on the acceleration response of a floor slab under a human-induced load is discussed in this paper. We took the point of excitation at the geometric center as an example, and assumed that the coordinate origin coincided with the geometric center of the floor. The peak acceleration response distribution of the hollow steel sandwich plate when the floor damping was varied between 2% and 8% is shown in Figure 9.

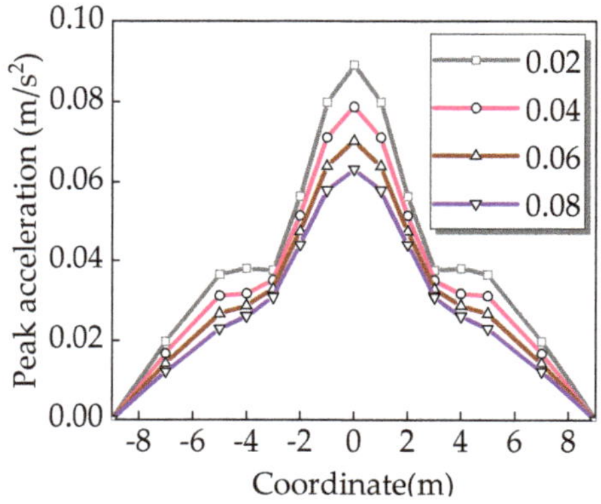

Figure 9. Effect of damping on peak acceleration response.

It can be seen in the above figure that when the damping was increased, the peak acceleration around the floor decreased significantly. This showed that increase in the structural damping had a significant effect on reducing the vibration response of the floor.

2.2.4. Effect of Load Parameters on Floor Acceleration

The main parameters that affect continuous walking excitation are walking frequency, self-weight, and other factors. Experimental research has shown that in general, the frequency of natural human walking is between 1.6 and 2.4 Hz. In order to study the influence of walking frequency on the structural response, this paper constructed pedestrian load curves under different frequencies, as shown in Figure 10, and the acceleration response of the floor under different pedestrian excitation frequencies was respectively calculated. The peak acceleration response distribution at different positions on the floor is shown in Figure 11 (assuming that the coordinate origin coincided with the geometric center of the floor). It can be seen in the figure that the distribution of peak acceleration on the floor was similar under different excitation frequencies. With an increase in the excitation frequency, the peak acceleration of the floor increased gradually.

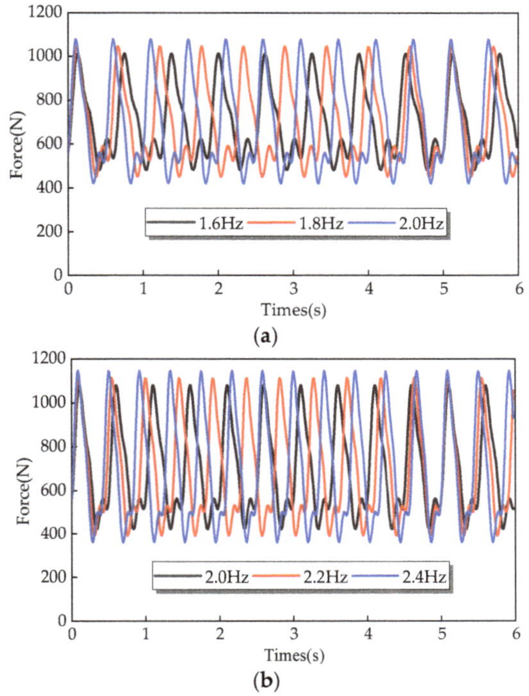

Figure 10. Pedestrian loads with different frequencies: (**a**) walking frequencies of 1.6 Hz, 1.8 Hz, and 2.0 Hz; (**b**) walking frequencies of 2.0 Hz, 2.2 Hz, and 2.4 Hz.

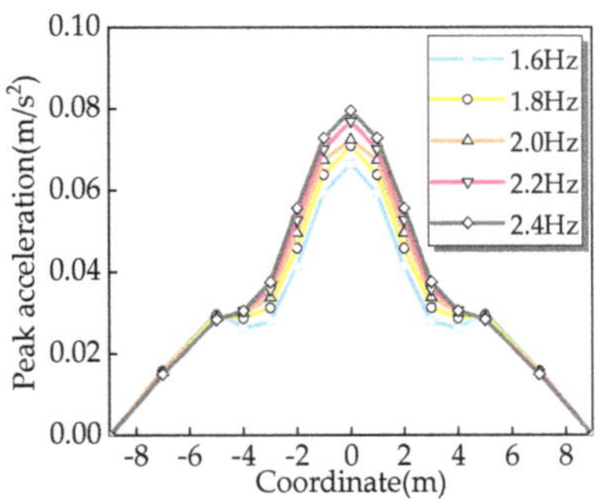

Figure 11. Effect of load frequency on peak acceleration response.

The peak value of the pedestrian load curve and the human-induced-load response of a floor are directly affected by human weight. The general value for a pedestrian load is 700 N according to [32], while the body weight of adults varies in the range of 700 ± 145 N according to [31]. In order to analyze the influence of human body weight on the peak response of the floor, load curves with adult self-weights of 555 N, 600 N, 650 N, 700 N, 750 N, 800 N, and 845 N were constructed (Figure 12), and then we analyzed the acceleration response. The peak acceleration distribution of the floor is shown in Figure 13 (assuming that the coordinate origin coincided with the geometric center of the floor). It can be seen in the figure that self-weight had a significant impact on the peak response of the floor caused by a human-induced load. With the increase in self-weight, the peak value of floor response increased significantly.

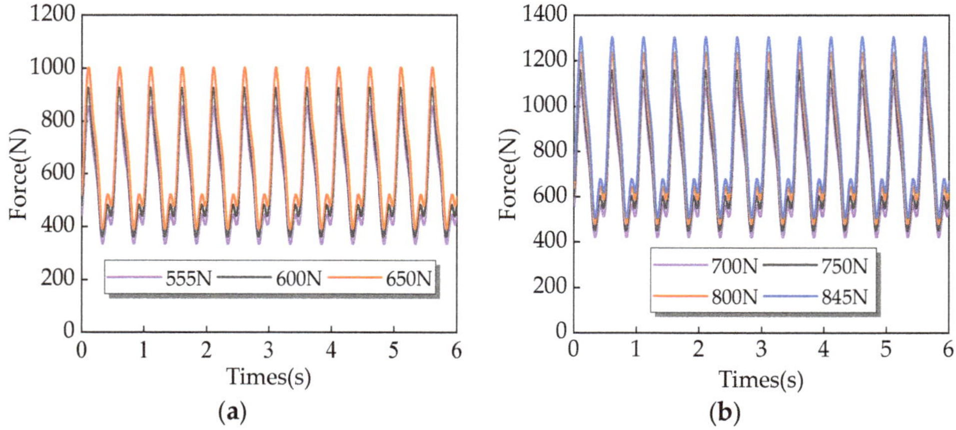

Figure 12. Pedestrian loads with different human weights: (**a**) human weights of 555 N, 600 N, and 650 N; (**b**) human weights of 700 N, 750 N, 800 N, and 845 N.

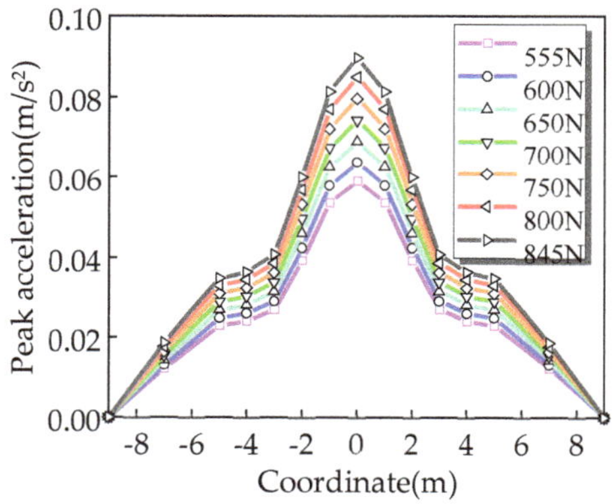

Figure 13. Effect of self-weight on peak acceleration response.

2.2.5. Effect of Structural Parameters on Floor Acceleration

According to [20], the main factors that affect the vertical dynamic characteristics of a Vierendeel sandwich plate are the floor span, grid size, and concrete slab thickness, while the secondary factors are chord height and shear connector thickness. In this paper, these five factors were selected to analyze the effects of structural parameters.

When analyzing the effects of structural parameters, the selected load mode was the same as given above. The human walking frequency was 2 Hz, the human weight was 700 N, and the point of excitation was the geometric center of the floor; that is, point A in Figure 6. When changing the structural parameters, only one variable was changed each time based on the basic model. The basic model parameters were the same as those given in Table 2, and the range of parameter variations is shown in Table 5. Assuming that the coordinate origin coincided with the geometric center of the floor, the damping ratio is 0.05. Through transient dynamic analysis, the peak acceleration profiles at different positions on the floor were obtained, as shown in Figure 14.

Table 5. Variation range of parameters.

Span (m)	Grid Size (mm)	Concrete Slab Thickness (mm)	Chord Height (mm)	Shear Connector Thickness (mm)
14	1500	60	160	5
16	1800	80	180	6
18	2000	100	200	7
20	2250	120	220	8
22	-	140	240	9
24	-	-	-	10

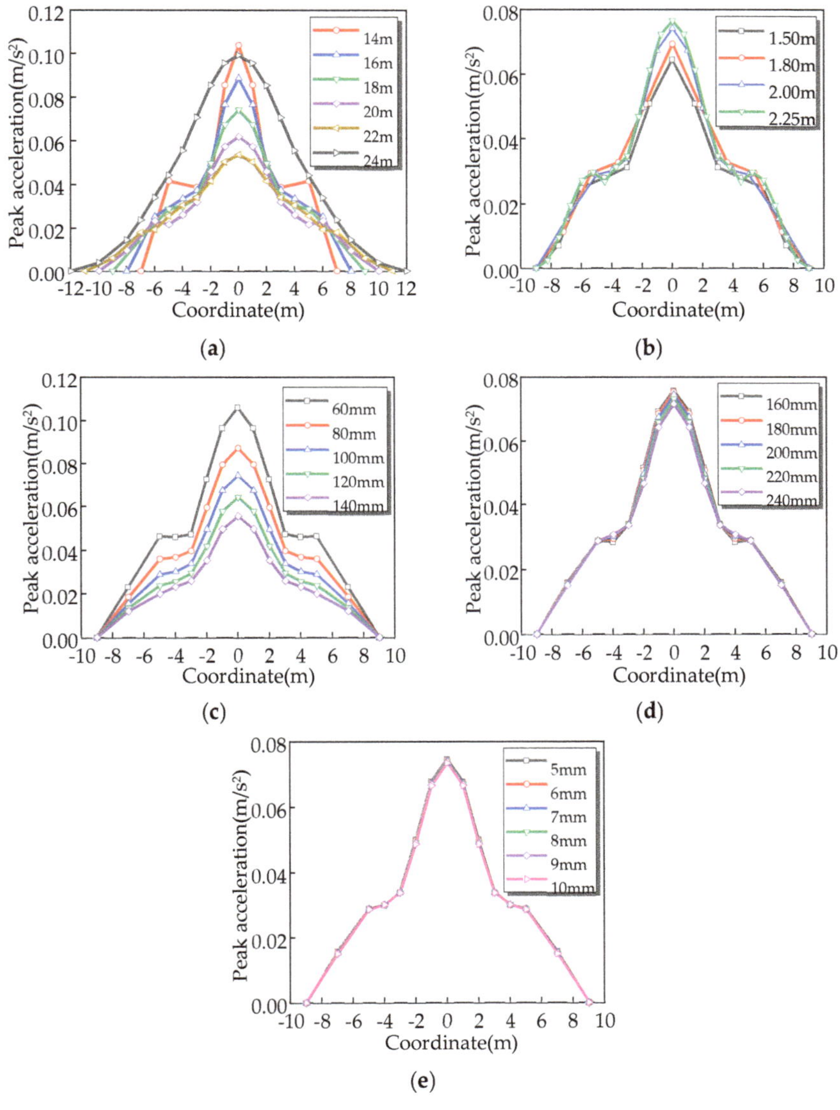

Figure 14. Effect of structural parameters on peak acceleration response: (**a**) effect of floor span; (**b**) effect of grid size; (**c**) effect of concrete slab thickness; (**d**) effect of chord height; (**e**) effect of shear connector thickness.

Figure 14 shows that the human-induced acceleration response was significantly affected by the span of the hollow sandwich plate and the thickness of the concrete plate, and the calculation results were affected by the grid size to some extent. The effects of the chord height and shear connector thickness were very small, so they could not be considered. With the increase of the span of the Vierendeel sandwich plate, the peak acceleration in the center of the span decreased. However, when the span was increased to 24 m (the fundamental frequency was 2 Hz), the peak acceleration response of the floor increased sharply, indicating that the floor resonated with people. With the increase in the thickness of the concrete slab, the peak acceleration response of the floor decreased obviously, and the peak acceleration response decreased with the decrease in the grid size.

3. Construction of Vibration Response Distribution Model

The response characteristics of the steel Vierendeel sandwich plate under human-induced fixed-point excitation were analyzed, and showed that the peak acceleration response of the floor was funnel-shaped in the plane, and the specific distribution shape was related to load parameters, damping, structural parameters, and resonance or lack thereof. In order to establish the distribution model of the acceleration response and position correlation of the Vierendeel sandwich plate under human-induced fixed-point excitation, based on the distribution characteristics of the acceleration response, this paper constructed the distribution model function, and different numerical results were fitted to determine the parameters of the distribution model.

3.1. Gaussian Distribution Model

Floor response can be roughly divided into two categories based on resonance or a lack thereof. When the floor response resonates due to human-induced excitation, the response distribution curve can be expressed by a Gaussian function; and when the curve is symmetrical about the x-axis, it can be expressed using Equation (5)—the curve is shown in Figure 15a. When the floor response does not resonate due to human-induced excitation, the response distribution curve can be expressed by a piecewise function (Equation (6)); the curve is shown in Figure 15b.

$$f(x) = Ae^{-\frac{x^2}{2w^2}} \tag{5}$$

where A is the height of the curve and w is a shape parameter.

$$f(x) = \begin{cases} \frac{2c}{L}x + c & (-\frac{L}{2} \le x < -\frac{L}{3}) \\ \frac{c}{3} + Ae^{-\frac{x^2}{2w^2}} & (-\frac{L}{3} \le x \le \frac{L}{3}) \\ -\frac{2c}{L}x + c & (\frac{L}{3} < x \le \frac{L}{2}) \end{cases} \tag{6}$$

where A is the height of Gaussian distribution curve, c is the parameter, and L is the span of the steel Vierendeel sandwich plate.

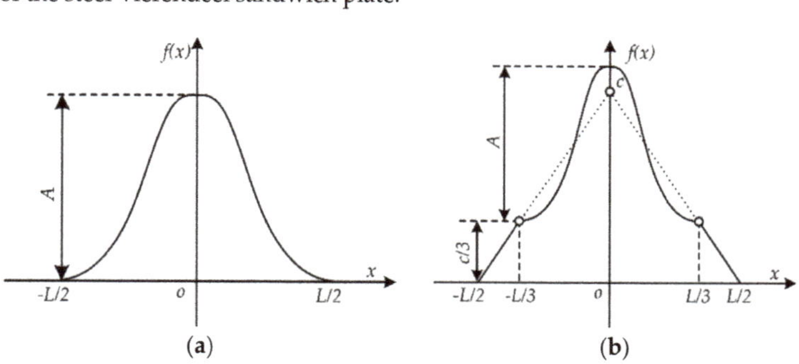

Figure 15. Acceleration response distribution model: (**a**) resonance; (**b**) nonresonance.

The above model is a plane model. For the square plane of the Vierendeel sandwich plate, according to Figure 15b, it can be assumed that the acceleration peak surface was centrosymmetric around the z-axis (x = 0, y = 0). Therefore, the plane Gaussian model could be extended to three-dimensional space, as shown in Equations (7) and (8):

$$f(x,y) = Ae^{-\frac{x^2+y^2}{2w^2}} \tag{7}$$

$$f(x,y) = \begin{cases} \frac{c}{3} + Ae^{-\frac{x^2+y^2}{2w^2}} & (\sqrt{x^2+y^2} \leq \frac{L}{3}) \\ -\frac{2c}{L}\sqrt{x^2+y^2} + c & (\frac{L}{3} < \sqrt{x^2+y^2} \leq \frac{L}{2}) \end{cases} \tag{8}$$

3.2. Model Parameter Calculation and Quality Evaluation

3.2.1. Parameter Calculation

It can be seen in the above analysis that the peak acceleration distribution was affected by factors, including damping, load frequency, pedestrian self-weight, floor span, grid size, concrete slab thickness, chord height, shear connector thickness, and other structural parameters. Among them, damping, pedestrian self-weight, floor span, and concrete slab thickness were the factors that had a greater influence. In order to simplify the parameter-estimation process, according to [33], the mean value of self-weight was 700 N, the damping ratio was 0.02, the mean value of human walking frequency was 2 Hz, the grid size was 2 m, the chord height was 150 mm, and the shear connector thickness was 5 mm. Different floor span and concrete slab thicknesses were considered, the acceleration response distribution of the square steel Vierendeel sandwich plate was supported by peripheral columns under single-person and fixed-point excitation, and the point of excitation was located in the middle of the span. The specific parameters of the analysis model are shown in Table 6, and the common parameters are shown in Table 7. The analysis model was divided into 2 groups with 5 in each group, for a total of 10.

Table 6. Model parameters.

Specimen	M1 Model					M2 Model				
	M1-1	M1-2	M1-3	M1-4	M1-5	M2-1	M2-2	M2-3	M2-4	M2-5
Span (m)	16	16	16	16	16	24	24	24	24	24
Concrete slab thickness (mm)	60	80	100	120	140	60	80	100	120	140

Table 7. Model common parameters.

Grid Size (mm)	Overall Height (mm)	Chord Height(mm)	Chord Width (mm)
2000	600	150	200
Chord flange thickness(mm)	Chord web thickness (mm)	Shear connector thickness (mm)	Ribbed stiffener thickness (mm)
10	8	5	100

The calculation results of each model were fitted by a nonlinear curve following the formulas of Equations (5) and (6); the fitting curve is shown in Figure 16.

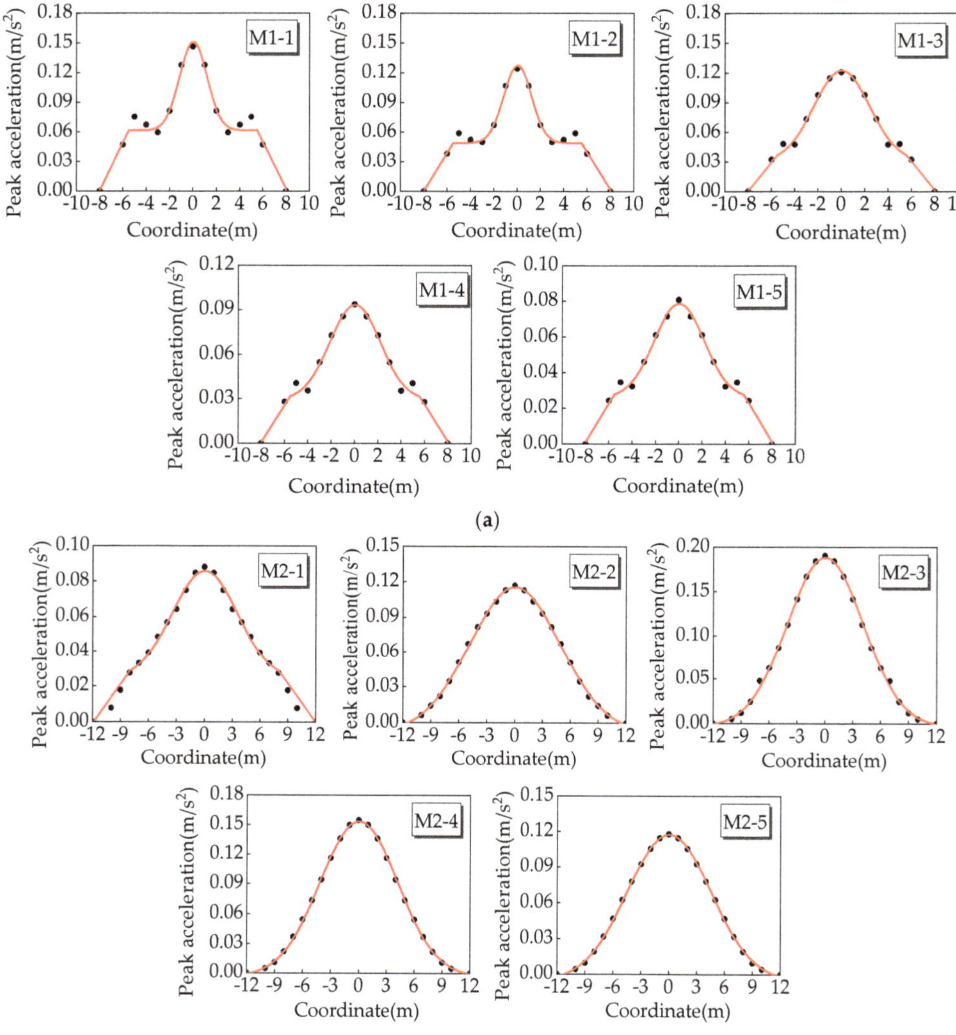

Figure 16. Nonlinear fitting curves: (**a**) model M1; (**b**) model M2.

3.2.2. Model Quality Assessment

Assuming that the numerical results were reliable, the accuracy of the regression model had to be evaluated [33]. In the regression analysis, the regression effect was characterized by the R^2 (coefficient of determination), and R was the ratio of the sum of regression square and the sum of total deviation square in the regression analysis. The larger the value, the more accurate the model was and the more significant the regression effect was. The R^2 can be calculated according to Equation (9):

$$R^2 = 1 - \frac{\Sigma(y - \hat{y})^2}{\Sigma(y - \overline{y})^2} \quad (9)$$

where $\hat{y}$ is the estimated value, $\overline{y}$ is the mean value, and y is the actual value.

In this paper, the estimated value $\hat{y}$, mean value $\overline{y}$, and actual value y corresponded to the calculated value of the finite element, the fitting value of the parametric model, and the mean value of the calculated value of the finite element model, respectively; the above parameters were brought into Equation (9). The calculated values are shown in Table 8. The natural frequencies of the models M1-1–M2-1 were 2.29–5.53 Hz, which were different from the human excitation frequency of 2 Hz, and since it was not easy to resonate, it was fitted according to the nonresonant model. The natural frequency of models M2-2–M2-5 was between 1.95 and 2.11 Hz, which was very close to the human induced excitation frequency of 2 Hz, and since it was very easy to resonate, it was fitted according to the resonant model.

Table 8. Parameter values of curve model.

Model		c	A	w	R^2	Adjusted R^2
M1	M1-1	0.186	0.089	1.18	0.92	0.89
	M1-2	0.147	0.079	1.2	0.95	0.93
	M1-3	0.093	0.092	2.39	0.98	0.97
	M1-4	0.087	0.064	2.17	0.97	0.96
	M1-5	0.078	0.053	2.15	0.98	0.97
M2	M2-1	0.072	0.062	3.54	0.99	0.99
	M2-2	-	0.123	4.85	0.99	0.99
	M2-3	-	0.19	4.08	0.99	0.99
	M2-4	-	0.157	4.18	0.99	0.99
	M2-5	-	0.123	4.52	0.99	0.99

It can be seen in Table 8 that the R^2 values for all models were between 0.89 and 0.99, indicating that the acceleration plane distribution model established in this paper had a very good fitting effect on the finite element calculation values.

3.3. Experimental Verification of Distribution Model

The New Museum of Guizhou Province is located in Guiyang City, Guizhou Province, China. The main structure was completed in September 2014, and various forms of long-span Vierendeel sandwich plates were adopted. In this paper, a steel–concrete composite Vierendeel sandwich plate with a 15.6 m × 17.5 m span orthogonal and upright grid was selected for analysis. Its location and a site photo are shown in Figure 17. The top (bottom) chords are T-shaped steel, the shear connectors are square steel pipe, one side of the floor is supported by the shear wall, and the other three sides are supported by the format frame wall.

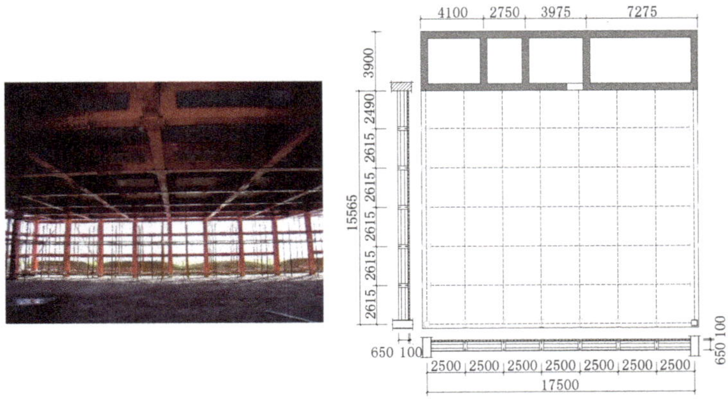

Figure 17. Structural diagram of the floor: (**a**) site photo; (**b**) plane layout.

There are five floors of the long-span floor, which are used as a conference hall and 5D cinema, respectively. Cinemas and conference rooms are large public buildings with high crowd densities. Under crowd excitation, their comfort degree affects the experience and satisfaction of tourists during operation. Because the evaluation of the floor comfort degree is mostly related to the structural dynamic response, the acceleration response distribution amplitude is taken as the index, and through the analysis of human-induced-vibration response of the floor, the dynamic characteristics of the steel Vierendeel sandwich plate can be deeply understood.

A 15.6 m × 17.5 m span Vierendeel sandwich plate in the New Museum of Guizhou Province was tested on site. The geometric center of the floor was selected as the point of excitation, and the excitation method was a single person standing in place. The experimental equipment mainly included five TST126V dynamic signal sensors and a TAISITE TST5912 dynamic signal acquisition and analysis system. The test site and five acceleration sensors (A~E) were arranged as shown in Figure 18. In the experiment, the tester was instructed to march on the spot at a step frequency of 2.0 Hz for 30 s at the center of the sandwich plate. The acceleration sensors were used to collect the response signals at different positions; the data acquisition time was 1 min and the frame rate was 100 Hz. After the collection, the data were preliminarily sorted, and the acceleration responses of different measuring points on the floor during this period were analyzed; thus, the peak acceleration of the floor at different positions could be obtained.

(a)

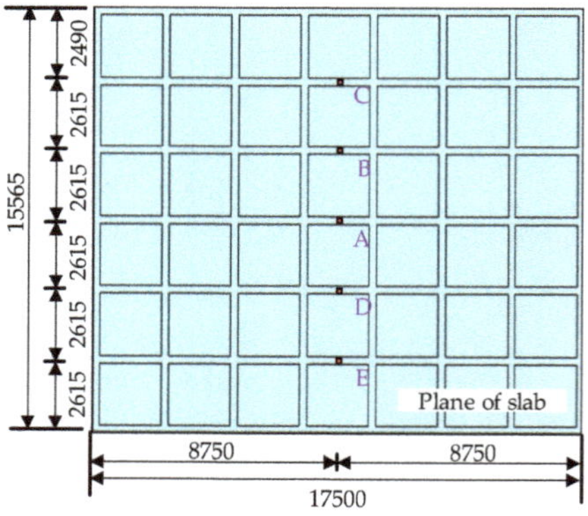

(b)

Figure 18. Dynamic response test: (**a**) test site; (**b**) layout position of acceleration sensors.

According to the actual measurement size, the finite element model of the 15.6 m × 17.5 m span open-web sandwich plate was established at a ratio of 1:1. The translational and rotational degrees of freedom of the grid intersections around the floor were constrained. We set the point of excitation at the geometric center of the floor (point A) for the transient analysis, and peak acceleration responses at different locations on the floor are recorded. The nonlinear curve was fitted to the finite element calculation results according to Equation (6) to obtain the model curve. A comparison between the model curve and the peak acceleration response of the measured points (A~E) on the project site is shown in Figure 19. It can be seen in Figure 19 that the model curve was consistent with the measured data. Table 9 shows a comparison between the fitting results and the measurement results for each measuring point. The maximum measurement error was less than 6%, indicating that the method proposed in this paper could effectively calculate the acceleration distribution characteristics of the sandwich plate.

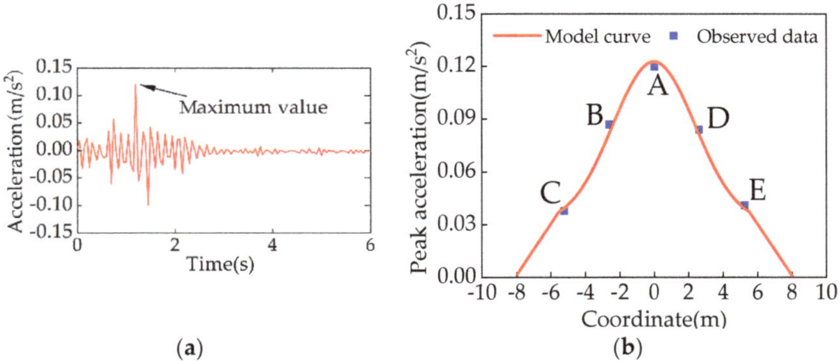

Figure 19. Peak acceleration response: (**a**) acceleration responses at position "A"; (**b**) measured acceleration value and model curve.

Table 9. Comparison of measurement results.

Sensor Location	Fitting Value (m/s²)	Measured Value (m/s²)	Error (%)
A	0.123	0.120	2.50%
B	0.082	0.087	5.74%
C	0.039	0.038	2.63%
D	0.082	0.084	2.30%
E	0.039	0.041	4.88%

4. Comfort Evaluation Method Based on Comfort Assurance Rate

At present, when evaluating the comfort degree, the peak response evaluation criterion is adopted in the codes of various countries; that is, the response of the floor under a human-induced load is not greater than the specified value. However, through the analysis conducted in this paper, it was found that for the large-span steel Vierendeel sandwich plate structure, the response peak distribution was funnel-shaped (Figure 7). The area with a large response only accounted for a small part of the total area of the floor. The maximum value was located in the center of the floor, and decayed sharply to the surrounding areas. In addition, the value at each point of the floor was also the maximum value on the acceleration response time history curve, and the duration of the maximum value accounted for a very small proportion of the entire response process, as shown in Figure 20. In addition, the maximum value on the acceleration response time history curve was taken as the value of each point of the floor, and the duration of the maximum value accounted for a very small proportion of the entire response process, as shown in Figure 20. Therefore, we found that the current peak acceleration evaluation scheme commonly used in engineering is too conservative. This paper attempted to establish a comfort evaluation method based on the floor area comfort assurance rate.

To reflect the proportion of the area with an acceleration response on the floor that was less than a certain value in the total floor area, we introduced coefficient λ, which we defined as the floor comfort assurance rate. According to the definition, it can be calculated according to Equation (10):

$$\lambda = 1 - \frac{\pi r^2}{L^2} \tag{10}$$

where r is the radius of the circular area when the peak acceleration response on the floor was greater than a and L is the side length (or span) of the floor, as shown in Figure 21.

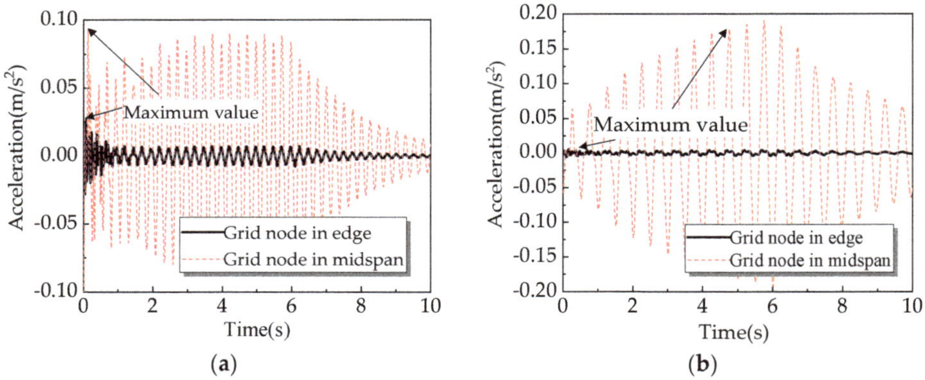

Figure 20. Response time history curves of two human-induced-vibration states: (**a**) resonance; (**b**) nonresonance.

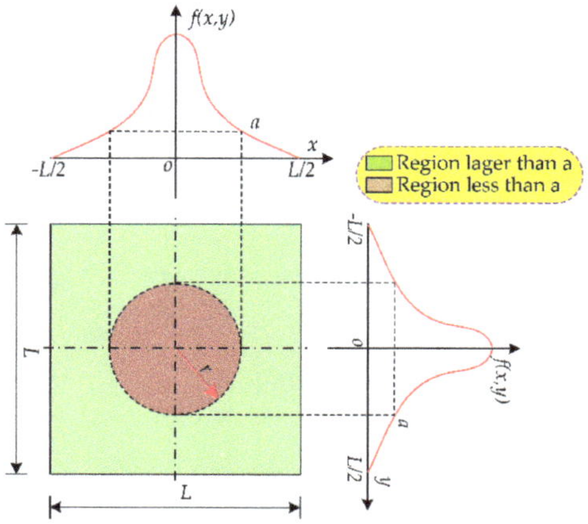

Figure 21. Schematic diagram of floor comfort assurance rate.

Referring to the unified standard reliability design of building structures (GB50068-2018) [34], we took the peak acceleration $a_{0.95}$ corresponding to λ = 95% as the representative value of the floor acceleration response and compared it with the specification limit (Equation (11)) used for comfort evaluation:

$$a_{0.95} \leq [a] \tag{11}$$

where a is the allowable value of the specification for human-induced-vibration acceleration; the value can be taken from Refs. [35,36].

We substituted λ = 95% into Equation (10) to obtain:

$$r = \sqrt{\frac{L^2}{20\pi}} \tag{12}$$

We substituted Equation (12) into Equations (5) and (6) to obtain the representative value of the acceleration response under resonance and nonresonance, which could be calculated according to Equations (13) and (14), respectively:

$$a_{0.95} = Ae^{-\frac{r^2}{2w^2}} = Ae^{-\frac{L^2}{40\pi w^2}} \tag{13}$$

$$a_{0.95} = \frac{c}{3} + Ae^{-\frac{x^2+y^2}{2w^2}} = \frac{c}{3} + Ae^{-\frac{L^2}{40\pi w^2}} \quad (14)$$

By substituting the fitting values of the parameters in Table 8 into Equations (13) and (14), the representative value of the acceleration response of the floor could be obtained. The comfort performance of the floor was evaluated using Equation (11).

For the numerical example in this study, the evaluation method used was compared with the maximum evaluation method; the relationship between the values of $a_{0.95}$ and $a_{\max}$ are shown in Table 10.

Table 10. Numerical size relationship between different models ($a_{0.95}$ and $a_{\max}$).

Model		$a_{0.95}$ (m/s^2)	$a_{\max}$ (m/s^2)	$a_{0.95}/a_{\max}$	Model		$a_{0.95}$ (m/s^2)	$a_{\max}$ (m/s^2)	$a_{0.95}/a_{\max}$
M1	M1-1	0.0826	0.1465	56.39%	M2	M2-1	0.067	0.0882	75.99%
	M1-2	0.0682	0.1246	54.73%		M2-2	0.1012	0.117	86.51%
	M1-3	0.0954	0.1211	78.78%		M2-3	0.1443	0.1904	75.77%
	M1-4	0.0705	0.0936	75.32%		M2-4	0.1208	0.1549	77.97%
	M1-5	0.0601	0.0809	74.35%		M2-5	0.0983	0.118	83.29%

It can be seen in Table 10 that the percentage of $a_{0.95}$ in $a_{\max}$ was affected by the span, concrete slab thickness, and whether resonance occurred. Overall, it was between 54.73% and 86.51%. In the case of nonresonance, the proportion of $a_{0.95}$ in $a_{\max}$ was between 54.73% and 78.78%. In the case of resonance, the proportion of $a_{0.95}$ in $a_{\max}$ was between 75.77% and 86.51%. It can be seen in the above analysis that the representative value of the evaluation could be greatly reduced by using $a_{0.95}$ to avoid being too conservative.

5. Conclusions

(1) Human-induced acceleration was affected by the span of the sandwich plate and the thickness of the concrete plate. The calculation results were affected by the grid size to a certain extent, and were less affected by the chord height and shear connector thickness;

(2) An acceleration response distribution model was established to accurately evaluate the dynamic response of a steel Vierendeel sandwich plate under human-induced fixed-point excitation;

(3) In view of the conservative peak acceleration evaluation scheme in engineering, this paper proposed a comfort evaluation method based on the floor area guarantee rate.

Author Contributions: Conceptualization, L.J. and R.C.; methodology, L.J.; software, H.Z.; validation, K.M.; formal analysis, R.C.; investigation, L.J.; resources, K.M.; data curation, L.J.; writing—original draft preparation, R.C.; writing—review and editing, L.J.; visualization, H.Z.; supervision, L.J.; project administration, K.M.; funding acquisition, L.J. All authors have read and agreed to the published version of the manuscript.

Funding: This research was funded by the Open Foundation Project of the State Key Laboratory for Safety Control and Simulation of Power System and Large Power Generation Equipment under grant number SKLD21KM11, as well as Key research and development projects in Hubei Province under grant number 2020BAB110.

Informed Consent Statement: Informed consent was obtained from all subjects involved in the study.

Data Availability Statement: Data will be shared upon request and consideration of the authors.

Conflicts of Interest: The authors declare no conflict of interest.

References

1. Wang, C.; Chang, W.S.; Yan, W.; Huang, H. Predicting the human-induced vibration of cross laminated timber floor under multi-person loadings. *Structures* **2021**, *29*, 65–78. [CrossRef]
2. Xie, Z.; Hu, X.; Du, H.; Zhang, X. Vibration behavior of timber-concrete composite floors under human-induced excitation. *J. Build. Eng.* **2020**, *32*, 101744. [CrossRef]
3. Cao, L.; Liu, J.; Li, J.; Zhang, R. Experimental and analytical studies on the vibration serviceability of long-span prestressed concrete floor. *Earthq. Eng. Eng. Vib.* **2018**, *17*, 417–428. [CrossRef]
4. Xiong, J.; Chen, J.; Caprani, C. Spectral analysis of human-structure interaction during crowd jumping. *Appl. Math. Model.* **2020**, *89*, 610–626. [CrossRef]
5. Jiang, L.; Ma, K.; Zhang, H.; Wu, Q.; Yang, Q. Seismic behavior of shear connectors of steel vierendeel sandwich plate. *Math. Probl. Eng.* **2019**, *2019*, 8047393. [CrossRef]
6. Harper, F.C. The mechanics of walking. *Res. Appl. Industry* **1962**, *15*, 23–28.
7. Kerr, S.C.; Bishop, N. Human induced loading on flexible staircases. *Eng. Struct.* **2001**, *23*, 37–45. [CrossRef]
8. Chen, J.; Wang, H.; Peng, Y. Experimental investigation on Fourier-series model of walking load and its coefficients. *J. Vib. Shock* **2014**, *33*, 11–15+28.
9. Ding, G.; Chen, J. Influences of walking load randomness on vibration responses of long-span floors. *J. Vib. Eng.* **2016**, *29*, 123–131.
10. *AISC811-97*; Steel Design Guide 11: Floor Vibrations due to Human Activity. American Institute of Steel Construction: Chicago, IL, USA, 1997.
11. Hakim, L. *PCI Design Handbook*, 7th ed.; Precast/Prestressed Concrete Institute: Chicago, IL, USA, 2010.
12. *Concrete Society Technical Report 43: Appendix G*, 2nd ed.; The Concrete Society: Surrey, UK, 2005.
13. *CCIP-016*; Standard for a Design Guide for Footfall Induced Vibration of Structures. The Concrete Centre: London, UK, 2007.
14. Parkhouse, J.G.; Ewins, D.J. Crowd-induced rhythmic loading. *Proc. Inst. Civil. Eng. Struct. Build.* **2006**, *159*, 247–259. [CrossRef]
15. Chen, J.; Peng, Y.; Wang, L. Experimental investigation and mathematical modeling of single footfall load using motion capture technology. *Chin. J. Eng. Des.* **2014**, *47*, 79–87. [CrossRef]
16. Zivanovic, S.; Pavic, A.; Reynolds, P. Vibration serviceability of footbridges under human-induced excitation: A literature review. *J. Sound Vibr.* **2005**, *279*, 1–74. [CrossRef]
17. Zhu, Q.; Liu, L.; Du, Y.; Chen, K. Human-induced vibration and control for cantilever steel bar truss deck slab based on pedestrain-structure interaction. *J. Build. Struct.* **2018**, *39*, 99–108.
18. Cao, L.; Li, A.; Zhang, Z.; Zhou, D.; Zhou, C. Human-induced vibration analysis and measurement of long-span composite floors. *J. Southwest Jiaotong Univ.* **2012**, *47*, 922–928.
19. Wang, Z.; Li, X.; Yi, J.; Li, Q. Human-induced vibration and optimal control of long-span glulam arch bridges. *China Civil. Eng. J.* **2021**, *54*, 79–94.
20. Jiang, L.; Ma, K.; Zhang, H.; Yang, Q.; Huang, J. Dynamic time-history analysis of large span open-web sandwich plate for vibration comfort based on walking route method. *Spat. Struct.* **2016**, *22*, 28–36+43.
21. *BS 5400*; Steel, Concrete and Composite Bridges-Part 2: Specification for Loads. British Standards Institution: London, UK, 2006.
22. Gulvanessian, H.; Calgaro, J.A.; Holický, M. *Designers' Guide to EN 1990 Eurocode: Basis of Structural Design*; Thomes Telford: London, UK, 2002; pp. 52–69.
23. *ISO10137*; Bases for Design of Structures-Serviceability of Buildings and Walkways against Vibrations. International Organization for Standard: Genève, Switzerland, 2007.
24. HiVoSS. *Human Induced Vibrations of Steel Structures: Design of Footbridges*; RFS2-CT-2007-00033, Footbridge_Guidelines_EN03; European Communities: Luxembourg, 2008.
25. *GB50010-2010*; Code for Design of Concrete Structures. China Construction Industry Press: Beijing, China, 2010.
26. *JGJ3-2010*; Technical Specification for Concrete Structures of Tall Building. China Construction Industry Press: Beijing, China, 2010.
27. Fiore, A.; Marano, G.C. Serviceability performance analysis of concrete box girder bridges under traffic-induced vibrations by structural health monitoring: A case study. *Int. J. Civ. Eng.* **2017**, *16*, 553–565. [CrossRef]
28. Jiang, L.; Ma, K.; Zhang, H.; Xu, X.; Wei, Y.; Hu, L. The dynamic property and comfort degree study on the steel-concrete composite vierendeel sandwich plate. *Earthq. Eng. Struct. Dyn.* **2017**, *6*, 122–131.
29. Blanchard, J.; Davies, B.L.; Smith, J.W. Design criteria and analysis for dynamic loading of footbridges. In Proceedings of the Symposium on Dynamic Behaviour of Bridges, Crowthorne, UK, 19 May 1977.
30. Matsumoto, Y.; Nishioka, T.; Shiojiri, H. *Dynamic Design of Footbridges*; Heft P-17; International Association for Bridge and Structural Engineering: Zürich, Switzerland, 1978.
31. Han, X.; Chen, X.; Mao, G.; Zheng, Y.; Ji, J. Research on analysis method for floor vibration and formula derivation of response spectra based on simulation of crowd walking. *Build. Sci.* **2009**, *5*, 4–9.
32. Lou, Y.; Huang, J.; Lv, Z. *Vibration Comfort Evaluation of Floor System*; Science Press: Beijing, China, 2012.
33. Xue, J.; Fiore, A.; Liu, Z.; Briseghella, B.; Marano, G.C. Prediction of ultimate load capacities of CFST columns with debonding by EPR. *Thin-Walled Struct.* **2021**, *164*, 107912. [CrossRef]

34. *GB50068-2018*; Unified Standard for Reliability Design of Building Structures. China Architecture & Building Press: Beijing, China, 2018.
35. *JGJ99-2015*; Technical Specification for Steel Structure of Tall Building. Ministry of Housing and Urban-Rural Development of the People's Republic of China. China Architecture & Building Press: Beijing, China, 2016.
36. *JGJ/T441-2019*; Technical Standard for Human Comfort of the Floor Vibration. China Architecture & Building Press: Beijing, China, 2019.

buildings

MDPI

Article

Evaluation of Inverted-Pendulum-with-Rigid-Legs Walking Locomotion Models for Civil Engineering Applications

Stana Živanović [1,*], Bintian Lin [1,2], Hiep Vu Dang [3], Sigong Zhang [1], Mladen Ćosić [4], Colin Caprani [5] and Qingwen Zhang [2]

1 College of Engineering, Mathematics and Physical Sciences, University of Exeter, Exeter EX4 4QF, UK
2 Key Lab of Structures Dynamic Behaviour and Control of the Ministry of Education, Harbin Institute of Technology, Harbin 150090, China
3 AmcoGiffen, Kensington House 3rd Floor, York Business Park, York YO26 6RW, UK
4 Institute for Testing of Materials—IMS, 11040 Belgrade, Serbia
5 Department of Civil Engineering, Monash University, Clayton, VIC 3800, Australia
* Correspondence: s.zivanovic@exeter.ac.uk

Abstract: Bipedal models for walkers, originally developed in the research field of biomechanics, have been identified as potential candidates for modelling pedestrians in structural engineering applications. These models provide insight into both the kinetics and kinematics of walking locomotion and are considered to have a significant potential to improve the vibration serviceability assessment of civil engineering structures. Despite this notion, the ability of the bipedal models to represent the key features of the walking gait and natural variability within the pedestrian population are still under-researched. This paper critically evaluates the performance of two bipedal models with rigid legs to realistically both reproduce key features of an individual pedestrian's walking gait and represent a wide range of individuals. The evaluation is performed for walking on a rigid, rather than vibrating, structure due to the availability of experimental data and expectation that successful modelling on rigid surfaces is a necessary condition for progressing towards modelling on the vibrating structures. Ready-to-use equations are provided and the ability of the models to represent the kinematics and kinetics of individual pedestrians as well as the inter-subject variability typical of the human population is critically evaluated. It was found that the two models could generate realistic combinations of the gait parameters and their correlations, but are less successful in reproducing genuine kinetic and kinematics profiles.

Keywords: walking locomotion; bipedal inverted pendulum; ground reaction force; walking kinematics

Citation: Živanović, S.; Lin, B.; Dang, H.V.; Zhang, S.; Ćosić, M.; Caprani, C.; Zhang, Q. Evaluation of Inverted-Pendulum-with-Rigid-Legs Walking Locomotion Models for Civil Engineering Applications. *Buildings* **2022**, *12*, 1216. https://doi.org/10.3390/buildings12081216

Academic Editor: Nerio Tullini

Received: 30 June 2022
Accepted: 4 August 2022
Published: 11 August 2022

1. Introduction

It is more than four decades since the first design guidance for the vibration serviceability assessment of footbridges, BS5400, was developed in 1978 [1] in recognition of the need to evaluate the structural vibration response to dynamic excitation induced by pedestrians. A pedestrian was modelled as a harmonic force moving across the bridge at a constant speed and at a "pacing" (also called "step") frequency matching a natural frequency of the structure. This single-pedestrian-exciting-the-resonance loading scenario has been at the heart of the vibration serviceability assessment of footbridges for almost three decades. A gradual introduction of high-strength and light-weight materials in contemporary structural design has resulted in new footbridges that are usually more slender, lighter, and less damped, and, therefore, more sensitive to dynamic loading than their older counterparts. As a consequence, there is a need for an improved modelling of pedestrian loading that exists. Some refinements, such as the inclusion of multi-pedestrian loading scenarios and the consideration of inter-subject variability in the pacing rate within a pedestrian crowd, have already been introduced in the new generation of design guidelines [2–4]. Research into stochastic models for pedestrian-induced force have been advanced from those that account for variations in the pacing rate within a pedestrian population [5] towards the inclusion of the probability distributions of pedestrian mass, dynamic force amplitude, and walking speed [6–8]. In addition, a detailed modelling of intra-subject (i.e., step-by-step) variations in pedestrian locomotion parameters is also available [6,9]. These relatively recent developments are underpinning the shift from a deterministic towards a probabilistic assessment of the vibration serviceability limit state of the structure. Nevertheless, the currently available models are limited to the structures on which pedestrians (and, therefore, the resulting dynamic forces) do not interact with the oscillating structure. The interaction term in this paper refers to the pedestrian–structure feedback loop, in which the structural vibration forces pedestrians to alter their walking locomotion which, in turn, alters the vibration response of the structure.

The excessive sway of the Millennium Bridge in London in June 2000, caused by a crowd of walkers, exposed a weakness in the design procedures that did not envisage the possibility of pedestrians interacting with the vibrating bridge deck [10]. The Millennium Bridge problem highlighted the need to understand the interaction mechanism and served as a motivation for developing more sophisticated models of pedestrians. An extensive experimental and theoretical line of research demonstrated that the structural behaviour could be explained by pedestrians continuously adjusting their placement of the foot on the moving deck to preserve their balance [11–13]. There is also some evidence that people might respond to deck vibration by synchronising with it [10,14]. In both cases, it has been acknowledged that pedestrians act as adaptable systems, highlighting the need for developing models that can genuinely represent pedestrian's walking on vibrating structures [15]. Similar detailed studies of walking gait are required for developing and understanding the human–structure interaction in the vertical direction of vibration [16–20]. The vertical component of the human-induced dynamic force will be the focus of this paper.

A number of researchers have identified that bipedal models for walkers, originally developed in the research fields of biomechanics and robotics, can qualitatively describe the pedestrian–structure interaction. These models include both a simple inverted pendulum model with rigid legs [13,16,17,21] and more complex bipedal models with deformable and damped legs [22–24]. While the initial studies show the potential of the bipedal models to be used in structural engineering applications, there is a need for a more detailed insight into the performance of these models, especially in terms of their ability to cover relevant parameter space and genuinely represent inherent variability in the pedestrian population. This is important for the accurate modelling of individuals to both study an emerging crowd behaviour [25] and the pedestrian–structure interaction. An overview of the governing equations for these models, relatively new to the structural engineering community, is also required to facilitate wider validation of the models.

The aim of this paper is to provide a reference source for the two simplest inverted pendulum models of the vertical force and evaluate their performance on rigid level ground surfaces. The evaluation will focus on the ability of the models to generate walking locomotion parameters observed in practice. The study concentrates on rigid level surfaces due to the availability of experimental data that can be used for the evaluation of the models, and due to the expectation that a satisfactory performance of the bipedal models on the rigid surface is a necessary condition for these models to be considered as good candidates for modelling walking on vibrating structures. To achieve the main aim, the paper also explains the human walking locomotion process. This is done by transferring knowledge from medical and biomechanics research areas and presenting it in the context relevant for civil engineering applications to enable the evaluation of the bipedal models not only in this paper, but also those to be published in the future.

The paper starts with describing the kinematic and kinetic features of the walking gait. It then proceeds towards a detailed evaluation of the performance of the two bipedal models with rigid legs: the inverted pendulum model (IPM) and inverted pendulum with rocker foot model (IPRFM). The discussion of the findings and their relevance to modelling pedestrians on vibrating structures are then presented and conclusions are briefly summarised.

2. Characteristics of Walking Locomotion

2.1. Kinematics and Kinetics of a Walking Gait Cycle

Gait analysis is a systematic study of the walking locomotion [26]. Traditionally, it is the ground reaction force (GRF) generated whilst walking that has been of most interest in structural engineering applications. This section, however, aims to describe not only the development of the force throughout a typical gait cycle (GC), but also the corresponding reference actions of the walker (e.g., heel-strike, toe-off, and other gait events) and the kinematics of the body centre of mass (BCoM). A GC is a cycle between two of the same nominal gait events of the same foot [27], such as the heel strike events of the right foot shown in Figure 1. Hence, a GC consists of two consecutive steps. Letters "R" and "L" in the figure refer to the right and left foot, respectively.

The human body in a walking posture is composed of a passenger unit and a locomotor unit [28]. The passenger unit consists of the upper body segments, which include head, arms, and trunk. These segments are carried by the lower body and they represent a passive contributor to the walking process. The locomotor unit, consisting of thighs, shanks, and feet, generates the body movement. Consequently, all gait events in a GC are usually described with reference to the positioning of the lower body parts, in particular, the two feet.

A GC consists of a stance and a swing phase of each leg (Figure 1). The stance phase is represented by the continuous contact of a leg with the ground, while the swing refers to the airborne phase of a leg. Besides the stance and swing phases, a GC can also be divided into a single support phase (SSP), when one foot only is in the contact with the ground, and double support phase (DSP), when both feet are simultaneously in the contact with the ground (Figure 1). SSP and DSP each occur twice in a GC.

Ayyappa [29], and Perry and Burnfield [28] have provided detailed descriptions of the walking phases and functions of body segments' joints. Key information from their work will be summarised here, with frequent cross-referencing to Figure 1. Assuming the heel strike of the right leg is adopted as the reference event, the GC can be said to start with the DSP and, at the same time, the right stance phase. The right foot makes an initial contact (IC, 0–2%, where the two percentage values refer to the typical start time and the end time, respectively, relative to the GC duration), during which the right knee is close to full extension and the leg is nearly straight. The abrupt impact of the foot generates a short-lasting peak in the time history of the GRF (the heel strike transient in Figure 1) and the body weight starts to be transferred to the ground through the right heel.

After the IC, the GRF time history enters the loading response phase (LR, 2–12%). During LR, the rapid transfer of the body weight to the right leg hinders the extended posture of the knee, and the knee starts to flex. As a result, the BCoM starts to descend, reaching the minimum elevation approximately half way through the LR/DSP (Figure 1). The bony segment between the heel and the ankle joint (also known as the heel rocker) of the right foot acts as a lever arm, in which the forward momentum of the leg drives the forefoot to the ground. The GRF builds up along with an increase of the contact area of the right foot. On the contrary, the left foot prepares to lift off by pushing against the ground and propelling the pedestrian forward. The push of the left foot increases the BCoM's elevation in the second half of the DSP (Figure 1). By the end of the LR, the GRF of the right leg reaches the first peak at F_1, while the force induced by the left leg decreases to zero, and the BCoM is at nominally the same elevation as at the beginning of the DSP. Meanwhile, the right knee is at the maximum flexion and the right foot is approximately in full contact with the ground. The DSP ends with the toe-off event of the left leg and the GC enters the SSP (Figure 1).

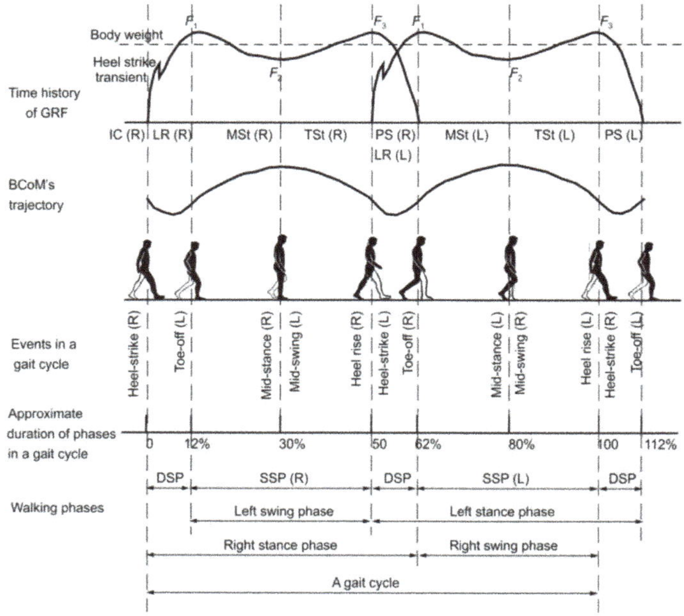

Figure 1. Ground reaction force, trajectory of body centre of mass, and key events in a gait cycle (adapted from [27]).

In the SSP (12–50%), only the right leg is in contact with the ground, while the left leg enters the swing phase. As soon as the left foot loses contact with the ground, the time history of the GRF only consists of the right leg's reaction (Figure 1). The first part of this SSP, during which the right foot is relatively stationary, is called the mid-stance (MSt, 12–30%). Due to the forward momentum, the body moves in front of the ankle joint axis (often referred to as the ankle rocker). At the same time, the knee-flexion keeps decreasing since the stability during the stance phase is at its optimum when the knee is in full extension. The swinging momentum of the left leg also plays a role in extending the right knee. As a consequence of the knee's extension, the BCoM's elevation continues to increase. On the contrary, the amplitude of the GRF decreases because of the upward momentum of the swinging leg. By the end of the MSt, the magnitude of GRF descends to its lowest point, F_2, while the BCoM reaches its maximum elevation (Figure 1). Once the swinging left leg becomes the leading leg (by overtaking the right leg), the left knee extends rapidly, allowing the pedestrian to achieve a certain step length.

After the MSt, the GC enters the terminal stance (TSt, 30–50%) that is initiated by the heel rise of the right foot. The TSt completes the SSP of the right leg and the swing phase of the left leg. The entire body makes a forward fall over the bony segment of the forefoot, acting as the forefoot rocker. The right knee reaches the state of full extension that makes the falling-forward movement of the BCoM relatively similar to the trajectory of an inverted pendulum. To prepare for the stance phase of the left leg, the left knee is in full extension. By the end of the TSt, the GRF of the right leg reaches the second peak at F_3. The TSt ends when the swing leg makes first contact with the ground. The SSP ends and the GC enters the second DSP.

Right after the TSt is the pre-swing (PS, 50–62%) phase of the right leg. This phase occurs at the same time as the IC and the LR of the left leg, and therefore, the BCoM experiences a similar trajectory pattern, as explained earlier in relation to the IC and LR phases of the right leg. The body weight is transferred from the right leg to the left leg, resulting in the rapid decrease of the GRF for the right and increase for the left leg (Figure 1).

To prepare for the swing phase, the right knee experiences significant flexion. Meanwhile, the right foot pushes the ground through metatarsal heads and toes (known as toe rocker) to progress the limb forward. The pre-swing phase ends with the toe-off event, which is also the terminal event of the stance phase of the right leg. The rest of the GC (62–100%) is the SSP of the left leg (and the swing phase of the right leg). The GC ends with the heel strike event of the right foot.

During a GC, the forward velocity is lowest when the BCoM is at its highest position, at around midstance [30], while it is at its maximum when the body descends to the lowest position during the DSP. In the walking process, therefore, the potential and kinetic energies are continuously interchanged, although it should be noted that the total energy is not conserved due to the damping effects of the human body [31]. To continue walking, the humans recover the lost energy by means of "external" work done by muscles [32].

The overview of the walking phases in a GC has shown three important aspects of a walking gait. First, the BCoM trajectory approximately resembles a series of arcs (Figure 1), especially during the SSPs. The transitions from one SSP to another are smoothened due to the influence of knee and ankle flexion [31]. The vertical excursion (i.e., the difference between maximum and minimum elevations of the BCoM) is observed to be 2–6 cm [33,34]. Second, during the stance phase, the foot that is in contact with the ground utilises four functional rockers: heel, ankle, forefoot, and toe rocker. These rockers maintain the stability of the gait and assist the forward progression of the limb [28]. Third, the time history of the GRF generated by one leg follows an M-shape with two peaks occurring approximately at the beginning and at the end of the SSP. At the normal walking speed, the peaks are about 110% of the body weight, while the trough, approximately in the middle of the SSP, is about 80% of the body weight [28]. These observations will be useful when evaluating the performance of the bipedal models.

2.2. Frequency Content of Ground Reaction Force

Modelling a continuous, multi-step GRF in civil engineering applications started with summing up a sequence of nominal single-step GRFs with an appropriate time overlap between two successive steps to achieve the intended pacing frequency (and therefore time period T), as shown in Figure 2a [35,36]. This created a periodic force consisting of distinct harmonics in the frequency domain. The typical continuous GRF, however, is of a narrow-band nature, as shown in Figure 2b. Its energy is concentrated not only in the main harmonics that occur at the pacing frequency and its integer multiples and less pronounced sub-harmonics that are consequence of the slight differences in forces generated by the left and right foot, but it also spills over to the neighbouring frequency lines [6,37]. The main harmonics are usually normalised by the weight of the pedestrian and expressed in the form of dynamic loading factors (DLFs).

The dynamic loading factor for the first harmonic, DLF_1, shows a strong dependency on the pacing frequency (Figure 3a), while the higher harmonics are frequency independent [38] (data for the 2nd harmonic only are shown in Figure 3b). It should be noted that DLFs are independent from the pedestrian's weight [9].

While DLF_1 can reach values as high as 0.7 (Figure 3a) for fast walking, DLF_2 (Figure 3b), DLF_3, and DLF_4 are characterised by maximum values around 0.22, 0.14, and 0.12, and mean values of 0.07, 0.06, and 0.05, respectively [38]. All four harmonics can have extremely low values approaching zero. In case of DLF_1, however, the values below 0.1 are only possible for extremely slow (and rarely seen in practice) walking frequencies below 1.3 Hz.

2.3. Pacing Frequency and Pedestrian's Forward Speed

Pacing frequency is one of the most important locomotion parameters in civil engineering applications [39]. When the pacing frequency (or one of its integer multiples) is close or equal to a natural frequency of a structure that is, at the same time, light and/or lightly damped, strong vibrations that might compromise the vibration serviceability state of the structure could develop [40]. While the pacing frequency for a population of structure users is known to follow a normal distribution, the distribution parameters (the mean and the standard deviation, STD) vary between user populations and structural purpose. Table 1 shows a summary of parameters observed on nine structures. Taking the mean ± 3STD as the boundaries of the possible pacing rates in each case suggests that the pacing rate ranges from 1.3 Hz to 2.5 Hz.

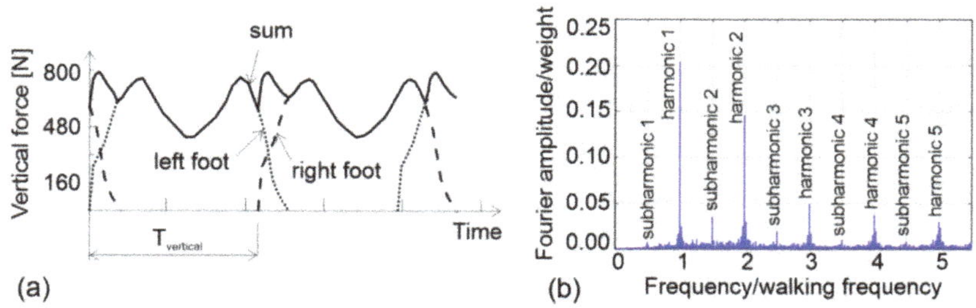

Figure 2. (**a**) Periodic GRF representation in the time domain and (**b**) actual GRF representation in the frequency domain (vertical component only).

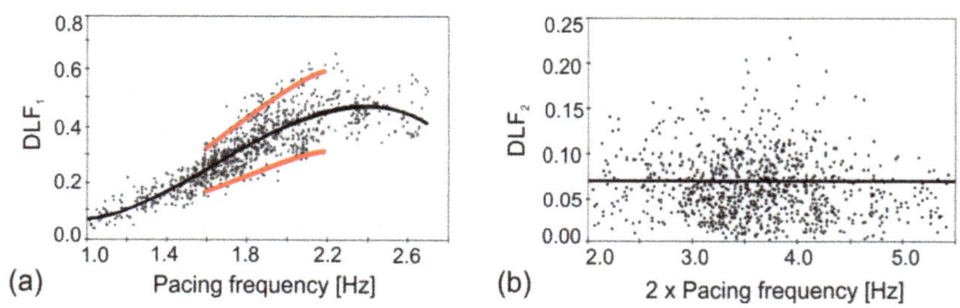

Figure 3. (**a**) Data points for DLF_1 versus pacing frequency. Solid black line—best data fit. Red solid lines—best fit ± 2 standard deviations. (**b**) Data points for DLF_2 versus pacing frequency. Solid black line—mean value. (after Kerr [38]).

Table 1. Statistics of pacing rate and walking speed on a range of structures.

Structure [Reference]	Country	Sample Size	Pacing Rate (Hz) Mean	Pacing Rate (Hz) STD	Walking Speed (m/s) Mean	Walking Speed (m/s) STD
Road [41]	Japan	505	1.99	0.17	-	-
Footbridge 1 [42]	UK	200	1.86	0.11	1.38	0.13
Footbridge 2 [42]	UK	200	1.80	0.10	1.23	0.09
Two shopping floors [42]	UK	400	2.00	0.13	1.41	0.13
Footbridge [43]	Germany	251	1.82	0.12	1.37	0.15
Walkway [44]	Italy	116	1.84	0.17	1.41	0.22
Indoor footbridge [45]	UK	939	1.94	0.19	1.47	0.23
Footbridge [46]	Montenegro	2019	1.87	0.19	1.39	0.20

Pedestrians' forward speed influences the amount of time a pedestrian requires to cross a structure, and therefore, the amount of time the structure is exposed to the dynamic excitation by the pedestrian. Assuming the possible speed values are within the mean ± 3STD of the data shown in Table 1, the speed could be as low as 0.7 m/s and as high as 2.1 m/s.

Some studies also report the mean and standard deviation for the step length: 0.75 m and 0.07 m [43], 0.77 m and 0.10 m [44], and 0.74 m and 0.08 m [46]. The range of step length values can be estimated to be from 0.47 m to 1.07 m.

For any individual pedestrian, the average pedestrian speed v, pacing rate f_p, and step length l_s are mutually dependent parameters ($v = f_p l_s$). Investigating the correlation of each pair of parameters on a population level suggests that the step length is relatively independent from the pacing frequency (Figure 4a), while the speed of walking increases with an increase in both pacing rate and step length (Figure 4b,c) [44,46].

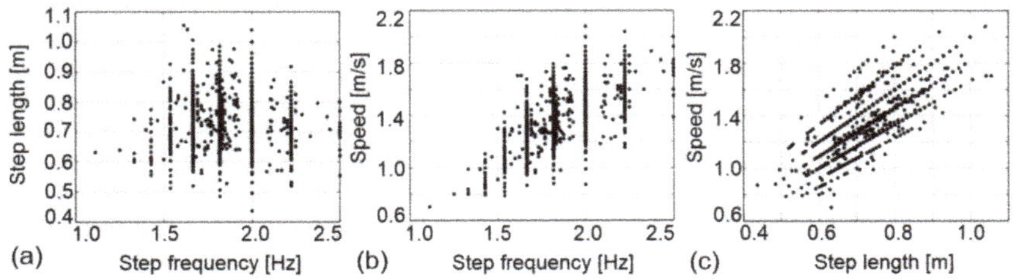

Figure 4. (**a**) Step length and (**b**) walking speed as functions of pacing frequency. (**c**) Pedestrian speed as a function of step length (adapted from [46]).

2.4. Criteria for Evaluation of Bipedal Models

When modelling pedestrian traffic in civil engineering applications, it is important to correctly reproduce the key time- and frequency-domain features of an individual pedestrian-induced dynamic force as well as the variability in the gait parameters within the studied population. The latter aspect requires the models to be able to generate the realistic range of gait parameters (e.g., DLF, pacing rate, and walking speed) as well as to correctly represent their correlation. The ability of the two simple bipedal models (IPM and IPRFM) to reproduce kinematics and kinetics of individual walkers as well as to cover the parameter space typical of a pedestrian population is investigated in the next section.

3. Inverted Pendulum Models with Rigid Legs

The research into walking locomotion was motivated by the curiosity about its mechanics [47], and it resulted in the development of a number of bipedal models. Most of the existing models have been developed based on the observations of major determinants of walking gait: (1) pelvic rotation, (2) pelvic tilt, (3) knee flexion, (4) the foot mechanism, (5) knee mechanisms, and (6) lateral displacement of the pelvis [31]. Since this study is concerned with the investigation of the walking gait in the sagittal plane (i.e., the vertical plane that includes the direction of progression), the first and sixth determinants could be excluded from considerations.

Bipedal models of walking locomotion were initially used to study normal and pathological gaits in medical applications [31]. Apart from medicine, the walking locomotion has been a subject of interest in research fields of biomechanics, animated image processing, and robotics. A number of models are shown in Figure 5 in order of increasing complexity. A detailed representation of the first two models will be presented in this paper; we hope this will inspire a similar type of analysis for other models in the future. The main feature of all these models is that the body mass is lumped into a single point representing the BCoM. As a result, the human kinematics is represented by the movement of this single point, as dictated by the geometry (and elasticity and damping, if included) of the legs.

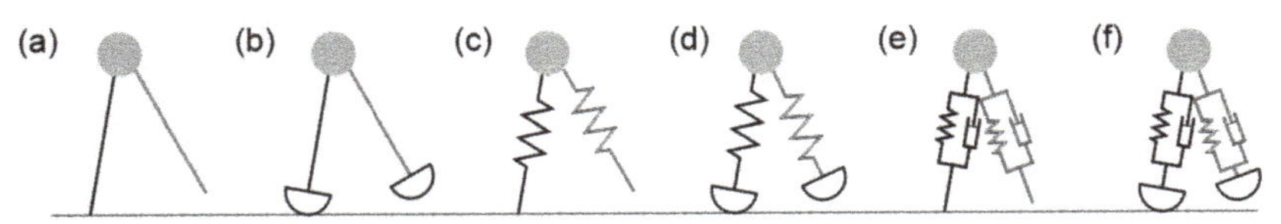

Figure 5. Bipedal walking locomotion models: (**a**) inverted pendulum [31], (**b**) inverted pendulum with rocker foot [30], (**c**) spring-loaded inverted pendulum, SLIP [48], (**d**) SLIP with rocker foot [49], (**e**) spring mass with damper [22], and (**f**) spring mass with rocker foot and damper [50].

3.1. Inverted Pendulum Model

The inverted pendulum model (IPM) is the simplest bipedal walking model (Figure 5a) developed by Saunders et al. [31]. The model consists of a point mass m_p and two rigid, straight, and massless legs (Figure 6). The human body is assumed to be symmetrical, i.e., the two legs have the same length. The foot is modelled as a point foot that does not slip during contact with the ground.

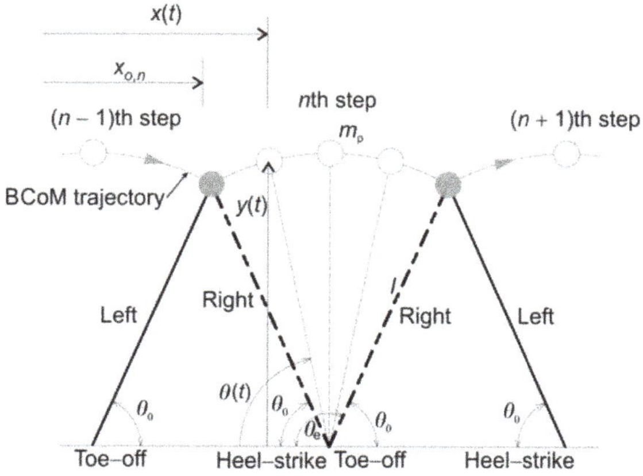

Figure 6. Inverted pendulum model.

The time instant when one leg touches the ground can be taken as a starting point of a step in the IPM. The model is set into motion by specifying the initial conditions for the attack angle θ_0 and angular speed $\dot{\theta}_0$. The resulting motion of the point mass is in the form of an arc, defined by the geometry of the supporting leg. A step is completed when the angle formed by the supporting leg, $\theta(t)$, (see Figure 6) becomes equal to the supplement of the attack angle θ_0. This angle is also called the end-of-step angle, denoted as θ_e in Figure 6.

At the time of step completion, the swinging leg touches the ground and the pedestrian's weight is instantaneously transferred from one foot to another. The next step is initiated by specifying a new set of initial conditions, which are usually assumed to be the same for all the steps.

Using the Lagrangian approach [51], the equation of motion for a single step can be written as:

$$\ddot{\theta}(t) = \frac{\cos\theta(t)}{l}g, \tag{1}$$

where $\ddot{\theta}(t)$ is the second derivative of $\theta(t)$ with respect to time, l is the distance from the BCoM to the foot (hereafter referred to as the pendulum length), and g is the acceleration of gravity ($g = -9.81$ m/s^2). Equation (1) describes the stance phase of the gait cycle only because the IPM neglects the DSP. The GRF generated within a single step $F_p(t)$ is:

$$F_p(t) = -m_p \sin\theta(t)\left(g\sin\theta(t) + l\dot{\theta}(t)^2\right), \tag{2}$$

where $\dot{\theta}(t)$ is the first derivative of $\theta(t)$ with respect to time, while all other variables are the same as before.

The step transition process requires redirecting the BCoM upwards, from the falling downwards momentum at the end of a step. During walking, this upwards momentum is provided by a foot pushing off the ground just before the toe-off event (Figure 1), and it can be simulated in the model by applying an upward impulse to the point mass. The amplitude of this vertical upward impulse I_n at the end of the nth step is [16]:

$$I_n = -m_p \dot{y}_{e,n} + m_p \dot{x}_{0,n+1} \cot \dot{\theta}_0, \tag{3}$$

where $\dot{y}_{e,n}$ and $\dot{x}_{0,n+1}$ are the vertical speed of the BCoM at the end of the nth step and the forward speed at the beginning of the (n + 1)th step, respectively. The first part of the impulse in Equation (3) cancels the falling effect at the end of nth step, while the second part pushes the mass upwards so to supply the initial forward speed $\dot{x}_{0,n+1}$. The initial forward speed $\dot{x}_0$ is assumed to be the same in all steps and it is linked to the initial angular speed $\dot{\theta}_0$ and the attack angle θ_0 through the equation:

$$\dot{x}_0 = l\dot{\theta}_0 \sin \theta_0. \tag{4}$$

3.1.1. Model Inputs

To perform simulations using the IPM, two sets of input parameters are required: model parameters and initial conditions. The model parameters consist of pedestrian mass and the pendulum length. The body mass of 77.5 ± 17.2 kg (mean ± STD) and height of 1.676 ± 0.097 m of an average person are chosen as representative values in this paper [52]. This body height corresponds to the physical leg length of 0.864 ± 0.050 m [53]. The physical leg length has to be increased by about 20% [54] to obtain the pendulum length parameter, which amounts to 1.037 ± 0.060 m.

Initial conditions required by the model are θ_0 and $\dot{\theta}_0$ (or $\dot{x}_0$ instead of $\dot{\theta}_0$, see Equation (4)). The attack angle θ_0 ranges between 65° and 80° [48]. The exact range of the initial forward speed $\dot{x}_0$ is not well known. After initial simulation trials, the range has been set to 1.0–2.5 m/s for the needs of the parametric study. The locomotion is assumed to be periodic, i.e., the initial conditions remain constant in each step.

3.1.2. Simulation Results

The solver *ode45* from the MATLAB library [55] that utilises the Runge–Kutta integration method with a variable step size was used to solve Equation (1). The maximum time step of the solver is set at 10^{-3} s. An example of the GRF generated by the model for the input values of $m_p = 77.5$ kg, $l = 1.037$ m, $\theta_0 = 69°$, and $\dot{x}_0 = 1.61$ m/s is shown in Figure 7. The initial conditions were chosen to generate walking at a pacing rate of 1.87 Hz and an average walking speed of 1.39 m/s, which correspond to the mean values observed on an as-built bridge (reported in [46], and included in Table 1). The arcs in the figure represent the inertia force of mass m_p while the vertical lines represent the externally applied impulses. The duration of each impulse for the numerical simulation is chosen to be such that the average of the total force is equal to the pedestrian weight.

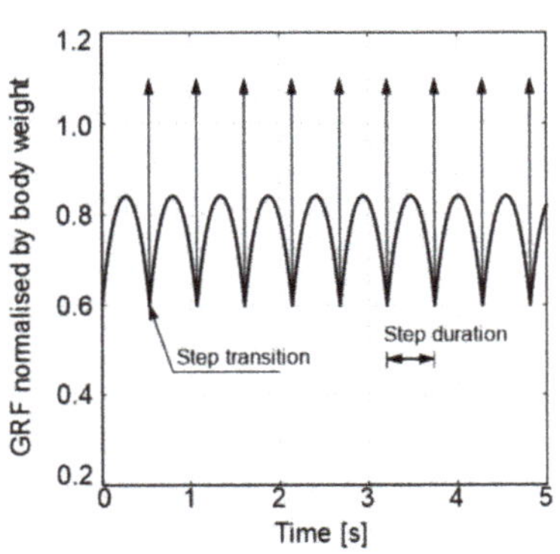

Figure 7. Normalised GRF as a function of time.

The amplitude of the first harmonic of this GRF is within the measured range shown in Figure 3. The DLFs for higher harmonics in the IPM are significantly overestimated due to the simplified modelling of the push-off phase using the impulses [17].

To investigate the ability of the IPM to reproduce realistic combinations of the pacing rate, DLF_1, step length, and average forward speed, the model has been exposed to a range of initial conditions: the initial forward speed varies between 1.0 m/s and 2.5 m/s (in discrete steps of 0.01 m/s) and the attack angle varies between 65° and 80° (in steps of 0.01°). The light grey area in Figure 8a shows all the combinations of the attack angle and initial forward speed that result in the physically relevant pacing rate between 1.3 and 2.5 Hz. The light grey in Figure 8b covers the parameter space which results in $DLF_1 \leq 0.7$, while the darker grey in the same figure represents the parameter combinations that produce realistic values of both the pacing frequency and DLF_1. It can be seen that unrealistically low values of DLF_1 (i.e., $DLF_1 \leq 0.1$) correspond to an unrealistically low pacing frequency, which is consistent with the data shown in Figure 3a. The light grey area in Figure 8c shows a parameter space that covers the realistic range of the walking speed (0.7–2.1 m/s) and the dark grey area in the same figure represents all the parameter combinations that result in a realistic pacing frequency, DLF_1, and pedestrian speed. Figure 8c reveals that, for the given pendulum parameters (i.e., $m_p = 77.5$ kg, $l = 1.037$ m), the most extreme pedestrian speeds (i.e., around 0.7 m/s and 2.1 m/s) are not achievable. To better represent extremely slow walkers, the pendulum length must be shortened, while the fastest walkers can be modelled by lengthening the pendulum length. Using a pendulum length between 0.9 m and 1.2 m (in agreement with the data reported by Hof et al. [54], it is possible to select initial conditions that result in realistic simulations of all of the pacing frequencies, DLF_1, and walking speeds.

To check the correlation between the gait parameters of interest, only those initial conditions that produced realistic ranges of the pacing rate, DLF_1, and average forward speed are utilised in Figure 9. The boundary of the parameter space that includes the points within the dark grey area in Figure 8c are shown as solid lines in Figure 9. Boundaries are also shown for a shorter pendulum length of 0.9 m (dashed line) and elongated pendulum length of 1.2 m (chain line). In addition, the experimental data points from Figure 4 are also shown in Figure 9a–c, while the boundaries for DLF_1 from Figure 3a are shown in Figure 9d. Figure 9 shows that the IPM is able to reproduce a (empirically observed) lack of correlation between the step length and the step frequency, as well as the positive correlations between the speed and step frequency, speed and step length, and DLF_1 and step frequency. It is also noticeable that the model covers a larger parameter space than that seen in practice (e.g., for a given step frequency of, say, 2 Hz, it can both underestimate and overestimate the DLF_1 value, depending on the choice of the pendulum length and the initial conditions).

The trajectory of the BCoM that is represented by a series of arcs in the IPM overestimates the vertical excursion by a factor of two or more [56]. The trajectory also lacks smoothness of the actual trajectory shown in Figure 1 due to an instantaneous transfer of the body weight from one foot to another and the inherent inability of the IPM to depict the double support phase of the gait.

The inability of the IPM to replicate the kinematics of the BCoM accurately is a drawback for the quantitative modelling of pedestrians on a vibrating structure as the kinematic state of the foot–structure interface cannot be genuinely represented. On the other hand, this simple model is a building block for understating more complex bipedal models and could offer some qualitative insight into the pedestrian–structure interaction [16,17].

3.1.3. Dimensional Analysis

Dimensional analysis is a convenient means of exploring how changing some parameters influences the others. Let us select mass m_p, pendulum length l, and acceleration of gravity g as the base quantities. For a given attack angle θ_0, it is possible to determine how changing the m_p or l (or gravity g, although this change is not of actual interest in this work) influences other parameters of interest. The new value of a parameter of interest can be found by preserving the dimensionless value, defined in Table 2, for the same parameter.

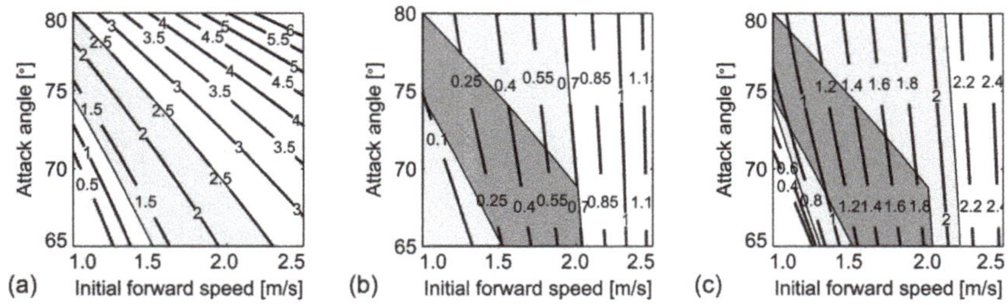

Figure 8. (**a**) Pacing rate, (**b**) DLF_1, and (**c**) average forward speed of a pedestrian resulting from different combinations of the initial conditions in the IPM ($m_p = 77.5$ kg, $l = 1.037$ m). Refer to the text for explanation of the colours.

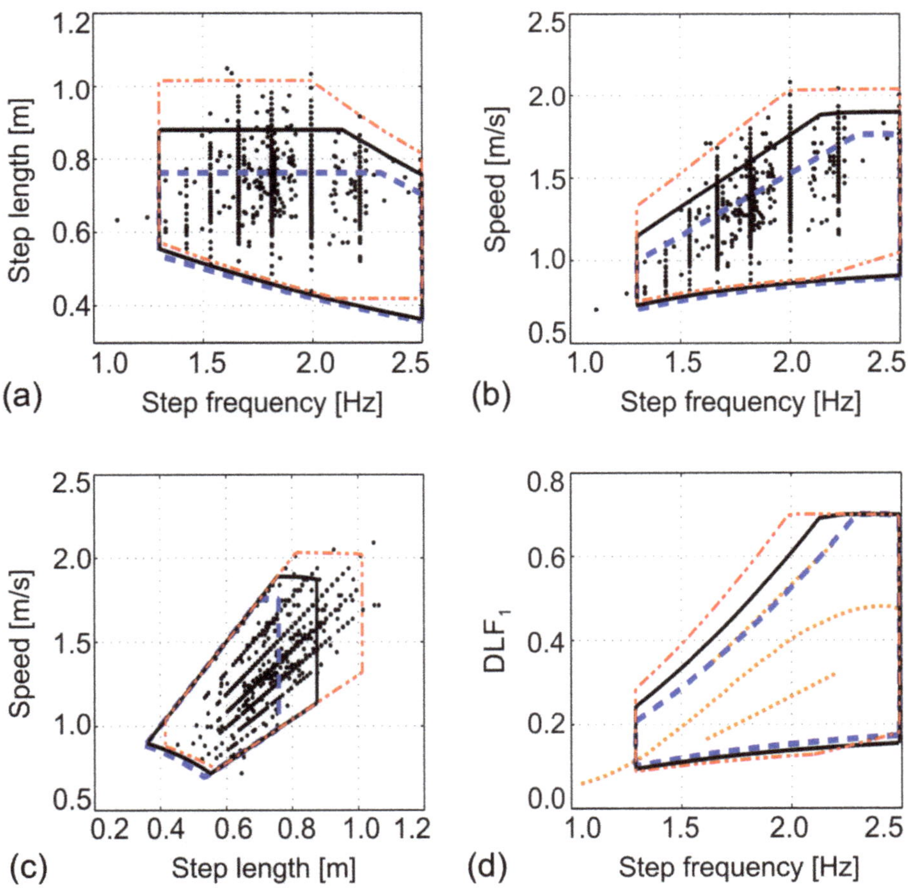

Figure 9. Relationships between (**a**) step length and step frequency, (**b**) speed and step frequency, (**c**) speed and step length, and (**d**) DLF_1 and step frequency, in the IPM. Blue dashed line: $l = 0.9$ m, black solid line: $l = 1.037$ m, red chain line: $l = 1.2$ m. Black dots: experimental data from Figure 4. Dotted lines: experimental data (mean and mean $\pm$ 2 STD) by Kerr [38].

Table 2. Parameters of interest.

Parameter	Dimensionless Parameter
Step frequency f_p	$f_p^* = f_p\sqrt{\frac{l}{g}}$
Step length d	$d^* = \frac{d}{l}$
Average speed v	$v^* = \frac{v}{\sqrt{gl}}$
Dynamic load factor DLF_1	-

This can be used, for example, to observe consequences of extending the pendulum length from, say, 1.037 m to 1.2 m. Let us assume the initial conditions: $\theta_0 = 70°$ and $\dot{x}_0 = 1.5$ m/s. These inputs (for $l = 1.037$ m) correspond to the following gait parameters: $f_p = 1.79$ Hz and $v = 1.27$ m/s (Figure 8). In addition, $DLF_1 = 0.31$. Extending the pendulum length to 1.2 m, keeping the same attack angle, and using dimensional analysis leads to an increase in the initial forward speed from $\dot{x}_0 = 1.5$ m/s to $\dot{x}_0\sqrt{1.2/1.037} = 1.61$ m/s. Table 2 indicates that the step frequency would reduce to $1.79\sqrt{1.037/1.2} = 1.66$ Hz, the average speed of walking would increase to $1.27\sqrt{1.2/1.037} = 1.36$ m/s, while DLF_1, being a dimensionless quantity, would remain 0.31. Moreover, the step length can now be calculated as a ratio between the average speed and step frequency. It has increased from 0.71 m (1.27 m/s/1.79 Hz) to 0.82 m (1.36 m/s/1.66 Hz), which corresponds to the expected increase by a multiplication factor of 1.2/1.037 (Table 2). A change in the pedestrian mass, on the other hand, does not influence the resulting parameters. However, an increase in the pedestrian mass will proportionally increase the harmonic force value, given that DLF_1 represents the force normalised by a pedestrian's weight.

3.2. Inverted Pendulum with Rocker Foot Model

The IPM overestimates the vertical excursion of the BCoM since it does not include all the relevant determinants of the walking gait in the modelling process [31]. While pelvic tilt and knee flexion (the second and third determinants) have little or no effect on the amount of the vertical excursion [57,58], the foot and ankle mechanisms (the fourth and fifth determinants) influence the excursion significantly. Thus, a way to improve the modelling of the vertical excursion is to replace point foot with a rocker. This intervention results in the inverted pendulum with the rocker foot model, IPRFM, shown in Figure 10 and explained in more detail by Hansen et al. [59], and Gard and Childress [30].

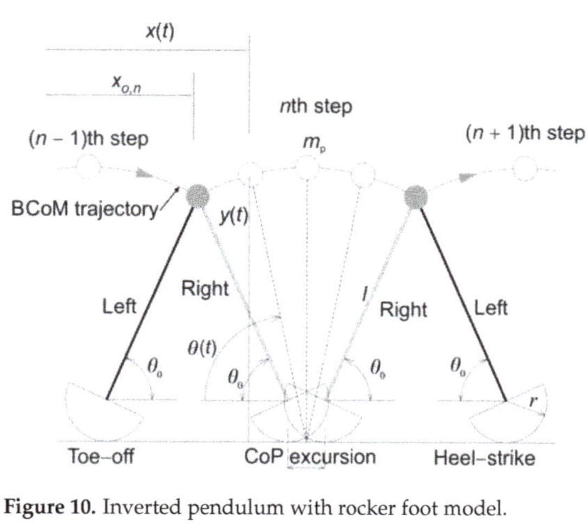

Figure 10. Inverted pendulum with rocker foot model.

The introduction of the rocker, in the form of a circular arc having radius r, is equivalent to lengthening the pendulum length of the IPM. In addition, it represents the walking mechanism during the stance phase more accurately as it enables the transition of the centre of pressure (CoP) from the "heel" to the "toe" as the foot rolls forward. Although the trajectory of the BCoM in the IPRFM still follows the pattern of a series of arcs, the effective lengthening of the leg and the forward progression of the CoP reduce the total excursion of the BCoM in comparison with the IPM [56,59].

Using the Lagrangian approach, the equation of motion for the IPRFM can be written as:

$$\ddot{\theta}(t) = \frac{l \cos\theta(t) \left(g - r\dot{\theta}^2(t)\right)}{r^2 + l^2 + 2rl\sin\theta(t)}. \tag{5}$$

The GRF generated by the IPRFM within a single step is:

$$F_p(t) = -m_p \left(g + l\dot{\theta}^2(t)\sin\theta(t) - \frac{l^2\cos^2\theta(t)\,(g - r\dot{\theta}^2(t))}{r^2 + l^2 + 2rl\sin\theta(t)} \right). \tag{6}$$

The end-of-the-step condition and the amplitude of applied impulses are calculated using the same approach as in the calculations for the IPM.

3.2.1. Model Inputs

The inputs for the IPRFM are pendulum length, $l + r$, and body mass, m_p. They are chosen in the same way as for the IPM. In addition, the proportion of r in the pendulum length has to be specified. McGeer [47] assumed that the supporting foot travels a distance of 20% of the leg length, and calculated that this requires a roller radius r that is equal to 30% of the leg length. Whittington and Thelen [49] reported that a roller radius of 0.3 m, which is approximately 30% of the limb length, resulted in the centre of pressure excursion that best agreed with experimental data at slow, preferred, and fast walking speeds. Adamczyk et al. [60] added that this size of the rolling feet appeared energetically advantageous, partially due to decreased work in step-to-step transitions. Hence, the rocker radius of 0.3 m is adopted in this paper.

3.2.2. Simulation Results

As before, *ode45* solver utilising the Runge–Kutta integration with a variable step size was implemented to solve the differential equation. Figure 11a shows how the introduction of a rocker in the IPRFM reduces the vertical excursion of the BCoM when compared with the IPM, while Figure 11b demonstrates that the progression of the CoP (i.e., excursion of the CoP in Figure 10) and the pacing frequency both increase with an increase in the rocker radius from 0.00 m to 0.45 m.

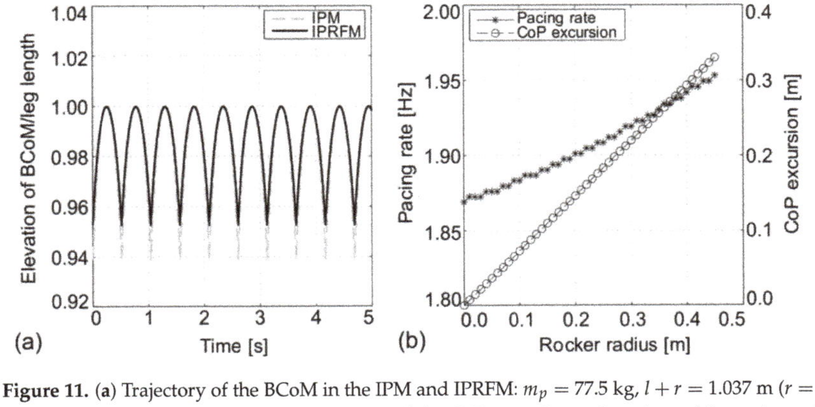

Figure 11. (**a**) Trajectory of the BCoM in the IPM and IPRFM: $m_p = 77.5$ kg, $l + r = 1.037$ m ($r = 0$ m in IPM, $r = 0.3$ m in IPRFM). (**b**) Pacing rate and the CoP excursion as functions of the rocker radius: $m_p = 77.5$ kg, $l + r = 1.037$ m. In all simulations: $\theta_0 = 69°$ and $\dot{x}_0 = 1.61$ m/s.

The results of a parametric scan of the IPRFM are shown in Figure 12. The rocker radius was set to $r = 0.3$ m and length $l + r = 1.037$ m. As before, the initial forward speed ranges from 1.0 m/s to 2.5 m/s, and the attack angle is between 65° and 80°. The light grey area in Figure 12a shows all the combinations of the initial conditions that result in the expected pacing rate between 1.3 and 2.5 Hz. Among the identified combinations, those that also result in a viable value of $DLF_1 \leq 0.7$ are shown as the dark grey area in Figure 12b. Furthermore, the dark grey area in Figure 12c denotes the parameter space that results in viable values of the pacing rate, DLF_1, and the average forward speed. In comparison with the IPM (Figure 8c), the IPRFM provides a slightly wider range of initial conditions that result in realistic walking parameters (Figure 12c). Varying the leg length and/or the rocker radius offers further flexibility in modelling different pedestrians.

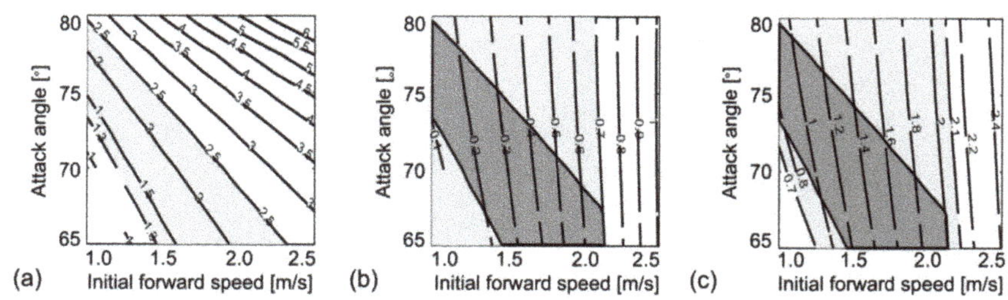

Figure 12. (**a**) Pacing rate, (**b**) DLF_1, and (**c**) average forward speed of a pedestrian resulting from different combinations of the initial conditions in the IPRFM ($l + r = 1.037$ m, $r = 0.3$ m). Refer to the text for explanation of the colours.

Figure 13 shows that the correlation graphs typical of the IPRFM are similar to those derived for the IPM in Figure 9. The solid lines represent the $l + r = 1.037$ m case, whereby $l = 0.737$ m and $r = 0.3$ m. The chain lines are derived for $l + r = 1.2$ m (achieved by increasing the original value for either l or r) and the dashed lines are for $l + r = 0.9$ m (achieved by decreasing either l or r). As before, increasing the pendulum length enables the modelling of pedestrians with the longest step lengths and fastest walking speeds, while a decrease in the pendulum length has the opposite effect. However, the pendulum length increase also increases the DLF_1 to values not easily encountered in practice. The lower boundary for DLFs underestimates the measured values for almost all but the slowest pacing frequencies. There is little difference between the cases having the same value of $l + r$ in Figure 13a–c, suggesting similar effects on the parameter space are achieved by changing either l or r. The effects on the DLF parameter are slightly more noticeable.

In summary, the IPRFM is a kinematic upgrade of the IPM as it reduces the vertical excursion of the BCoM and enables the progression of the CoP. However, it still models the single support phase of the gait only and requires applying the impulses as an external source of power to keep the model in motion. The latter means that the generated GRF has the same shortcomings as the IPM in terms of overestimating the higher harmonics.

3.2.3. Dimensional Analysis

As in the case of the IPM, the three base quantities (m_p, l and g) are used to determine how changing m_p or l influences other parameters of interest for a given θ_0. Both r and l have a dimension of length, and therefore, multiplying l by the factor l^* would require multiplying r by the same factor. The dimensionless parameters of interest are the same as those presented in Table 2.

Let us assume the initial conditions: $\theta_0 = 70^{\circ}$ and $\dot{x}_0 = 1.5$ m/s. These inputs correspond to $f_p = 1.85$ Hz and $v = 1.32$ m/s (Figure 12). In addition, $DLF_1 = 0.30$. Let us observe how extending length $l + r = 1.037$ m (whereby $l = 0.737$ m and $r = 0.3$ m) to $l^*(l + r) = 1.2$ m influences the other derived parameters. This corresponds to $l^* = 1.2/1.037 = 1.16$, and therefore, the new lengths $l = 0.853$ m and $r = 0.347$ m. Dimensional analysis (Table 2) suggests that the observed lengthening of the pendulum results in the initial forward speed, which increases from $\dot{x}_0$= 1.5 m/s to $\dot{x}_0\sqrt{1.16} = 1.61$ m/s. The step frequency will reduce to $\dot{x}_0\sqrt{1/1.16} = 1.72$ Hz, the average speed of walking will increase to $1.32\sqrt{1.16} = 1.42$ m/s, while DLF_1 will remain 0.30. The step length (a ratio between the average speed and step frequency) has increased from 0.71 m (1.32 m/s/1.85 Hz) to 0.82 m (1.42 m/s/1.72 Hz), which corresponds to the expected increase by a multiplication factor of 1.16 (Table 2). The change in the pedestrian mass does not influence any quantity apart from the harmonic force amplitude.

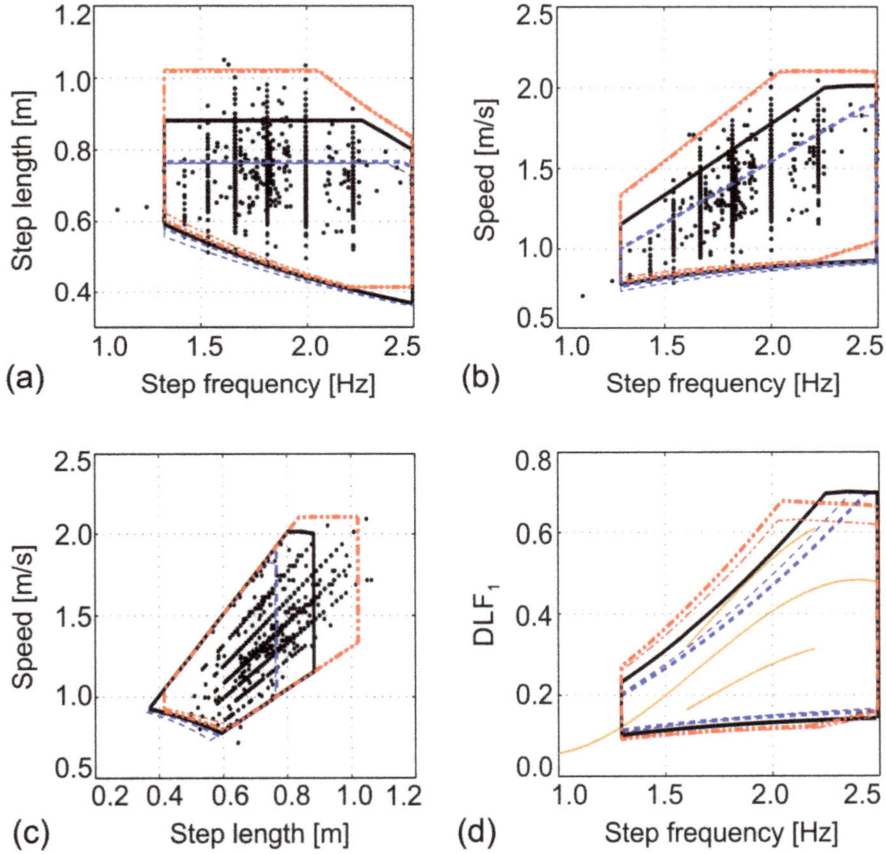

Figure 13. Relationships between (**a**) step length and step frequency, (**b**) speed and step frequency, (**c**) speed and step length, and (**d**) DLF1 and step frequency, in the IPRFM. Blue dashed lines: $l + r = 0.9$ m (thick $l = 0.6$ m and $r = 0.3$ m, thin $l = 0.737$ m and $r = 0.163$ m), black solid line: $l + r = 1.037$ m (thick $l = 0.737$ m and $r = 0.3$ m), red chain lines: $l + r = 1.2$ m (thick $l = 0.9$ m and $r = 0.3$ m, thin $l = 0.737$ m and $r = 0.463$ m). Black dots: experimental data from Figure 4, yellow thin solid lines in (**d**): experimental data (mean ± 2 STD) by Kerr [38].

4. Discussion and Conclusions

Both the IPM and IPRFM are simplifications of human gait that have no capability of modelling the double support phase of walking locomotion. As a result, the kinematics of the body's centre of mass consists of a series of arcs with a non-smooth transition between the successive steps. The vertical excursion of the centre of mass is exaggerated in the IPM and reduced to a more realistic level by enabling the forward progression of the centre of pressure in the IPRFM. However, due to a lack of the double support phase, the kinematics in the two models are unrepresentative of actual human walking. As the kinematic compatibility of the leg and vibrating surface is likely to play an important role in the realistic simulation of actual walking on lively structures, it is unrealistic to expect that these models can be used in a quantitative vibration serviceability assessment of civil engineering structures. Initial research into using the IPM on lively structures confirms this notion. Namely, simulations by Bocian et al. [16] demonstrated that the IPM produces reasonable estimates of additional damping that pedestrians could add to the structures under a specific vibration frequency, pacing frequency, and vibration amplitude conditions, as observed in practice. However, the IPM also predicts negative damping effects for some parameter combinations—a phenomenon that has not been observed in practice so far. In addition, Dang [17] demonstrated that the IPM, on average, generates an improved prediction of the measured vibration response compared with the classical harmonic force model that neglects the pedestrian–structure interaction. At the same time, however, the IPM tends to react to the oscillating structure by reducing the pacing rate, especially when the pacing rate is close to the natural frequency of the structure. This is in contradiction with the slight increase in the pacing rate observed in experiments, and it makes the model erroneous in the most important frequency region. The authors are unaware of the research that investigated the performance of the IPRFM on lively structures.

Parametric analysis focused on the gait parameters for the two models showed that they are able to generate realistic combinations of the gait parameters and their correlations. In addition, they include some parameter combinations that are not seen in pedestrian populations. Namely, a realistic value for one parameter, say DLF_1, is not necessarily associated with the realistic values of the other parameters, e.g., pacing frequency. The introduction of the rocker foot has a kinematic effect similar to extending the pendulum length in the IPM. However, the ranges of the locomotion parameters that the two models can produce are similar, with no obvious advantage of using one model over another. Finally, it should be noticed that the two models have the ability to generate a realistic value of the DLF_1, while higher harmonics are overestimated due to presence of impulse components in the dynamic force.

The next natural step in the evolution of pedestrian modelling is to replace the rigid leg with a compliant (i.e., deformable) leg (with or without a rocker foot), which is a feature of the spring-loaded inverted pendulum, SLIP, models (Figure 5c,d). These models are known to improve the kinematics of the body centre of mass and replicate a typical M-shaped GRF pattern [48]. However, these models, together with the two models analysed in this paper, neglect the role of the muscles in generating walking locomotion by relying on either artificially generated external impulses (as in the IPM and IPRFM) or energy preservation (in SLIP models). They also neglect the energy dissipation that characterises the walking process [32]. In response to these drawbacks, models that account for leg damping and positive work done by muscles have been developed (Figure 5e,f). Both constant [50] and time-variant damping models [22,23,61] exist. They propose different modelling mechanisms for compensating the energy lost due to damping. This brief model overview illustrates the richness of modelling approaches within the inverted pendulum model family. Unfortunately, the verification of these models, especially on lively structures, lags behind the theoretical developments. The adoption of one or more models in the vibration serviceability field could be accelerated by: (1) the development of an open-access database of experimental data related to walking on lively structures (that preferably should include kinematic and kinetic data for the human, as well as vibration data for the structure) and (2) providing detailed insight into the performance of the inverted pendulum models on both rigid and lively surfaces.

To conclude, the IPM and IPRFM are the simplest models from the bipedal family of models of pedestrians. This paper provides the necessary background for the application of these models by other researchers and information about their ability to represent kinematic and kinetic features that characterise human walking locomotion on rigid level surfaces. A limited amount of research into the performance of the IPM on lively structures suggests that, while the model can qualitatively describe some effects observed on as-built structures, it cannot be relied upon for the accurate assessment of structural vibration, especially in the most interesting case of (near) resonance excitation. Further research should investigate whether more complex bipedal models could provide a better representation of pedestrian locomotion on both rigid and lively surfaces.

Author Contributions: Conceptualisation, S.Ž. and C.C.; methodology, S.Ž., H.V.D. and C.C.; software, B.L. and H.V.D.; validation, M.Ć. and S.Z.; formal analysis, B.L. and H.V.D.; writing—original draft preparation, H.V.D. and S.Ž.; writing—review and editing, B.L., S.Z., C.C. and Q.Z.; supervision, S.Ž. and Q.Z.; funding acquisition, S.Ž., S.Z. and B.L. All authors have read and agreed to the published version of the manuscript.

Funding: This research was funded by the UK Engineering and Physical Sciences Research Council, grant numbers EP/I03839X/1 (Pedestrian interaction with lively low-frequency structures) and EP/M021505/1 (Characterising the dynamic performance of fibre-reinforced polymer structures for resilience and sustainability), EU funding from the Horizon 2020 MSCA-IF project vPERFORM, grant number 898216 (Developing advanced vibration performance assessment for a new generation of lightweight pedestrian structures using motion platforms and virtual reality environments), and the China Scholarship Council (CSC) scholarship No. 202006120341 for B.L.

Institutional Review Board Statement: Not applicable.

Informed Consent Statement: Not applicable.

Data Availability Statement: Data are contained within the article.

Conflicts of Interest: The authors declare no conflict of interest.

References

1. *BS 5400*; Steel, Concrete and Composite Bridges-Part 2: Specification for Loads; Appendix C: Vibration Serviceability Requirements for Foot and Cycle Track Bridges. British Standards Association: London, UK, 1978.
2. Sétra. *Footbridges: Assessment of Vibrational Behaviour of Footbridges under Pedestrian Loading: Technical Guide*; Service d'Études Techniques des Routes et Autoroutes: Paris, France, 2006.
3. *NA to BS EN 1991-2:2003*; UK National Annex to Eurocode 1: Actions on Structures-Part 2: Traffic Loads on Bridges. British Standards Institution: London, UK, 2008.
4. Research Fund for Coal and Steel. Human Induced Vibrations of Steel Structures: Design of Footbridges—Guideline. RFS2-CT-2007-00033. 2009. Available online: https://www.coursehero.com/file/59133890/Footbridge-Backgroundpdf/ (accessed on 5 April 2013).
5. Brownjohn, J.M.W.; Pavic, A.; Omenzetter, P. A spectral density approach for modelling continuous vertical forces on pedestrian structures due to walking. *Can. J. Civ. Eng.* **2004**, *31*, 65–77. [CrossRef]
6. Živanović, S.; Pavic, A.; Reynolds, P. Probability based prediction of multi-mode vibration response to walking excitation. *Eng. Struct.* **2007**, *29*, 942–954. [CrossRef]
7. Caprani, C.C.; Keogh, J.; Archbold, P.; Fanning, P. Enhancement factors for the vertical response of footbridges subjected to stochastic crowd loading. *Comput. Struct.* **2012**, *102–103*, 87–96. [CrossRef]
8. Piccardo, G.; Tubino, F. Equivalent spectral model and maximum dynamic response for the serviceability analysis of footbridges. *Eng. Struct.* **2012**, *40*, 445–456. [CrossRef]
9. Racic, V.; Brownjohn, J.M.W. Stochastic model of near periodic vertical loads due to humans walking. *Adv. Eng. Inform.* **2011**, *25*, 259–275. [CrossRef]
10. Dallard, P.; Fitzpatrick, A.J.; Flint, A.; Le Bourva, S.; Low, A.; Ridsdill-Smith, R.M.; Willford, M. The London Millennium Footbridge. *Struct. Eng.* **2001**, *79*, 17–33.
11. Macdonald, J.H.G. Lateral excitation of bridges by balancing pedestrians. *Proc. R. Soc. Lond. Ser. A* **2009**, *465*, 1055–1073. [CrossRef]
12. Ingólfsson, E.T.; Georgakis, C.T.; Ricciardelli, F.; Jönsson, J. Experimental identification of pedestrian-induced lateral forces on footbridges. *J. Sound Vib.* **2011**, *330*, 1265–1284. [CrossRef]
13. Carroll, S.P.; Owen, J.S.; Hussein, M.F.M. Experimental identification of the lateral human-structure interaction mechanism and assessment of the inverted-pendulum biomechanical model. *J. Sound Vib.* **2014**, *333*, 5865–5884. [CrossRef]
14. McRobie, A.; Morghental, G.; Lasenby, J.; Ringer, M. Section model tests on human-structure lock-in. *Proc. ICE-Bridge Eng.* **2003**, *156*, 71–79. [CrossRef]
15. Ricciardelli, F.; Mafrici MIngólfsson, E.T. Lateral Pedestrian-Induced Vibrations of Footbridges: Characteristics of Walking Forces. *J. Bridge Eng.* **2014**, *19*, 04014035. [CrossRef]

16. Bocian, M.; Macdonald, J.H.G.; Burn, J.F. Biomechanically-inspired modelling of pedestrian-induced vertical self-excited forces. *J. Bridge Eng.* **2013**, *18*, 1336–1346. [CrossRef]
17. Dang, H.V. Experimental and Numerical Modelling of Walking Locomotion on Vertically Vibrating Low-Frequency Structures. Ph.D. Thesis, University of Warwick, Coventry, UK, 2014.
18. Ahmadi, E.; Caprani, C.; Živanović, S.; Heidarpour, A. Vertical ground reaction forces on rigid and vibrating surfaces for vibration serviceability assessment of structures. *Eng. Struct.* **2018**, *172*, 723–738. [CrossRef]
19. Ahmadi, E.; Caprani, C.; Živanović, S.; Heidarpour, A. Assessment of human-structure interaction on a lively lightweight GFRP footbridge. *Eng. Struct.* **2019**, *199*, 109687. [CrossRef]
20. Caprani, C.C.; Ahmadi, E. Formulation of human-structure interaction system models for vertical vibration. *J. Sound Vib.* **2016**, *377*, 346–367. [CrossRef]
21. Bocian, M.; Macdonald, J.H.G.; Burn, J.F. Biomechanically inspired modelling of pedestrian-induced forces on laterally oscillating structures. *J. Sound Vib.* **2012**, *331*, 3914–3929. [CrossRef]
22. Qin, J.W.; Law, S.S.; Yang, Q.S.; Yang, N. Pedestrian-bridge dynamic interaction, including human participation. *J. Sound Vib.* **2013**, *332*, 1107–1124. [CrossRef]
23. Lin, B.; Zhang, Q.; Fan, F.; Shen, S. A damped bipedal inverted pendulum for human-structure interaction analysis. *Appl. Math. Model.* **2020**, *87*, 606–624. [CrossRef]
24. Vega Ruiz, D.; Magluta, C.; Roitman, N. Experimental verification of biomechanical model of bipedal walking to simulate vertical loads induced by humans. *Mech. Syst. Signal Process.* **2022**, *167*, 108513. [CrossRef]
25. Carroll, S.P.; Owen, J.S.; Hussein, M.F.M. Modelling crowd-bridge dynamic interaction with a discretely defined crowd. *J. Sound Vib.* **2012**, *331*, 2685–2709. [CrossRef]
26. Whittle, M.W. *Gait Analysis: An Introduction*, 3rd ed.; Butterworth-Heinemann Elsevier: Oxford, UK, 2001.
27. Inman, V.T.; Ralston, H.; Todd, F. *Human Walking*; Edwin Mellen Press Ltd.: Lewiston, NY, USA, 1989.
28. Perry, J.; Burnfield, J.M. *Gait Analysis: Normal and Pathological Function*, 2nd ed.; SLACK Incorporated: Thorofare, NI, USA, 2010.
29. Ayyappa, E. Normal human locomotion, Part 2: Motion, ground-reaction force and muscle activity. *Prosthet. Orthot.* **1997**, *9*, 4957. [CrossRef]
30. Gard, S.A.; Childress, D.S. What determines the vertical displacement of the body during normal walking? *J. Prosthet. Orthot.* **2001**, *13*, 6467. [CrossRef]
31. Saunders, J.B.D.M.; Inman, V.T.; Eberhart, H.D. The major determinants in normal and pathological gait. *J. Bone Jt. Surg.* **1953**, *53*, 543–558. [CrossRef]
32. Cavagna, G.A.; Thys, H.; Zamboni, A. The sources of external work in level walking and running. *J. Physiol.* **1976**, *262*, 639657. [CrossRef]
33. Whittle, M.W. Three-dimensional motion of the center of gravity of the body during walking. *Hum. Mov. Sci.* **1997**, *16*, 347–355. [CrossRef]
34. Gard, S.A.; Miff, S.C.; Kuo, A.D. Comparison of kinematic and kinetic methods for computing the vertical motion of the body center of mass during walking. *Hum. Mov. Sci.* **2004**, *22*, 597610. [CrossRef]
35. Blanchard, J.; Davies, B.L.; Smith, J.W. Design criteria and analysis for dynamic loading of footbridges. In Proceedings of the Symposium on Dynamic Behaviour of Bridges at the Transport and Road Research Laboratory, Crowthorne, UK, 19 May 1977; pp. 90–100.
36. Rainer, J.H.; Pernica, G. Vertical dynamic forces from footsteps. *Can. Acoust.* **1986**, *14*, 12–21.
37. Racic, V.; Pavic, A.; Brownjohn, J.M.W. Experimental identification and analytical modelling of human walking forces: Literature review. *J. Sound Vib.* **2009**, *326*, 1–49. [CrossRef]
38. Kerr, S.C. Human Induced Loading on Staircases. Ph.D. Thesis, University College London, London, UK, 1998.
39. Pedersen, L.; Frier, C. Sensitivity of footbridge vibrations to stochastic walking parameters. *J. Sound Vib.* **2010**, *29*, 2683–2701. [CrossRef]
40. Živanović, S.; Pavic, A.; Reynolds, P. Vibration serviceability of footbridges under human-induced excitation: A literature review. *J. Sound Vib.* **2005**, *279*, 1–74. [CrossRef]
41. Matsumoto, Y.; Nishioka, T.; Shiojiri, H.; Matsuzaki, K. Dynamic design of footbridges. *IABSE Proc.* **1978**, *P-17/78*, 1–15.
42. Pachi, A.; Ji, T. Frequency and velocity of people walking. *Struct. Eng.* **2005**, *83*, 36–40.
43. Kasperski, K.; Sahnaci, C. Serviceability of Pedestrian Structures. In Proceedings of the IMAC-XXV, Orlando, FL, USA, 19–22 February 2007.
44. Ricciardelli, F.; Briatico, C.; Ingólfsson, E.T.; Georgakis, C.T. Experimental validation and calibration of pedestrian loading models for footbridges. In Proceedings of the International Conference of Experimental Vibration Analysis for Civil Engineering Structures, Porto, Portugal, 24–26 October 2007.
45. Živanović, S.; Racic, V.; El-Bahnasy, I.; Pavic, A. Statistical characterization of parameters defining human walking as observed on an indoor passerelle. In Proceedings of the International Conference of Experimental Vibration Analysis for Civil Engineering Structures, Porto, Portugal, 24–26 October 2007.
46. Živanović, S. Benchmark footbridge for vibration serviceability assessment under vertical component of pedestrian load. *J. Struct. Eng.* **2012**, *138*, 1193–1202. [CrossRef]
47. McGeer, T. Passive dynamic walking. *Int. J. Robot. Res.* **1990**, *9*, 6282. [CrossRef]

48. Geyer, H. Simple Models of Legged Locomotion Based on Compliant Limb Behavior. Ph.D. Thesis, Friedrich-Schiller-Universität, Jena, Germany, 2005.
49. Whittington, B.R.; Thelen, D.G. A simple mass-spring model with roller feet can induce the ground reactions observed in human walking. *J. Biomech. Eng.* **2009**, *131*, 011013. [CrossRef]
50. Kim, S.; Park, S. Leg stiffness increases with speed to modulate gait frequency and propulsion energy. *J. Biomech.* **2011**, *44*, 1253–1258. [CrossRef]
51. Goldstein, H. *Classical Mechanics, 2nd edition*; Addison-Wesley: San Francisco, CA, USA, 1980.
52. National Health Service (NHS). Health Survey for England, United Kingdom. 2010. Available online: https://digital.nhs.uk/data-and-information/publications/statistical/health-survey-for-england/health-survey-for-england-2009-trend-tables (accessed on 5 April 2013).
53. Pheasant, S.T. Anthropometric estimates for British civilian adults. *Ergonomics* **1982**, *25*, 993–1001. [CrossRef]
54. Hof, A.L.; Gazendam, M.G.J.; Sinke, W.E. The condition for dynamic stability. *J. Biomech.* **2005**, *38*, 1–8. [CrossRef]
55. Mathworks Inc. *MATLAB*, version 9.11.0.1769968 (R2021b); Matworks: Natick, MA, USA, 2021.
56. Lee, C.R.; Farley, C.T. Determinants of the center of mass trajectory in human walking and running. *J. Exp. Biol.* **1998**, *201*, 2935–2944. [CrossRef]
57. Gard, S.A.; Childress, D.S. The effect of pelvic list on the vertical displacement of the trunk during normal walking. *Gait Posture* **1997**, *5*, 233–238. [CrossRef]
58. Gard, S.A.; Childress, D.S. The influence of stance-phase knee flexion on the vertical displacement of the trunk during normal walking. *Arch. Phys. Med. Rehabil.* **1999**, *80*, 26–32. [CrossRef]
59. Hansen, A.; Gard, S.; Childress, D. The determination of foot/ankle roll-over shape: Clinical and research applications. In Proceedings of the Pediatric Gait: A New Millennium in Clinical Care and Motion Analysis Technology, Chicago, IL, USA, 22 July 2000; pp. 159–165. [CrossRef]
60. Adamczyk, P.G.; Collins, S.H.; Kuo, A.D. The advantages of a rolling foot in human walking. *J. Exp. Biol.* **2006**, *209*, 3953–3963. [CrossRef]
61. Gao, Y.; Wang, J.; Liu, M. The Vertical Dynamic Properties of Flexible Footbridges under Bipedal Crowd Induced Excitation. *Appl. Sci.* **2017**, *7*, 677. [CrossRef]

buildings

MDPI

Article

Equivalent Dynamic Load Factor of Different Non-Exceedance Probability for Crowd Jumping Loads

Jiecheng Xiong [1], Shuqian Duan [1,2,*], Hui Qian [1] and Ziye Pan [1]

[1] School of Civil Engineering, Zhengzhou University, Zhengzhou 450001, China; xiongjc2019@zzu.edu.cn (J.X.); qianhui@zzu.edu.cn (H.Q.); lizzie.feehily@foxmail.com (Z.P.)
[2] Henan Urban Planning Institute and Corporation, Zhengzhou 450044, China
* Correspondence: shuqianduan@zzu.edu.cn

Abstract: Existing dynamic load factors (DLF) for crowd jumping loads are modeled by identifying the peaks and energies from the corresponding Fourier amplitude spectrum, which may underestimate and overestimate structural responses, respectively. Based on the principle of equal structural responses, this study herein develops an equivalent DLF, using the frequency response function to weight and integrate the power spectral density (PSD) matrix of crowd jumping loads. Firstly, massive PSD matrices of different crowd sizes and different metronome frequencies are simulated using a random field model of crowd jumping loads. Thereafter, the equivalent DLF of different structural damping ratios, crowd sizes, metronome frequencies, and non-exceedance probabilities are investigated, and a model of the equivalent DLF is established accordingly. It is believed that this model could provide an efficient and accurate way to conduct reliability design for structures subjected to crowd jumping loads.

Keywords: crowd jumping load; dynamic load factor (DLF); vibration serviceability; human-induced vibration

Citation: Xiong, J.; Duan, S.; Qian, H.; Pan, Z. Equivalent Dynamic Load Factor of Different Non-Exceedance Probability for Crowd Jumping Loads. *Buildings* **2022**, *12*, 450. https://doi.org/10.3390/buildings12040450

Academic Editor: Luca Pelà

Received: 16 February 2022
Accepted: 31 March 2022
Published: 6 April 2022

1. Introduction

Currently, the modern assembly structures (e.g., large-span floor slabs, footbridges, grandstands, etc.), gradually exhibit the characteristics of low natural frequencies and low damping ratios, due to the extensive use of lightweight, high-strength building materials [1–5]. Accordingly, structural vibration serviceability issues [6,7] as well as safety risks [8,9] caused by crowd activities have been increasingly emerging, which may further lead to huge economic losses and even heavy casualties in extreme cases. One typical example of this was the London Millennium Bridge in 2000, in which severe lateral vibrations of the bridge were induced by pedestrian use [10]. The bridge was therefore closed to traffic for three days after the grand opening, and it cost GBP 5 million to solve the vibration problem. Additionally, at a sports concert given in Vigo, Spain in 2018, the wooden stands collapsed due to the audience's rhythmic activities, with over 300 people injured [11]. Jumping is usually considered to generate the most significant vertical loads in contrast to other common activities [12,13], in particular when the crowd is coordinated by an external beat. This will be the focus of this research.

Load models are necessary, both in the design stage and the as-built stage to evaluate the vibration performance of structures subjected to crowd jumping activities [14]. Considering that jumping activities are approximate periodic, Fourier series models are mostly used to represent jumping loads [13,15–19], where the dynamic load factor (DLF) is the key parameter to characterize the load magnitude and crowd synchronization. When crowds jump to an external beat, it is unlikely to achieve perfect synchronization because of both inter- and intra-subject variability [20], and the crowd behave differently to the beat of different dominant frequencies [17]. Therefore, the DLF is a stochastic variable and related to the crowd size and guidance frequency, where the guidance frequency denotes the dominant frequency of the metronome or music that guides the crowd to jump. Massive investigations on jumping loads have been conducted [13–19], and DLF models have been established accordingly. The DLF of these models is identified by the peaks [16] or energies [17] from the corresponding Fourier amplitude spectrum. The intra-subject variability of a jumping activity caused the load energy to spread around the jumping frequency and its multiplier [21], where the jumping frequency denotes the dominant frequency of a person's jumping load time history. When a crowd jumps at a given guidance frequency, most people can follow the beat, and their jumping frequency is equal to the guidance frequency. As for a few people who cannot follow the beat, their jumping frequency is unequal to the guidance frequency. The DLF obtained by identifying the peaks can underestimate structural responses because it overlooks the energy near the jumping frequency and its multiplier [22]. The DLF obtained by identifying the energy can overestimate structural responses because it assumes that the energy is all concentrated at the jumping frequency and its multiplier. When using the resonance assumption to calculate structural responses, the energy at the jumping frequency and its multiplier contributes more to structural responses than the energy around them. Moreover, most of the DLF models lack the data support of large crowd jumping tests. For one thing, the DLF model is developed based on the individual jumping load records aligned by the time instant of the beat [17,19], in which the influence of human–human interaction is not involved. For another, the DLF is modeled through crowd jumping loads identified by structural responses [15], in which the jumping load of each person could not be recorded. Moreover, owing to the limitation of test conditions, the model ignores the change of the DLF with the guidance frequency. In addition, only the mean value or a certain quantile value of the DLF is provided, which is not applicable for the structural design of different reliabilities.

The accurate calculation of structural responses is the main purpose of load modeling. Hence, the modeling parameters could be obtained based on the principle of equal structural responses. This idea has been applied to the modeling of crowd walking loads, in which the number of equivalent synchronized pedestrians is obtained using this principle [23,24]. Similarly, the DLF can be identified with the equal structural response, hereafter termed as the equivalent DLF. This is the main highlight of this study. It can overcome the shortcomings of the existing DLF modeled by identifying the peaks or energies from the corresponding Fourier amplitude spectrum. Moreover, structural response could be calculated in the form of Fourier series using the equivalent DLF, which is further more convenient for application to structural design and evaluation than other complicated stochastic models of jumping loads [12,25–29]. In addition, to further apply the model to structural reliability design, the equivalent DLF of different non-exceedance probability is established. This is another main highlight of this study. It can overcome the shortcomings of the existing DLF, which only provides its mean value or a certain quantile value. The paper begins with the formula of the equivalent DLF calculation, which adopts the frequency response function (FRF) to weight and integrate the power spectral density (PSD) matrix of crowd jumping loads. This approach makes the root mean square value of structural acceleration responses equal. The formula of simulating PSD matrices of crowd jumping loads is then provided. This is followed by the modeling of the equivalent DLF with four parameters, i.e., structural damping ratio, guidance frequency, crowd size, and non-exceedance probabilities. The Engineering application of the equivalent DLF is followed. Afterwards, discussions on this work are provided.

2. Formula of the Equivalent Dynamic Load Factor

The Fourier series-based model for crowd jumping loads, $F(t)$, is usually expressed as [15]:

$$F(t) = W\left[1 + \sum_{j=1}^{n_h} r_j \sin(2\pi j f_{gu} t + \varphi_j)\right] \quad (1)$$

where W is the total body weight of the jumping crowd, r_j and φ_j are the DLF and phase angle of the jth harmonic, f_{gu} is the guidance frequency, and n_{h} is the number of load harmonics considered.

The values of r_j for jumping loads are commonly determined by two methods. In the first method, hereafter termed as the peak method, the maximum value of the jth harmonic region in the Fourier amplitude spectrum is adopted as r_j, as depicted in Figure 1, and the values are listed in Table 1 for further comparison. In the above, the jth harmonic region is $((j-0.5)f_{\mathrm{gu}}, (j+0.5)f_{\mathrm{gu}})$. As for the second method, hereafter termed as the energy method, the total energy of the jth harmonic region in the Fourier amplitude spectrum is adopted as r_j, as calculated by:

$$r_j = \sqrt{\sum_{h=1}^{n_{\mathrm{s}}} A_{jh}^2} \tag{2}$$

where A_{jh} is the amplitude of the hth sinusoid in the jth harmonic region and n_{s} is the number of the sinusoids. The values of r_j calculated by the energy method for the Fourier amplitude spectrum in Figure 1 are provided in Table 1. The peak method only considers the energy at the jumping frequency and its multiplier, and the energy in between is ignored, which can underestimate structural responses. The energy method assumes that all the energy of a jumping load is concentrated at the jumping frequency and its multiplier, which overestimates structural responses when the resonance assumption is made.

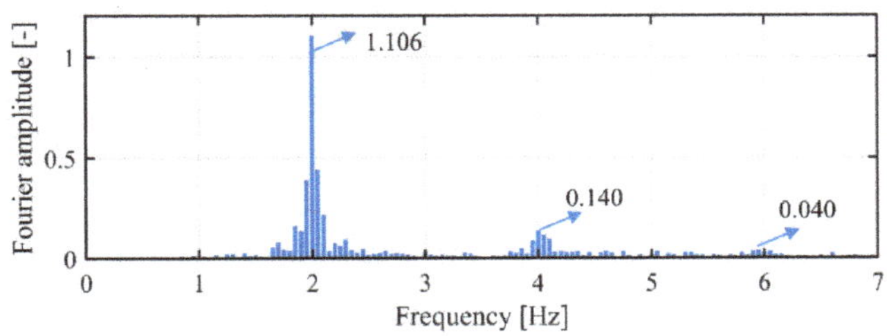

Figure 1. Fourier amplitude spectrum of a 2.0 Hz individual jumping load.

Table 1. Values of the DLF obtained by different methods.

Harmonic Order	Peak Method	Energy Method	FRF-Weighting Method	
			ζ_{s} = 0.005	ζ_{s} = 0.05
r_1	1.106	1.313	1.113	1.248
r_2	0.140	0.276	0.152	0.236
r_3	0.040	0.126	0.047	0.090

A novel DLF calculating method that can accurately estimate the root mean square value of structural acceleration responses is therefore proposed. Considering a single degree-of-freedom system with a unit mass subjected to crowd jumping loads, as shown in Figure 2, the equation of motion of the system is expressed as:

$$\ddot{u}_{\mathrm{s}}(t) + 2\zeta_{\mathrm{s}}\omega_{\mathrm{s}}\dot{u}_{\mathrm{s}}(t) + \omega_{\mathrm{s}}^2 u_{\mathrm{s}}(t) = \sum_{l=1}^{n_{\mathrm{p}}} W_l x_l(t) \tag{3}$$

where ζ_{s}, ω_{s}, and $u_{\mathrm{s}}(t)$ are the damping ratio, circular frequency, and displacement of the system, respectively, W_l is the body weight of the lth jumping person, and $x_l(t)$ is the ground reaction force time history of the lth jumping person divided by W_l, which is dimensionless. Therefore, $x_l(t)$ reflects the change of center when an individual jumps, and it is the most direct variable to reflect the jumping pattern. The FRF $H(f; f_{\mathrm{s}}, \zeta_{\mathrm{s}})$ related to the input, $\sum_{l=1}^{n_{\mathrm{p}}} W_l x_l(t)$, and output, $\ddot{u}_{\mathrm{s}}(t)$, is deduced from Equation (3):

$$H(f;f_s,\zeta_s) = \frac{-f^2}{(-f^2+f_s^2+2\zeta_s\mathrm{i}ff_s)} \tag{4}$$

where $f_s = 2\pi\omega_s$ is the cyclic frequency of the structure and i is the imaginary unit. Then, the auto-PSD [21] $G(f)$ of $\ddot{u}_s(t)$ can be obtained by the stochastic vibration theory [30]:

$$G(f) = |H(f;f_s,\zeta_s)|^2 \sum_{l=1}^{n_p}\sum_{k=1}^{n_p} W_l W_k G_{lk}(f) \tag{5}$$

where $G_{lk}(f)$ is the lth row and kth column element of the PSD matrix of crowd jumping loads. If $l \neq k$, $G_{lk}(f)$ is the cross-PSD of $x_l(t)$ and $x_k(t)$. If $l = k$, $G_{ll}(f)$ is the auto-PSD of $x_l(t)$. Integrating $G(f)$ within the frequency range yields the root mean square value a_{rms} of $\ddot{u}_s(t)$:

$$a_{rms}^2 = \int_0^\infty G(f)\mathrm{d}f = \int_0^\infty |H(f;f_s,\zeta_s)|^2 \sum_{l=1}^{n_p}\sum_{k=1}^{n_p} W_l W_k G_{lk}(f)\mathrm{d}f \tag{6}$$

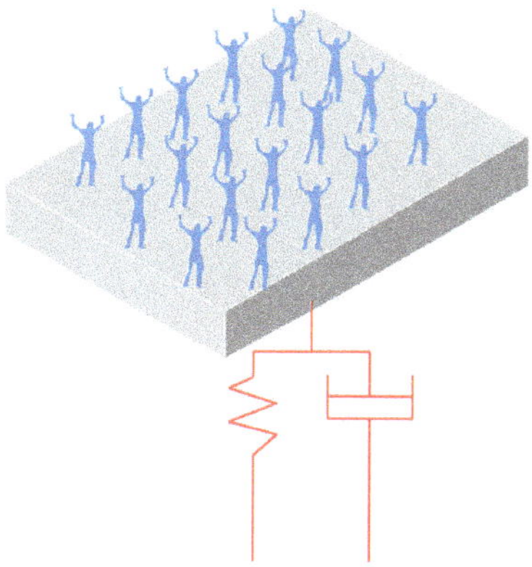

Figure 2. Model of single degree-of-freedom system subjected crowd jumping loads.

If the crowd jumping load, i.e., the right term of Equation (3), is modeled by the Fourier series, as expressed by Equation (1), the auto-PSD $G_F(f)$ of $F(t)$ is:

$$G_F(f) = W^2\left[\delta(f) + \sum_{j=1}^{n_h} r_j^2 \delta(f - jf_{gu})\right] \tag{7}$$

where $\delta()$ is the Dirac delta function. Then, as the derivation procedure of Equations (3)–(5), a_{rms} of $\ddot{u}_s(t)$ can be calculated by:

$$\begin{aligned} a_{rms}^2 &= \int_0^\infty G(f)\mathrm{d}f \\ &= \int_0^\infty |H(f;f_s,\zeta_s)|^2 W^2\left[\delta(f) + \sum_{j=1}^{n_h} r_j^2\delta(f-jf_{gu})\right]\mathrm{d}f \\ &= W^2\left[|H(0;f_s,\zeta_s)|^2 + \sum_{j=1}^{n_h} r_j^2 |H(jf_{gu};f_s,\zeta_s)|^2\right] \\ &= W^2\sum_{j=1}^{n_h} r_j^2 |H(jf_{gu};f_s,\zeta_s)|^2 \end{aligned} \tag{8}$$

The resonance scenario is commonly assumed for structural response estimation to consider the severest case [31], where structural frequency is assumed to be equal to the guidance frequency or its multiplier. Consequently, structural responses induced by the jth harmonic of jumping load is much larger than the rest of the harmonics:

$$a_{\mathrm{rms}}^2 = W^2 \sum_{j=1}^{n_{\mathrm{h}}} r_j^2 \left| H\left(jf_{\mathrm{gu}}; f_{\mathrm{s}}, \zeta_{\mathrm{s}}\right) \right|^2 \approx W^2 r_j^2 \left| H\left(jf_{\mathrm{gu}}; jf_{\mathrm{gu}}, \zeta_{\mathrm{s}}\right) \right|^2 \tag{9}$$

Making a_{rms} calculated by Equations (6) and (9) equal and assuming a resonance scenario in Equation (6) yields:

$$W^2 r_j^2 \left| H\left(jf_{\mathrm{gu}}; jf_{\mathrm{gu}}, \zeta_{\mathrm{s}}\right) \right|^2 = \int_0^\infty \left| H\left(f; jf_{\mathrm{gu}}, \zeta_{\mathrm{s}}\right) \right|^2 \sum_{l=1}^{n_{\mathrm{p}}} \sum_{k=1}^{n_{\mathrm{p}}} W_l W_k G_{lk}(f) \mathrm{d}f \tag{10}$$

If the weight of every jumping person is assumed to be equal, Equation (10) will be further simplified as:

$$r_j = \frac{\sqrt{\int_0^\infty \left| H\left(f; jf_{\mathrm{gu}}, \zeta_{\mathrm{s}}\right) \right|^2 \sum\limits_{l=1}^{n_{\mathrm{p}}} \sum\limits_{k=1}^{n_{\mathrm{p}}} G_{lk}(f) \mathrm{d}f}}{\left| H\left(jf_{\mathrm{gu}}; jf_{\mathrm{gu}}, \zeta_{\mathrm{s}}\right) \right|} \tag{11}$$

In the above, $H(jf_{\mathrm{gu}}; jf_{\mathrm{gu}}, \zeta_{\mathrm{s}})$ is the maximum of $H(f; jf_{\mathrm{gu}}, \zeta_{\mathrm{s}})$, so it is indicated that the equivalent DLF for crowd jumping loads can be identified using the FRF to weight and integrate the PSD matrix of crowd jumping loads, hereafter termed as the FRF-weighting method. The equivalent DLF is related to the structural damping ratio because Equation (11) is derived from the principle of equal structural responses. Then, the equivalent DLF of the Fourier amplitude spectrum in Figure 1 calculated by Equation (11) is provided in Table 1. It is observed from Table 1 that the DLF obtained by the peak method and energy method are quite different, especially for high-order harmonics, because the degree of energy diffusion increases as the harmonic order increases [21]. The maximum-normalized FRF is adopted in the FRF-weighting method to weight the contribution of different sinusoids to structural responses, and the equivalent DLF obtained by this method is between the DLF obtained by the peak method and the DLF obtained by the energy method. Moreover, the equivalent DLF obtained by the FRF-weighting method is strongly related to structural damping ratio. When the structural damping ratio is small, the bandwidth of the FRF is narrow, so the energy around the jumping frequency and its multiplier contributes less to structural responses. Therefore, the equivalent DLF is small and close to the DLF obtained by the peak method. When the structural damping ratio is large, the bandwidth of the FRF is wide and the energy around the jumping frequency and its multiplier contributes more to structural responses. Therefore, the equivalent DLF is large and close to the DLF obtained by the energy method.

3. Simulation of the Equivalent Dynamic Load Factor

The PSD matrix for crowd jumping loads is fundamental to calculating the equivalent DLF according to Equation (11). In this section, a random field model for crowd jumping loads proposed by the authors [32] is adopted for the simulation of the PSD matrix.

3.1. Random Field Model for Crowd Jumping Loads

The auto-PSD of individual jumping loads, i.e., the diagonal element of the random field model, is established according to the individual jumping load records collected by force plates. The cross-PSD of crowd jumping load, i.e., the off-diagonal element of the random field model, is established according to the crowd jumping records collected by three-dimensional motion capture technology. It is proved by spectral analysis that the energy after the second harmonic of the load is weak, and the test error is large for high-frequency components due to the high-frequency trembling of the skin when three-dimensional motion capture technology is used. Therefore, the first two harmonics of crowd jumping loads are considered in the random field model. Spectral analysis of the individual load records demonstrates that the energy of the jumping load is mainly distributed around the fundamental jumping frequency and its multiple integers (harmonics), and the energy in between is almost equal to zero. Therefore, the auto-PSD can be modeled separately for each harmonic. The auto-PSD sample is then normalized by the jumping frequency and the harmonic order, and a two-term Gaussian function is used to model the shape of the normalized auto-PSD. Finally, the lth diagonal element of the PSD matrix, i.e., the auto-PSD $G_{ll}(f)$ of the lth person's jumping load, is modeled by:

$$\begin{aligned} &G_{ll}(f;f_{\text{ju}}) = \sum_{j=1}^{2} G_{ll,j}(f;f_{\text{ju}}) = \sum_{j=1}^{2} \frac{\rho(f_{\text{ju}})s_j(f_{\text{ju}})G_{\text{no}}(f/jf_{\text{ju}})}{jf_{\text{ju}}} \quad && f \in [0.9jf_{\text{ju}}, 1.1jf_{\text{ju}}] \\ &G_{ll}(f;f_{\text{ju}}) = 0 && \text{otherwise} \end{aligned} \tag{12}$$

where $G_{ll,j}(f;f_{\text{ju}})$ is the jth harmonic of the auto-PSD and $s_j(f_{\text{ju}})$ is the energy for the jth harmonic, which can be calculated by Equation (13); the coefficients identified by the test records are listed in Table 2. $\rho(f_{\text{ju}})$ is used to reflect the energy truncation effect and can be calculated by Equation (14). $G_{\text{no}}(f/jf_{\text{ju}})$ is used to describe the shape of the normalized-PSD, which can be calculated by Equation (15), and the coefficients are provided in Table 2.

$$s_j(f_{\text{ju}})(s_{\text{T}}(f_{\text{ju}})) = p_1 f_{\text{ju}}^3 + p_2 f_{\text{ju}}^2 + p_3 f_{\text{ju}} + p_4 \tag{13}$$

$$\rho(f_{\text{ju}}) = \frac{s_{\text{T}}(f_{\text{ju}})}{\sum_{j=1}^{2} s_j(f_{\text{ju}})} \tag{14}$$

$$G_{\text{no}}(f/jf_{\text{ju}}) = p_5 \exp\left[-\left(\frac{f/jf_{\text{ju}}-1}{p_6}\right)^2\right] + p_7 \exp\left[-\left(\frac{f/jf_{\text{ju}}-1}{p_8}\right)^2\right] \tag{15}$$

Table 2. Model coefficients for the auto-PSD.

Parameter	p_1 [s]	p_2 [s]	p_3 [s]	p_4 [s]	p_5 [-]	p_6 [-]	p_7 [-]	p_8 [-]
$s_{\text{T}}(f_{\text{ju}})$	0.210	−1.919	5.621	−3.958	/	/	/	/
$s_1(f_{\text{ju}})$	0.211	−1.842	5.231	−3.769	/	/	/	/
$s_2(f_{\text{ju}})$	0.035	−0.363	1.166	−0.993	/	/	/	/
$G_{\text{no}}(f/jf_{\text{ju}})$	/	/	/	/	2.804	0.079	29.27	0.012

The coherence function is the key to the modeling of the cross-PSD, because the auto-PSD has been well modeled by Equation (12). First, the coherence function is modeled by filtering the records with a dominant frequency equal to the metronome frequency. The magnitude of the coherence function is then simplified as an energy scaling factor, because the real part and imaginary part of cross-PSD have the same curve shape, as auto-PSD. The phase angle of the coherence function is simplified as a set of time lag shifts, because different load records have the same dominant frequency. Finally, the off-diagonal element of the PSD matrix, i.e., the cross-PSD $G_{lk}(f)$ of the lth person's jumping load and kth person's jumping load is modeled by:

$$\begin{aligned} &G_{lk}(f;f_{\text{ju},l},f_{\text{ju},k}) = \exp\left[\mathrm{i}2\pi f\left(\Delta t_k(f_{\text{ju},k}) - \Delta t_l(f_{\text{ju},l})\right)\right] \times \\ &\left[\sum_{j=1}^{2}\sqrt{\left|\gamma_{lk}(j;f_{\text{ju},l})\right|\left|\gamma_{lk}(j;f_{\text{ju},k})\right|G_{ll,j}(f;f_{\text{ju},l})G_{kk,j}(f;f_{\text{ju},k})}\right] \quad && f \in [0.9jf_{\text{ju}}, 1.1jf_{\text{ju}}] \\ &G_{lk}(f;f_{\text{ju},l},f_{\text{ju},k}) = 0 && \text{otherwise} \end{aligned} \tag{16}$$

where $f_{\text{ju},l}$ is the lth person's jumping frequency, $G_{ll,j}(f;f_{\text{ju},l})$ is defined by Equation (12), and $|\gamma_{lk}(j;f_{\text{ju},l})|$ is the magnitude of the coherence function at the jth harmonic, as provided in Table 3. $\Delta t_l(f_{\text{ju},l})$ denotes the time lag shift, which follows a normal distribution $N(0, \sigma^2)$ [s], and the value of σ is provided in Table 3. When a crowd jumps at a given guidance frequency, some people can follow the beat while others cannot. Therefore, the jumping frequency of each person, i.e., $f_{\text{ju},l}$, is modelled as a random variable, and Table 4 in the literature [32] shows its probability distribution.

Table 3. Model coefficients for the cross-PSD.

f_{ju} [Hz]	σ [-]	$\mid\gamma_{lk}(1)\mid$ [-]	$\mid\gamma_{lk}(2)\mid$ [-]
1.5	0.131	0.628	0.301
1.6	0.084	0.620	0.242
1.7	0.098	0.702	0.325
1.8	0.084	0.632	0.311
1.9	0.069	0.725	0.381
2.0	0.071	0.762	0.405
2.1	0.059	0.785	0.442
2.2	0.054	0.798	0.463
2.3	0.057	0.785	0.449
2.4	0.053	0.795	0.452
2.5	0.042	0.809	0.509
2.6	0.053	0.733	0.407
2.7	0.054	0.722	0.401
2.8	0.053	0.718	0.377
2.9	0.047	0.727	0.369
3.0	0.049	0.743	0.389
3.1	0.045	0.722	0.368
3.2	0.042	0.760	0.417
3.3	0.042	0.691	0.345
3.4	0.037	0.705	0.342
3.5	0.041	0.665	0.296

3.2. Simulation Procedure

Because the first two harmonics of crowd jumping loads are considered in the random field model, the equivalent DLF for the first two harmonics is simulated in this section. As detailed in Table 1, the equivalent DLF is relevant to the structural damping ratio, and the structural damping ratio of common structures subjected to crowd jumping loads, such as gymnasia, grandstands, and concert halls, is in the range of 0.005–0.05 [31,33]. Therefore, the structural damping ratio is taken from 0.005 to 0.05 with an interval of 0.005. Moreover, a crowd reacts differently to different guidance frequencies, so it is crucial to investigate the relationship between the equivalent DLF and the guidance frequency. The guidance frequency is taken from 1.5 to 3.5 Hz, with an interval of 0.1 Hz, because it is hard for a crowd to keep jumping outside this range [32]. The DLF and crowd size exhibit a nonlinear relationship because of the inter-subject variability. In addition, crowd coordination tends to be a constant when the crowd size exceeds fifty [15,18], so the crowd size is taken from 1 to 50 with an interval of 1. To further apply the model to structural reliability design, 1000 Monte Carlo simulations are performed for each working condition; to investigate the equivalent DLF of different non-exceedance probability, the adequacy of the number of Monte Carlo simulations will be discussed in Section 4.4. Details of the simulation algorithm can be seen in Algorithm 1.

Algorithm 1: Algorithm for the equivalent DLF simulation

```
for  i1 = 1.5, 1.6, ... , 3.5 do                         # Set the loop range of the guidance frequency
  f_gu ← i1.
  for  i2 = 1, 2, ... , 50 do                             # Set the loop range of the crowd size
    n_p  ← i2.
    for  i3 = 1, 2, ... , 1000 do                         # Set the number of the Monte Carlo simulations
      Calculate G_lk(f) (l = k) using Equation (12).      # Simulate the PSD matrix of crowd jumping loads
      Calculate G_lk(f) (l ≠ k) using Equation (16).
      for  i4 = 0.005, 0.005, ... , 0.05 do               # Set the loop range of the structural damping ratio
        ζ_s ← i4.
        Calculate r_j using Equation (11).                # Calculate the equivalent DLF
      end for
    end for
  end for
end for
return  r_j
```

4. Modeling of the Equivalent Dynamic Load Factor

The equivalent DLF is related to the structural damping ratio, guidance frequency, crowd size, and non-exceedance probability, so the modeling of the equivalent DLF is a multivariable regression problem. The working condition, where the crowd size is equal to 50, the structural damping ratio is equal to 0.005, and the non-exceedance probability is equal to 0.5, is defined as the standard working condition. First, the equivalent DLF model of the standard working condition is built. Afterwards, the conversion models for other crowd size, structural damping ratio, and non-exceedance probability are established.

4.1. Modeling of the Equivalent DLF for the Standard Working Condition

According to the definition of the standard working condition, the median of 1000 simulation data with a structural damping ratio equal to 0.005 and crowd size equal to 50 are adopted to build the model. The crowd reacts differently to different beats. This is also proven by the probability distribution of the jumping frequency, the probability distribution of the time lag shift, and the values of the coherency function magnitude, all of which are related to the guidance frequency. Therefore, the equivalent DLF of the standard working condition is related to the guidance frequency, as depicted in Figure 3, from which it is observed that the equivalent DLF increases first and then decreases with the guidance frequency. This is because it is hard for a crowd to keep up with the beat, and the jumping activity is exhausted when the guidance frequency is high or low. Particularly, the equivalent DLF at the low guidance frequency is only about one-third of the maximum value at the moderate guidance frequency. A fourth-order polynomial is used to model the equivalent DLF for the standard working condition, as depicted in Figure 3, in which Equation (17) is used for the first harmonic and Equation (18) is used for the second harmonic.

$$r_{\text{sta},1}(f_{\text{gu}}) = 3.673f_{\text{gu}}^4 - 29.4f_{\text{gu}}^3 + 71.42f_{\text{gu}}^2 - 38.31f_{\text{gu}} - 16.83 \tag{17}$$

$$r_{\text{sta},2}(f_{\text{gu}}) = 0.934f_{\text{gu}}^4 - 8.258f_{\text{gu}}^3 + 24.62f_{\text{gu}}^2 - 27.12f_{\text{gu}} + 9.135 \tag{18}$$

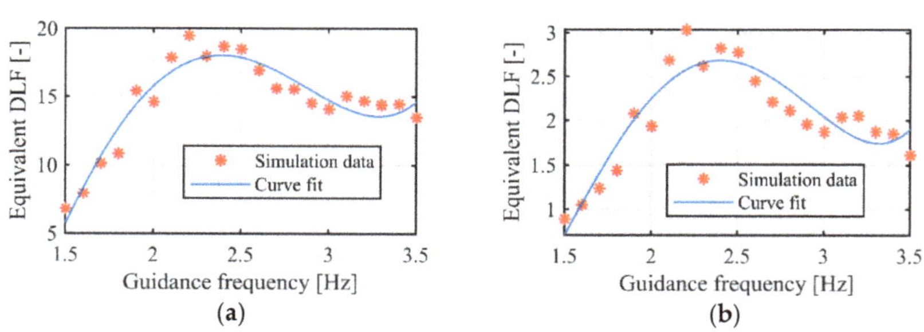

Figure 3. Modeling of the equivalent DLF for the standard working condition. (**a**) Equivalent DLF for the first harmonic, (**b**) Equivalent DLF for the second harmonic.

4.2. Conversion Model for Structural Damping Ratio

According to Equation (11), the structural damping ratio affects the equivalent DLF by changing the magnitude of the FRF. The bandwidth of the FRF increases as the structural damping ratio increases, so the contribution for the energy between different harmonics to structural responses increases. As a result, the equivalent DLF will increase with the structural damping ratio. Moreover, the influence degree of the structural damping ratio on the equivalent DLF is related to the PSD matrix. In particular, the impact degree increases as the bandwidth of the sum of the PSD matrix's elements increases, because more energies are distributed between the jumping frequency and its multiplier, and these energies are weighted by the FRF to contribute for structural responses. First, the bandwidth increases as the jumping frequency increases, which is illustrated by the expression of the normalized PSD, i.e., Equation (15). Moreover, this bandwidth increases as the dispersion degree of the jumping frequency increases according to Equation (16). The dispersion degree of the jumping frequency is large at the low guidance frequency [32], because many people cannot keep up with the beat. Consequently, the influence degree of the structural damping ratio on the equivalent DLF for different guidance frequencies is complicated. To further yield the influence degree, the median of the 1000 simulations with a crowd size equal to 50 is used to plot Figure 4, where the equivalent DLF is normalized by the equivalent DLF with a structural damping ration equal to 0.005.

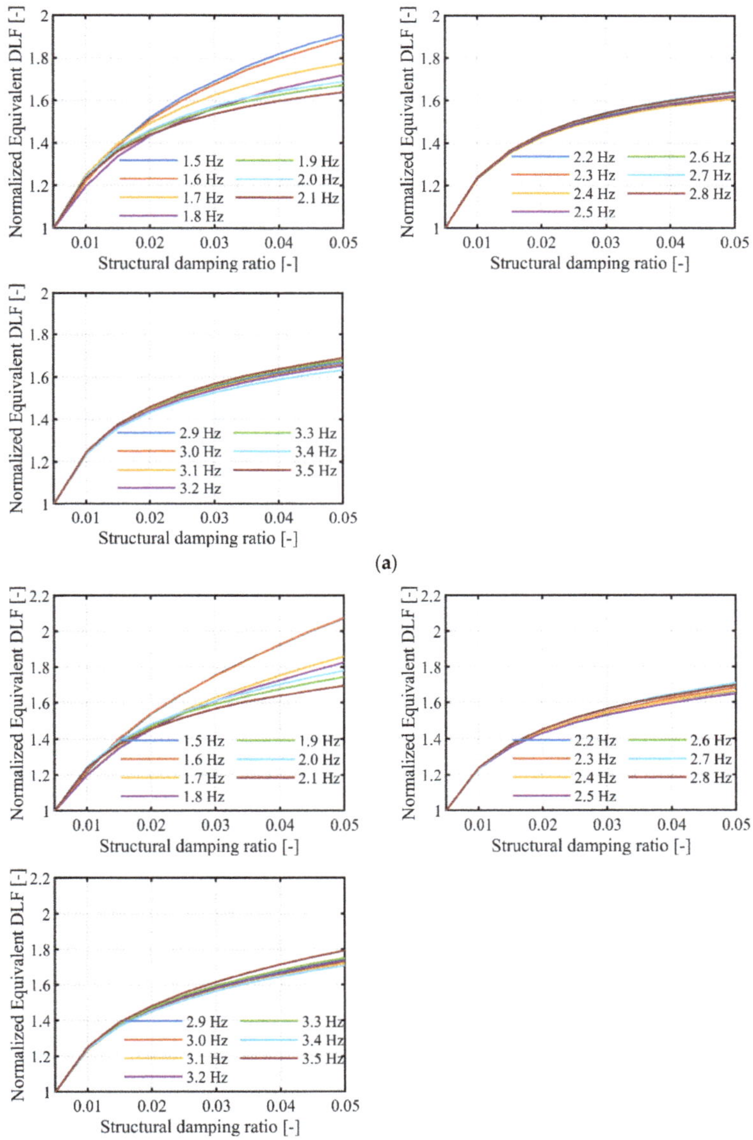

Figure 4. Relationship between structural damping ratio and normalized equivalent DLF. (**a**) Relationship for the first harmonic, (**b**) Relationship for the second harmonic.

It is observed from Figure 4 that the influence of the structural damping ratio on the equivalent DLF is not greatly affected by the guidance frequency in general. Only at the low guidance frequency, such as 1.5 and 1.6 Hz, the influence is different and is slightly greater than the other guidance frequencies. Therefore, the influence of the guidance frequency is overlooked in the conversion model of Equations (19) and (20) for the structural damping ratio. In addition, the crowd size and non-exceedance probability are also overlooked in the conversion model of Equations (19) and (20), because these two variables are not related to the influence of the structural damping ratio on the equivalent DLF according to Equation (11). Then, by fitting the average curve of different guidance frequencies in Figure 4, power functions, as depicted in Figure 5, are adopted to establish the conversion model, as shown in Equations (19) and (20).

$$c_1(\zeta_s) = -0.5241{\zeta_s}^{-0.2253} + 2.728 \tag{19}$$

$$c_2(\zeta_s) = -5.702{\zeta_s}^{-0.0483} + 8.367 \tag{20}$$

where $c_1(\zeta_s)$ is the conversion model for the first harmonic and $c_2(\zeta_s)$ is the conversion model for the second $c_1(\zeta_s) = -0.5241{\zeta_s}^{-0.2253} + 2.728$ harmonic.

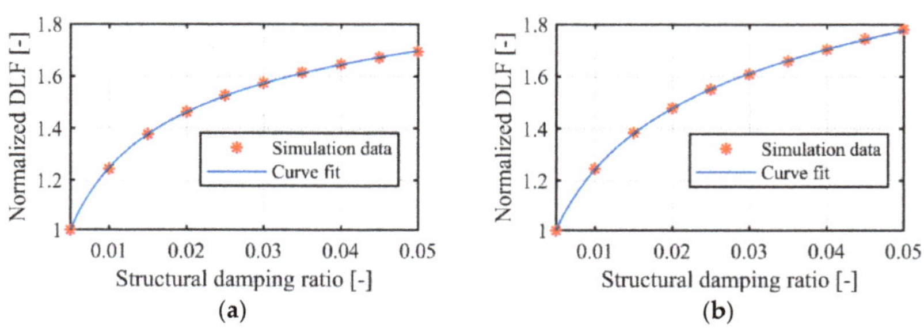

Figure 5. Conversion model for structural damping ratio. (**a**) Model for the first harmonic, (**b**) Model for the second harmonic.

4.3. Conversion Model for Crowd Size

Owing to the inter-subject variability, the equivalent DLF is not proportional to the crowd size. In existing models, the crowd DLF divided by the crowd size is defined as the coordination factor to quantify the crowd synchronization [15], because the coordination factor tends to be a constant as the crowd size increases, which makes the modeling of the crowd DLF easy. Learning from this idea, the coordination factor, co_i, in this section is defined as:

$$co_j(n_p) = \frac{c_j(n_p)}{n_p} = \frac{r_j(n_p)}{n_p r_j(50)} \tag{21}$$

where the subscript j denotes the harmonic order, c_j is the conversion model for the crowd size, and $r_j(n_p)$ is the equivalent DLF with crowd size n_p. To further investigate the law of the coordination factor, the median of the 1000 simulations with structural damping ratio equal to 0.005 is used to plot Figure 6, where the ordinate is obtained by Equation (21).

It is observed from Figure 6 that the influence of the crowd size on the coordination factor is affected by the guidance frequency when the crowd size is small. Nevertheless, as the crowd size increases, the relationships between the coordination factor and crowd size for different guidance frequencies tend to coincide, and practical structures are commonly subjected to jumping loads with large crowd. Therefore, the influence of the guidance frequency is overlooked in the modeling of the coordination factor. In addition, the structural damping ratio and non-exceedance probability are also overlooked in the coordination factor model, because these two variables are not related to crowd synchronization. Then, by fitting the average curve of different guidance frequencies in Figure 6, power functions, as depicted in Figure 7, are adopted to establish the model of coordination factor:

$$co_1(n_{\mathrm{p}}) = 0.01993 n_{\mathrm{p}}{}^{-1.033} + 0.01973 \tag{22}$$

$$co_2(n_{\mathrm{p}}) = 0.1012 n_{\mathrm{p}}{}^{-0.6797} + 0.01308 \tag{23}$$

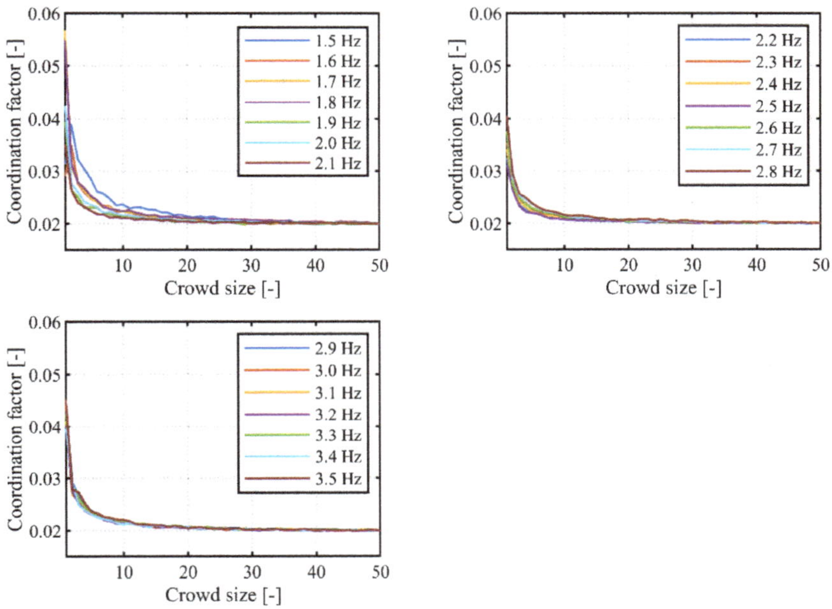

(a)

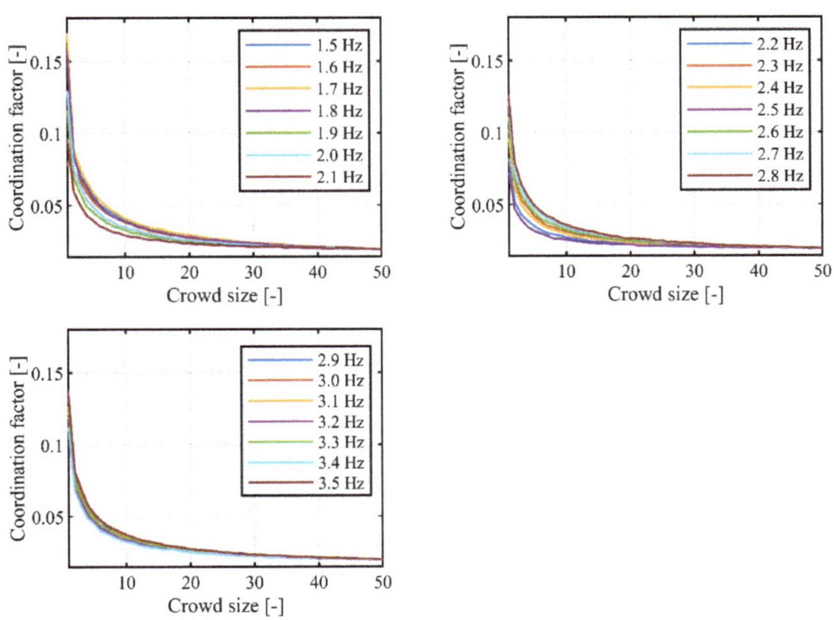

(b)

Figure 6. Relationship between coordination factor and crowd size. (**a**) Relationship for the first harmonic, (**b**) Relationship for the second harmonic.

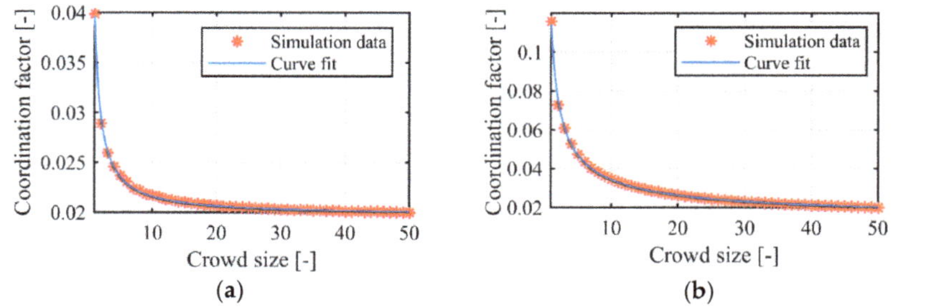

Figure 7. Conversion model for the coordination factor. (**a**) Model for the first harmonic, (**b**) Model for the second harmonic.

According to Equation (21), the conversion model for the crowd size can be obtained by:

$$c_1(n_\mathrm{p}) = 0.01993{n_\mathrm{p}}^{-0.033} + 0.01973n_\mathrm{p} \tag{24}$$

$$c_2(n_\mathrm{p}) = 0.1012{n_\mathrm{p}}^{0.3203} + 0.01308n_\mathrm{p} \tag{25}$$

4.4. Conversion Model for Non-Exceedance Probability

Due to the randomness of crowd jumping loads, the equivalent DLF is a random variable. Therefore, the probability distribution of the equivalent DLF is the key to building the conversion model for non-exceedance probability. Some histograms with crowd size equal to 50 and structural damping ratio equal to 0.005 are provided in Figures 8 and 9. It is observed that the equivalent DLF for both the first harmonic and second harmonic follows a normal distribution, and similar histograms are found for other crowd sizes and structural damping ratios.

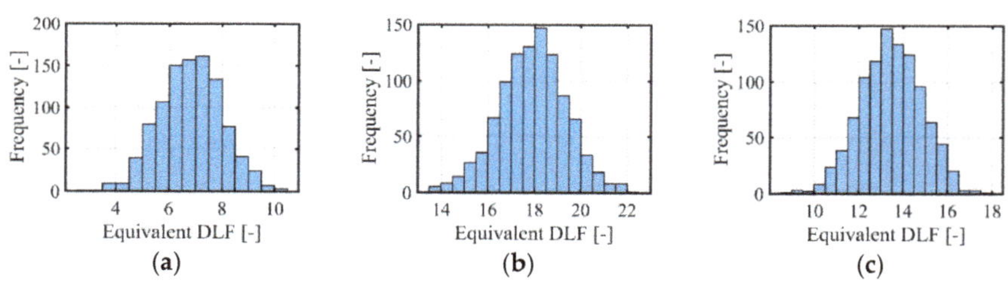

Figure 8. Equivalent DLF histogram of the first harmonic. (**a**) 1.5 Hz guidance frequency, (**b**) 2.5 Hz guidance frequency, (**c**) 3.5 Hz guidance frequency.

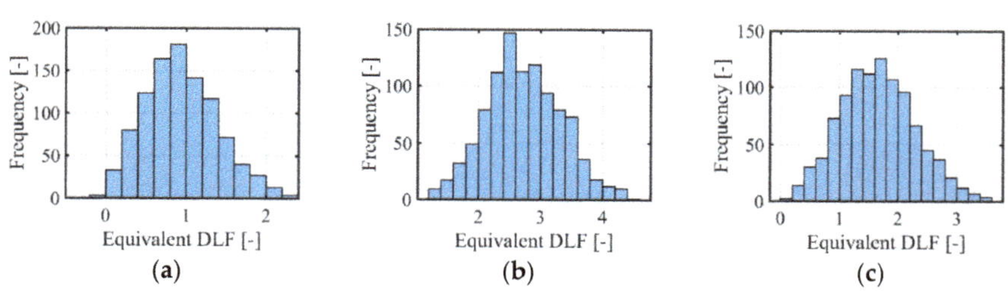

Figure 9. Equivalent DLF histogram of the second harmonic. (**a**) 1.5 Hz guidance frequency, (**b**) 2.5 Hz guidance frequency, (**c**) 3.5 Hz guidance frequency.

The equivalent DLF r_j of a certain non-exceedance probability p can be deduced by:

$$F^{-1}(p) = \frac{r_j - \bar{r}_j}{\bar{r}_j c_{\mathrm{v}j}} \tag{26}$$

where $F^{-1}()$ is the inverse function of the cumulative distribution function for the standard normal distribution, $\bar{r}_j$ is the average of the equivalent DLF, and c_{vj} is the coefficient of variation of the equivalent DLF. According to Equations (17)–(20) and Equations (24) and (25), the average of the equivalent DLF can be calculated by:

$$\bar{r}_j = r_{\mathrm{sta},j}(f_{\mathrm{gu}})c_j(\zeta_{\mathrm{s}})c_j(n_{\mathrm{p}}) \tag{27}$$

Introducing Equation (27) into Equation 26) and reorganizing the equation yields:

$$r_j = r_{\mathrm{sta},j}(f_{\mathrm{gu}})c_j(\zeta_{\mathrm{s}})c_j(n_{\mathrm{p}})\left[1 + F^{-1}(p)c_{\mathrm{v}j}\right] \tag{28}$$

Therefore, the coefficient of variation of the equivalent DLF is the key to building the conversion model for the non-exceedance probability. First, the simulations with crowd size equal to 50 and structural damping ratio equal to 0.005 are taken to investigate the c_{vj} of different guidance frequencies. To further determine the number of Monte Carlo simulations, the value of c_{vj} of a different number of Monte Carlo simulations is calculated, as shown in Tables 4 and 5. It can be observed that when the number of Monte Carlo simulations exceeds 600, c_{vj} tends to be stabilized. Therefore, 1000 is adopted as the number of Monte Carlo simulation in Section 3.2, and the simulated results are depicted in Figure 10. It is observed from Figure 10a that the trend of the coefficient of variation for the first harmonic is inverse to the trend in Figure 3a, because the randomness of the crowd jumping load is weak when the crowd synchronization is good. A fourth-order polynomial is then used to fit the coefficient of variation for the first harmonic of different guidance frequencies:

$$c_{\mathrm{v}1}(f_{\mathrm{gu}}) = -0.00884f_{\mathrm{gu}}^4 + 0.01899f_{\mathrm{gu}}^3 + 0.2506f_{\mathrm{gu}}^2 - 1.028f_{\mathrm{gu}} + 1.132 \tag{29}$$

As for the second harmonic in Figure 10b, the difference of the coefficient of variation for different guidance frequency is small, so the average of the coefficient of variation for different guidance frequency is adopted:

$$c_{\mathrm{v}2}(f_{\mathrm{gu}}) = 0.214 \tag{30}$$

Table 4. The coefficient of variation for the first harmonic of different numbers of MC simulations.

Guidance Frequency [Hz]	1.5	1.7	1.9	2.1	2.3	2.5	2.7	2.9	3.1	3.3	3.5
1000 MC simulation	0.134	0.116	0.077	0.072	0.074	0.056	0.090	0.096	0.092	0.088	0.092
800 MC simulation	0.135	0.117	0.077	0.072	0.075	0.056	0.090	0.097	0.092	0.089	0.091
600 MC simulation	0.133	0.116	0.077	0.072	0.075	0.056	0.090	0.096	0.091	0.088	0.091
400 MC simulation	0.125	0.112	0.073	0.070	0.073	0.055	0.093	0.090	0.093	0.084	0.091
200 MC simulation	0.142	0.110	0.078	0.075	0.073	0.053	0.089	0.088	0.094	0.079	0.094

Table 5. The coefficient of variation for the second harmonic of different numbers of MC simulations.

Guidance Frequency [Hz]	1.5	1.7	1.9	2.1	2.3	2.5	2.7	2.9	3.1	3.3	3.5
1000 MC simulation	0.175	0.153	0.193	0.201	0.212	0.179	0.211	0.201	0.198	0.193	0.165
800 MC simulation	0.176	0.154	0.193	0.202	0.213	0.180	0.211	0.202	0.200	0.195	0.164
600 MC simulation	0.174	0.154	0.193	0.202	0.214	0.180	0.211	0.199	0.199	0.194	0.166
400 MC simulation	0.167	0.150	0.190	0.197	0.210	0.175	0.211	0.189	0.198	0.188	0.163
200 MC simulation	0.194	0.147	0.192	0.205	0.212	0.167	0.215	0.184	0.204	0.173	0.170

Simulations with a structural damping ratio equal to 0.005 are taken to investigate the relationship between the coefficient of variation and crowd size, and the relationships of some guidance frequencies are illustrated in Figure 11. It is observed from Figure 11 that the coefficient of variation tends to be a constant as the crowd size increases. Because practical structures are commonly subjected to large crowd jumping loads, the influence of the crowd size on the coefficient of variation is overlooked. Then, the simulations with a crowd size equal to 50 are taken to investigate the relationship between coefficient of variation and structural damping ratio, and the relationships of some guidance frequencies are illustrated in Figure 12. It is observed from Figure 12 that the coefficient of variation is only slightly affected by the structural damping ratio, so the influence of the structural damping ratio on the coefficient of variation is ignored.

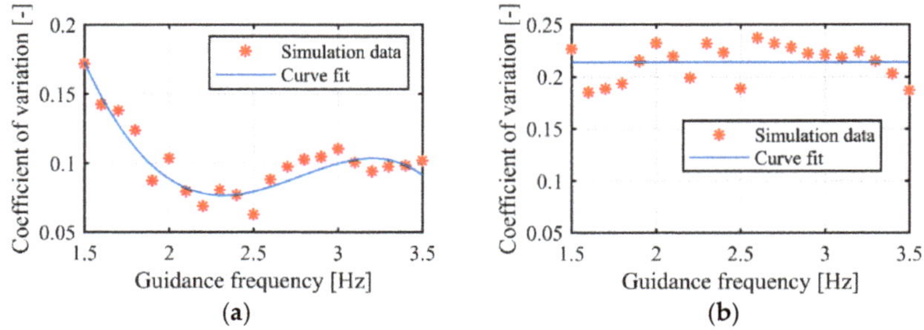

Figure 10. Modeling of the coefficient of variation of different guidance frequencies. (**a**) Modeling for the first harmonic, (**b**) Modeling for the second harmonic.

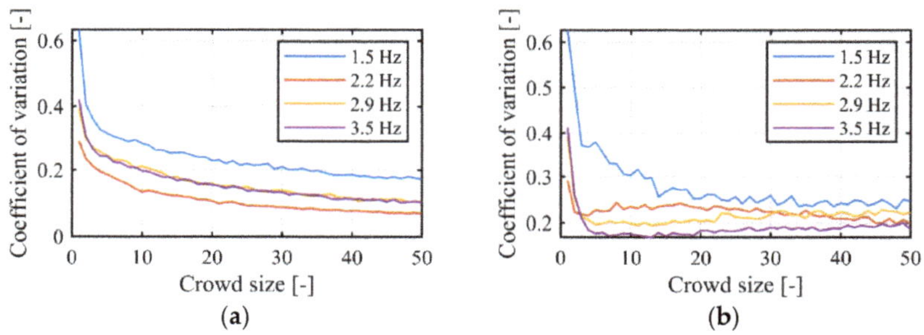

Figure 11. Relationship between coefficient of variation and crowd size. (**a**) Relationship for the first harmonic, (**b**) Relationship for the second harmonic.

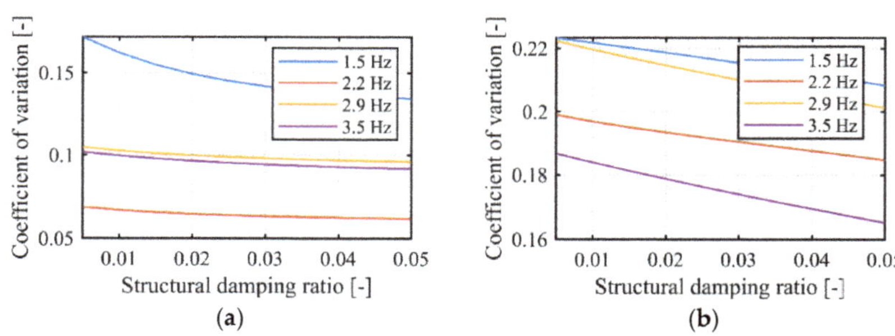

Figure 12. Relationship between coefficient of variation and structural damping ratio. (**a**) Relationship for the first harmonic, (**b**) Relationship for the second harmonic.

4.5. Model of the Equivalent DLF

To date, the model of the equivalent DLF for the standard working condition (Equations (17) and (18)) and the conversion model for structural damping ratio (Equations (19) and (20)), crowd size (Equations (24) and (25)), and non-exceedance probability (Equations (28)–(30)) have been established. Consequently, the equivalent DLF can be calculated by:

$$
\begin{aligned}
r_1 &= r_{\mathrm{sta},1}(f_{\mathrm{gu}})c_1(\zeta_{\mathrm{s}})c_1(n_{\mathrm{p}})\left[1+F^{-1}(p)c_{\mathrm{v}1}\right] \\
&= \left(3.673f_{\mathrm{gu}}^4-29.4f_{\mathrm{gu}}^3+71.42f_{\mathrm{gu}}^2-38.31f_{\mathrm{gu}}-16.83\right) \\
&\times\left(-0.5241\zeta_{\mathrm{s}}{}^{-0.2253}+2.728\right)\times\left(0.01993n_{\mathrm{p}}{}^{-0.033}+0.01973n_{\mathrm{p}}\right) \\
&\times\left[1+F^{-1}(p)\times\left(-0.00884f_{\mathrm{gu}}^4+0.01899f_{\mathrm{gu}}^3+0.2506f_{\mathrm{gu}}^2-1.028f_{\mathrm{gu}}+1.132\right)\right]
\end{aligned} \tag{31}
$$

$$
\begin{aligned}
r_2 &= r_{\mathrm{sta},2}(f_{\mathrm{gu}})c_2(\zeta_{\mathrm{s}})c_2(n_{\mathrm{p}})\left[1+F^{-1}(p)c_{\mathrm{v}2}\right] \\
&= \left(0.934f_{\mathrm{gu}}^4-8.258f_{\mathrm{gu}}^3+24.62f_{\mathrm{gu}}^2-27.12f_{\mathrm{gu}}+9.135\right) \\
&\times\left(-5.702\zeta_{\mathrm{s}}{}^{-0.0483}+8.367\right)\times\left(0.1012n_{\mathrm{p}}{}^{0.3203}+0.01308n_{\mathrm{p}}\right) \\
&\times\left[1+F^{-1}(p)\times 0.214\right]
\end{aligned} \tag{32}
$$

5. Engineering Application of the Equivalent DLF

The obtained equivalent DLF r_j can be used to estimate structural responses under crowd jumping activities in the design stage or as-built stage. First, structural dynamic properties can be obtained by dynamic tests or finite element models. Normally, structural responses are controlled by the resonant mode, where it is assumed the sth vibration mode is resonant, i.e., $f_{\mathrm{s}} = jf_{\mathrm{gu}}$. The crowd size, n_{p}, can be determined by structural functional requirements, and the non-exceedance probability, p, can be determined by structural reliability requirements. Then, introducing the values of $j, f_{\mathrm{gu}}, n_{\mathrm{p}}$, and p and the damping ratio ζ_{s}, r_j can be calculated by Equations (31) and (32). The root mean square value of structural acceleration a_{rms} can then be calculated by:

$$
a_{\mathrm{rms}} = \frac{Wr_j}{2M_{\mathrm{s}}\zeta_s} \tag{33}
$$

where M_{s} is the modal mass of the resonant mode, normalized by the maximum modal coordinate.

6. Discussions

The principal contribution of this paper is to develop an equivalent DLF model for crowd jumping loads. The FRF is applied to weight and integrate the PSD matrix to calculate the equivalent DLF. The structural damping ratio is introduced as an independent variable in the model, so that the root mean square of the structural responses can be accurately calculated, which overcomes the shortcomings of the existing DLF [13,15–19] overestimating or underestimating structural responses. Moreover, the Fourier series model based on the developed equivalent DLF is more convenient and efficient to calculate structural responses than complicated stochastic models [12,25–29]. In addition, by introducing the non-exceedance probability as an independent variable, a normal distribution-based model is built to obtain the equivalent DLF of different reliabilities, which could avoid using massive Monte Carlo simulations for structural reliability assessment and design. However, to better predict structural responses in real scenes, some limitations and extensions can be described as follows:

(1) Practical engineering structures, such as concert halls and grandstands, are commonly occupied by crowds of a large size. A crowd that is in constant contact with the structure, such as a standing crowd or a sitting crowd, can influence structural dynamic properties [34–36], causing an increase of the structural damping ratio in particular. Therefore, when Equations (31) and (32) are adopted to obtain the equivalent DLF, the influence of crowd on the structural damping ratio should be considered.

(2) Crowd jumping loads are multi-point excitations, because different people are at different positions in the structure. The Fourier series model based on the developed equivalent DLF simplifies multi-point excitation to single-point excitation, which cannot consider the modal value of each person. As a result, the modal value of each jumping person is assumed to be 1 to predict structural responses, as shown in Equation (33), which indicates that every jumping person is located at the maximum vibration mode, and it can

overestimate structural responses. To reduce the errors in structural response calculation, a representative value to characterize each jumping person's modal value needs to be further investigated.

7. Conclusions

This study builds on an equivalent DLF model for crowd jumping loads based on the massive simulated PSD matrices using a random field model. The equivalent DLF is identified from the principle of equaling the root mean square of structural responses. Four independent variables, the structural damping ratio, guidance frequency, crowd size, and non-exceedance probability, are introduced in the equivalent DLF model. A fourth-order polynomial is used to fit the equivalent DLF of different guidance frequencies. Power functions are used to convert the equivalent DLF of different damping ratios and crowd sizes. A normal distribution-based model is used to convert the equivalent DLF of different non-exceedance probabilities.

Author Contributions: Conceptualization, J.X.; methodology, J.X. and Z.P.; software, J.X.; validation, S.D.; writing—original draft preparation, J.X.; writing—review and editing, S.D.; supervision, H.Q.; funding acquisition, J.X. and S.D. All authors have read and agreed to the published version of the manuscript.

Funding: This work was supported by the National Natural Science Foundation of China (Grant numbers: 52008376 and 51909241), the Postdoctoral Research Foundation of China (Grant number: 2020M682350 and 2021M701111), and Zhengzhou University Youth Enterprise Cooperation Innovation Team (Grant number: 32320405).

Informed Consent Statement: Informed consent was obtained from all subjects involved in the study.

Data Availability Statement: New data were created or analyzed in this research. Data will be shared upon request and the consideration of the authors.

Conflicts of Interest: The authors declare no conflict of interest.

References

1. Wang, H.; Zhang, Z.; Chen, J. Prediction of structural responses induced by single-person jumping through a physical principle based on transfer functions. *Adv. Struct. Eng.* **2022**, *25*, 146–157. [CrossRef]
2. van Nimmen, K.; Lombaert, G.; de Roeck, G.; van den Broeck, P. The impact of vertical human-structure interaction on the response of footbridges to pedestrian excitation. *J. Sound Vib.* **2017**, *402*, 104–121. [CrossRef]
3. Chen, J.; Wang, P.; Wang, H. Pedestrian-induced load identification from structural responses using genetic algorithm with numerical and experimental validation. *J. Bridge Eng.* **2021**, *26*, 4021001. [CrossRef]
4. Živanović, S.; Pavic, A.; Reynolds, P. Vibration serviceability of footbridges under human-induced excitation: A literature review. *J. Sound Vib.* **2005**, *279*, 1–74. [CrossRef]
5. Xiong, J.; Chen, J. Open access and updated human-induced load data set. *J. Struct. Eng.* **2021**, *147*, 4720003. [CrossRef]
6. Wang, J.; Chen, J.; Yokoyama, Y.; Xiong, J. Spectral model for crowd walking load. *J. Struct. Eng.* **2020**, *146*, 04019220. [CrossRef]
7. Lee, S.H.; Lee, K.K.; Woo, S.S.; Cho, S.H. Global vertical mode vibrations due to human group rhythmic movement in a 39 story building structure. *Eng. Struct.* **2013**, *57*, 296–305. [CrossRef]
8. Wolmuth, B.; Surtees, J. Crowd-related failure of bridges. *Proc. ICE Civ. Eng.* **2003**, *156*, 116–123. [CrossRef]
9. de Brito, V.L.; Pimentel, R.L. Cases of collapse of demountable grandstands. *J. Perform. Constr. Facil.* **2009**, *23*, 151–159. [CrossRef]
10. Dallard, P.; Flint, A.; le Bourva, S.; Low, A.; Smith, R.M.R.; Willford, M. The London Millenium Footbridge. *Struct. Eng.* **2001**, *79*, 17–35. [CrossRef]
11. Sky News. UK. 13 August 2018. Available online: https://news.sky.com/story/300-injured-as-wooden-pier-collapses-at-o-marisquino-music-festival-in-spain-11471557 (accessed on 5 December 2021).
12. Sim, J.; Blakeborough, A.; Williams, M.S.; Parkhouse, G. Statistical model of crowd jumping loads. *J. Struct. Eng.* **2008**, *134*, 1852–1861. [CrossRef]
13. Ji, T.; Ellis, B.R. Floor vibration induced by dance-type loads: Theory. *Struct. Eng.* **1994**, *72*, 37–44.
14. Jones, C.A.; Reynolds, P.; Pavic, A. Vibration serviceability of stadia structures subjected to dynamic crowd loads: A literature review. *J. Sound Vib.* **2011**, *330*, 1531–1566. [CrossRef]
15. Ellis, B.R.; Ji, T. Loads generated by jumping crowds: Numerical modelling. *Struct. Eng.* **2004**, *82*, 35–40.
16. Li, G.; Ji, T.; Chen, J. Determination of the dynamic load factors for crowd jumping using motion capture technique. *Eng. Struct.* **2018**, *174*, 1–9. [CrossRef]

17. Parkhouse, J.G.; Ewins, D.J. Crowd-induced rhythmic loading. *Struct. Build.* **2006**, *159*, 247–259. [CrossRef]
18. International Organization for Standardization. *Bases for Design of Structures—Serviceability of Buildings and Walkways against Vibration*; ISO: Geneva, Switzerland, 2007.
19. Kasperski, M.; Agu, E. Prediction of crowd-induced vibrations via simulation. In Proceedings of the 23rd International Modal Analysis Conference, Orlando, FL, USA, 31 January–3 February 2005.
20. Živanović, S.; Pavić, A.; Reynolds, P. Probability-based prediction of multi-mode vibration response to walking excitation. *Eng. Struct.* **2007**, *29*, 942–954. [CrossRef]
21. Xiong, J.; Chen, J. Power spectral density function for individual jumping load. *Int. J. Struct. Stab. Dyn.* **2018**, *18*, 1850023. [CrossRef]
22. Brownjohn, J.M.W.; Pavic, A.; Omenzetter, P. A spectral density approach for modelling continuous vertical forces on pedestrian structures due to walking. *Can. J. Civ. Eng.* **2004**, *31*, 65–77. [CrossRef]
23. Ingólfsson, E.T.; Georgakis, C.T.; Ricciardelli, F.; Jönsson, J. Experimental identification of pedestrian-induced lateral forces on footbridges. *J. Sound Vib.* **2011**, *330*, 1265–1284. [CrossRef]
24. Sétra, Assessment of vibrational behaviour of footbridges under pedestrian loading. In *Technical Guide*; French Association of Civil Engineering: Paris, France, 2006.
25. Racic, V.; Pavic, A. Stochastic approach to modelling of near-periodic jumping loads. *Mech. Syst. Signal Process.* **2010**, *24*, 3037–3059. [CrossRef]
26. Martínez, J.F.; Hermanns, L.; de Lerma, A.F.; Álvarez, E.A. Jumping load models applied on a gymnasium floor. *Eng. Struct.* **2016**, *125*, 26–38. [CrossRef]
27. Xiong, J.; Chen, J. A generative adversarial network model for simulating various types of human-induced loads. *Int. J. Struct. Stab. Dyn.* **2019**, *19*, 1950092. [CrossRef]
28. Duan, S.; Xiong, J.; Qian, H. Generative adversarial networks-based stochastic approach to the modeling of individual jumping loads. *Int. J. Struct. Stab. Dyn.* **2021**, *21*, 2150047. [CrossRef]
29. Chen, J.; Li, G.; Racic, V. A data-driven wavelet-based approach for generating jumping loads. *Mech. Syst. Signal Process.* **2018**, *106*, 49–61. [CrossRef]
30. Wirsching, P.H.; Paez, T.L.; Ortiz, K. *Random Vibrations: Theory and Practice*; Courier Corporation: New York, NY, USA, 2006.
31. Chen, J.; Li, G.; Racic, V. Acceleration response spectrum for predicting floor vibration due to occupants jumping. *Eng. Struct.* **2016**, *112*, 71–80. [CrossRef]
32. Xiong, J.; Chen, J. Random field model for crowd jumping loads. *Struct. Saf.* **2019**, *76*, 197–209. [CrossRef]
33. Chen, J.; Wang, L.; Racic, V.; Lou, J. Acceleration response spectrum for prediction of structural vibration due to individual bouncing. *Mech. Syst. Signal Process.* **2016**, *76*, 394–408. [CrossRef]
34. Xiong, J.; Chen, J.; Caprani, C. Spectral analysis of human-structure interaction during crowd jumping. *Appl. Math. Model.* **2021**, *89*, 610–626. [CrossRef]
35. Tubino, F. Probabilistic assessment of the dynamic interaction between multiple pedestrians and vertical vibrations of footbridges. *J. Sound Vib.* **2018**, *417*, 80–96. [CrossRef]
36. Caprani, C.C.; Ahmadi, E. Formulation of human–structure interaction system models for vertical vibration. *J. Sound Vib.* **2016**, *377*, 346–367. [CrossRef]

buildings

Article

Maximal Multivariable Coefficient Analysis between Vibration Limits and Relevant Factors in General Buildings

Lei Cao [1] and Jun Chen [2,*]

[1] College of Engineering, Fujian Jiangxia University, Fuzhou 350108, China; 2021020@fjjxu.edu.cn
[2] College of Civil Engineering, Tongji University, Shanghai 200092, China
* Correspondence: cejchen@tongji.edu.cn

Abstract: The vibration limit is an essential prerequisite for building vibration serviceability assessment, and various biological/environmental factors affect it deeply. Yet quantitative relationships between vibration limits and these factors in general buildings, such as the human weight, height and number of stories, stay unknown. Based on data collected by an investigation conducted on a cell phone application, this paper proposed a novel approach for quantifying correlations between common relevant factors in general buildings and limits by maximal information coefficient (MIC). Vibration serviceability was thoroughly proved to be a multivariable system and crest factor/BMI had a higher correlation than other factors. A functional relationship and 95% confidence intervals between vibration limits and crest factor/BMI were proposed, respectively. Lilliefors test and normal probability plot show that residuals between fitted values of limits and measured ones follow a normal distribution. Finally, estimation of vibration serviceability based on probability was suggested when the crest factor/BMI and vibration magnitude were known.

Keywords: relevant factors; vibration limits; MIC; functional relationship; normal distribution; prediction of vibration serviceability

Citation: Cao, L.; Chen, J. Maximal Multivariable Coefficient Analysis between Vibration Limits and Relevant Factors in General Buildings. *Buildings* **2022**, *12*, 807. https://doi.org/10.3390/buildings12060807

Academic Editor: Vitor Silva

Received: 4 May 2022
Accepted: 8 June 2022
Published: 11 June 2022

1. Introduction

A building vibration serviceability issue refers to discomfort or disturbance of occupants and impediment of sensitive operations caused by structural vibration. Vibration limits are the key issue of vibration serviceability research. Up to the early years of the 19th century, researchers began to observe building vibration serviceability problems. However, not until the 20th century did researchers suggest any vibration limits. With the development of material and construction techniques, it became impossible to ignore the vibration serviceability problem for the reasons of larger span and weaker stiffness emerging. In 1931, Reiher and Meister [1] conducted a milestone experiment to obtain vibration limitations by using semantic labels such as 'easily perceptible' and 'strongly perceptible' to describe volunteers' feelings of simulated vibrations. From then on, it is a common practice for researchers to employ a few volunteers and collect their judgments through vibration tests to determine vibration limits.

A growing body of literature [2–6] shows that vibration serviceability is a multivariable issue. Factors such as the biological characteristics of humans [2,4,5] and environmental characteristics [3] have an effect on vibration serviceability. Researchers have already reported vibration limits labeled by gender [4], and gesture of body [5]. Relationships between these factors and limits are also unlikely to be found as linear [7,8]. However, there is no research showing which one of the relevant factors is more important to vibration serviceability. The functional relationship between the most relative factors and limits also stays unknown.

Two reasons lead to the dilemma of vibration serviceability research. One is that the data scale in earlier research was too small (with dozens or a few hundred volunteers and several buildings) to contain a statistically significant range of various factors' values. The other is that traditional correlation analysis tools are not able to quantify normalized nonlinear relationships efficiently on a large scale [8]. For example, the Pearson correlation coefficient and Spearman correlation coefficient have problems with estimating nonlinear relationships, K–nearest neighbor (KNN) and kernel density estimation (KDE) fail in normalizing the correlation coefficient, and the distance coefficient is inefficient when the data scale is large [9,10].

To overcome the obstacles in vibration serviceability research, researchers should carry out vibration serviceability investigations on a much larger scale and with smaller costs than the traditional method. A new correlation coefficient that is capable of quantifying correlation in various relationships efficiently and in a normalized way is also essential.

With the help of rapidly developing technologies, such as smart mobile phones and internet cloud calculation, it is possible to obtain reliable large-scale data at a relatively low cost. Many researchers [11,12] found that sensors integrated into smartphones, such as tri-axial accelerometers, GPS and gyroscopes, could be successfully applied for structural monitoring. In the authors' earlier research [13], a smartphone-based application (App) was designed and spread through the network to volunteers to collect information including biological and environmental factors.

Reshef et al. [9] proposed a novel correlation analysis tool in 2011 and proved that the maximum information coefficient (MIC) has many advantages, such as being normalized, general to all kinds of functional relationships, equitable to any level of noise and low in calculation complexity. Hence the MIC is suitable for estimating the correlation coefficient between vibration limits and relevant factors.

This paper demonstrates the methodology of data collection and shows statistics of relevant factors briefly at first. By comparing the MICs of different relevant factors with vibration limits, the factor with the biggest MIC is chosen to obtain the fitting function with vibration limits. The normal distribution of residuals between fitted limits and actual ones is checked by the Lilliefors test and the normal probability plot. In the case of residuals following normal distribution, 95% confidence intervals of vibration limits are proposed. Finally, a probability-based method is suggested to predict vibration serviceability.

2. Methodology and Statistics of Factors

In the previous study [13], we conducted a smartphone-based data collection campaign, as shown in Figure 1. The data collection procedure, data cleaning principle and representativeness verification are briefly described herein for completeness. The scheme of data collection consists of three parts (Figure 1): App design; App promotion; Self-help investigation and upload.

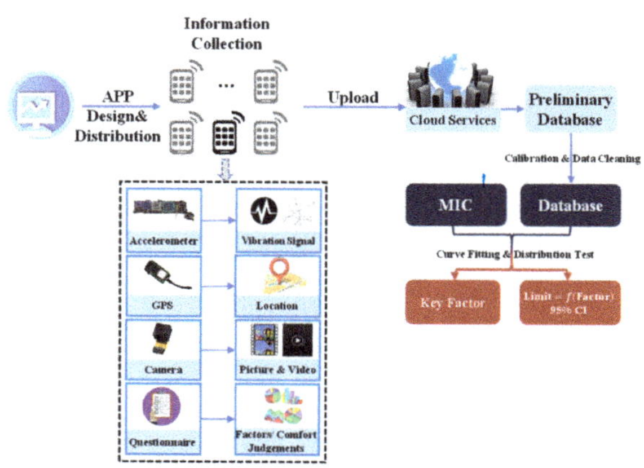

Figure 1. Proposed research scheme for correlation between vibration limits and relevant factors.

After data cleaning and representativeness analysis, the data set was then used as a database for correlation analysis. The procedure was carried out for more than 2 years and 8521 pieces of data were collected.

2.1. Data Collection of Relevant Factors

Since the natural frequencies of human organs fall in the range of 1–80 Hz [14], with most organs in the range of 4–8 Hz, it is feasible to measure vibration which may cause vibration serviceability problems by smartphone [11,13]. As a reason for resonance, the sensitive region of vibration comfort falls in the same range. Earlier researchers [11] found that the measurement accuracy of micro-electro-mechanical systems (MEMS) integrated into a smart mobile phone complies with request of vibration signal test in vibration serviceability research. A program named 'VCheck' (vibration check) was finally designed and the interface of vibration measurement and questionnaire are shown in Figure 2. VCheck is designed to collect system time and the reading of built-in accelerometer of smartphone at the same time. The sampling frequency of VCheck falls in the range of 100–1000 Hz, depending on the type of smartphone. In case some mistakes happened and no data were collected for some time, a corresponding gap would emerge in the series of total data.

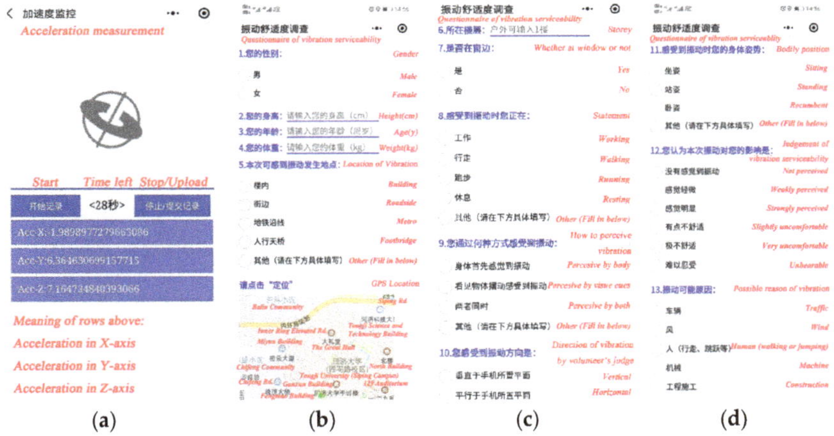

Figure 2. Vibration measurement (**a**) and questionnaire (**b**–**d**) in VCheck (Menu in Chinese, English translation in italic).

To obtain as much data as possible and make sure the data are representative, VCheck program was promoted through internet and bonuses were used to encourage smartphone users to participate in the investigation or introduce VCheck to others.

Although more than 8000 pieces of data were collected, many invalid data existed due to some smartphones being of poor quality or volunteers' misunderstandings. Redundancy of data was also a problem in that some volunteers uploaded too much data just for more bonuses. Hence, authors carried out data cleaning by following several elaborate rules. Finally, 3319 pieces of data were recognized as valid. Further analysis of these data showed that the sample was representative and rational [13].

Figure 3 shows all data collected around the world.

Figure 3. Distribution of data (by December 2021).

2.2. Statistics of Relevant Factors

Table 1 shows the factors which were studied before and found to be related to vibration serviceability in daily life. The range of these factors was also listed according to the statistics of the database after data cleaning.

Table 1. Factors investigated in this research and the corresponding ranges.

Relevant Factor	Range
Gender	Male/Female
Age	12–72 (y)
Height	132–189 (cm)
Weight	37–95 (kg)
Statement	Resting/Working/Walking/Running/Other
Gesture	Sitting/Standing/Recumbent/Other
Vibration magnitude	0.0037–4.38 (rms, m/s^2)
Crest factor	4–64
Direction of vibration	Vertical/Horizontal/Other
Vibration source	Human activity/ Traffic/Wind/Machine/Construction/Other
Longitude	−157.81–139.78 (°)
Latitude	19.77–53.47 (°)
Site	Building/Roadside/Metro/Footbridge/Other
Storey	1–58
Near window	Yes/No
Visual cues	Firstly perceive vibration by: body/visual cues/both/other

Two types of relevant factors, which were proved to have an influence on vibration serviceability, were investigated in this research. Biological factors: gender [4], age [2], height, weight [7], statement, gesture [15]; Environmental factors: magnitude, form [3], direction [16] and source of vibration [7,17], type of site [18], storey [13], whether near a window, visual cues [19,20]. Previous researchers [3] found that the crest factor Equation (1) influenced vibration serviceability greatly, which was also a common indicator used to distinguish different types of vibration, such as constant vibration and shock.

$$\text{CrestFactor} = \frac{\text{peak}}{\text{rms}} \quad (1)$$

Among these factors, vibration magnitude is obviously a key issue in vibration serviceability. Earlier research showed that different types of vibration magnitude indicators performed quite differently from each other, especially when the crest factor Equation (1) of vibration changes was large than six [21]. To make sure conclusions were exact, five most commonly used indicators were chosen to assess the magnitudes of vibration (Table 2).

Table 2. Different indicators of vibration magnitude.

Vibration Indice	Abbreviation	Formula
peak of acceleration	peak	$\text{peak} = \max(\lvert a_{\text{w}}(t)\rvert)$ *
root-mean-square of acceleration	rms	$\text{rms} = \left(\frac{1}{T}\int_0^T a_{\text{w}}^2(t)\text{d}t\right)^{\frac{1}{2}}$
vibration dose value	VDV	$\text{VDV} = \left(\int_0^T a_{\text{w}}^4(t)\text{d}t\right)^{\frac{1}{4}}$
root-mean-quad of acceleration	rmq	$\text{rmq} = \left(\frac{1}{T}\int_0^T a_{\text{w}}^4(t)\text{d}t\right)^{\frac{1}{4}}$
maximal transient vibration value	MTVV	$\text{Running-rms} = \left\{\frac{1}{\tau}\int_{t_0-\tau}^{t_0} [a_{\text{w}}(t)]^2\text{d}t\right\}^{\frac{1}{2}}$ $\text{MTVV} = \max(\text{Running-rms})$

* $a_{\text{w}}(t)$ is the weighted acceleration time history; T is the duration of the entire signal; t_0 is a certain time in the range of 0–T s; τ is the MTVV duration, where 1 s MTVV corresponds to τ = 1 s, and 0.5 s MTVV corresponds to τ = 0.5 s.

At the end of each questionnaire, six semantic labels were listed for volunteers to describe their subject senses of the vibration: 'not perceived', 'weakly perceived', 'strongly perceived', 'slightly uncomfortable', 'very uncomfortable', or 'unbearable'. By analyzing vibration magnitudes corresponding to different factors, authors were able to obtain vibration limits corresponding to various factors.

2.2.1. Biological Factors

Statistics of biological factors (Figures 4–6) show that the biological characteristics of the sample are abundant in diversification.

Since the proportion of 'Other' in both statistics (Figure 7) is very small, statistics of volunteers' statements and gestures show that the sample covers almost all conditions where people may daily encounter vibration serviceability problems.

2.2.2. Environmental Factors

Figure 8 shows the types of vibration sources in daily life and the proportion of them. Over 98% of vibrations were caused by 5 types of commonly seen sources: humans (36.52%), traffic (25.31%), wind (17.63%), machines (13.35%) and construction (5.75%).

Figure 9 shows that almost all (except for vehicles) perceptible vibrations (the semantic label for vibration is at least 'weakly perceived') happened in buildings or roadside. In 76.84% of data uploaded, people sensed a perceptible or stronger vibration in a building, with a range of 1–58th floor.

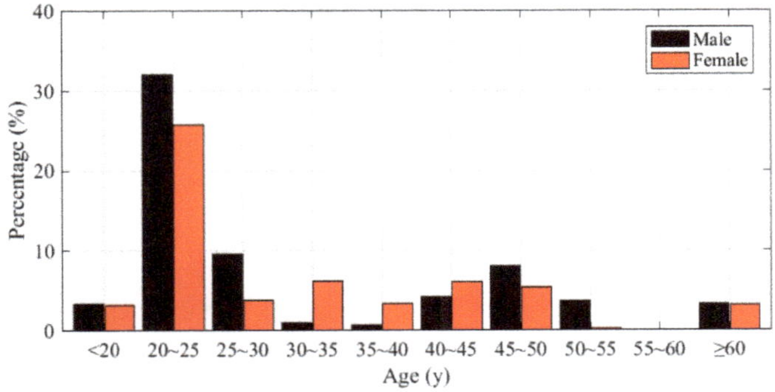

Figure 4. Age distribution of male and female.

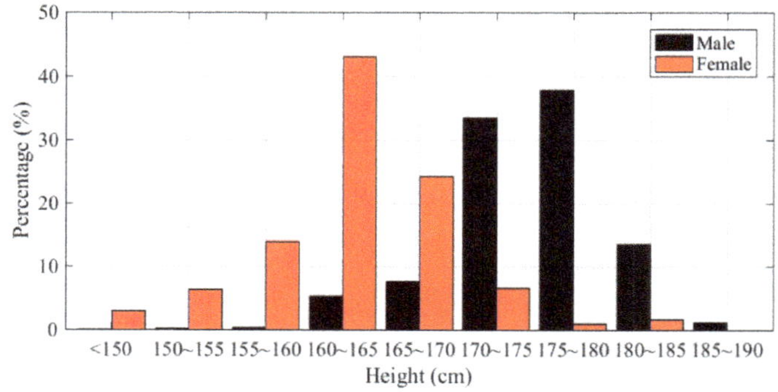

Figure 5. Height distribution of male and female.

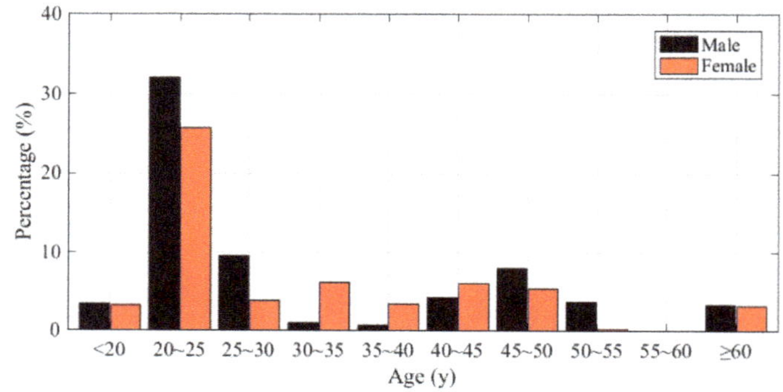

Figure 6. Weight distribution of male and female.

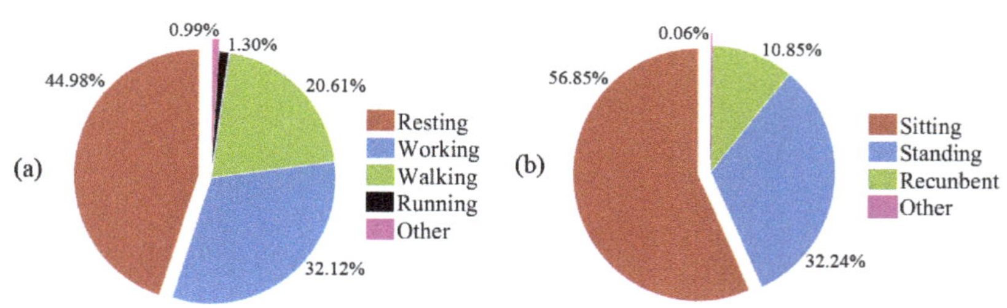

Figure 7. Statistics of volunteers' statements (**a**) and gestures (**b**).

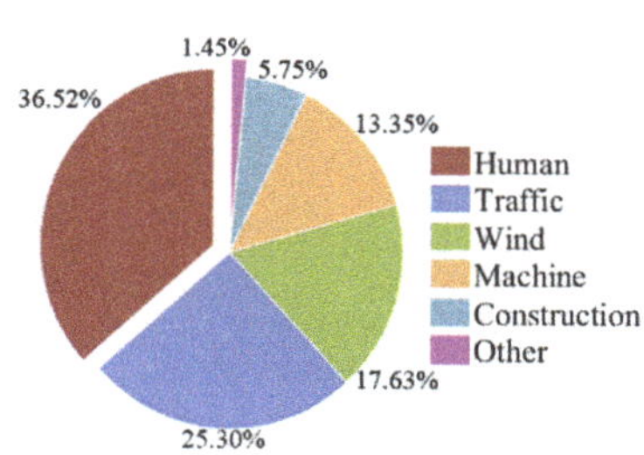

Figure 8. Statistic of vibration source.

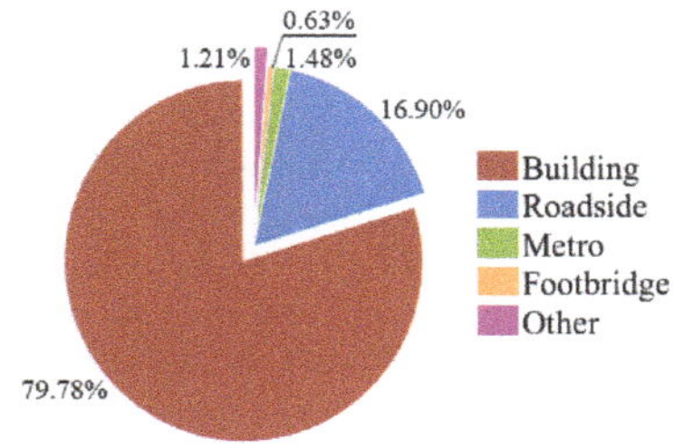

Figure 9. Statistic of perceptible vibration site.

The magnitudes of vibration investigated range from 0.0037–4.38 m/s^2 (rms), and the crest factors of vibration range from 4–64. By contrast, in previous research and criteria of vibration serviceability, perception limits of vibration range from 0.005–0.015 m/s^2 [22–24] while comfort limits of vibration range from 0.315–0.5 m/s^2 [21,25]. Crest factor in sinusoidal vibration is 1.414 and some criteria suggest different methods of signal analysis when the crest factor is higher than 6 [26] or 9 [21]. It should be emphasized that there is seldom sinusoidal vibration in a real environment and the crest factors of most forms of actual vibration are bigger than 1.414.

The statistics of biological and environmental factors show that the data collected cover a wide range of various factors and the data set is big enough for the following analysis.

3. Correlation Analysis by MIC

The MIC (maximal information coefficient) evolves from MI (mutual information). Equation (2) shows the definition of MI between two variables of x and y:

$$\mathrm{MI}[x;y] = \int \mathrm{p}(x,y)\log_2 \frac{\mathrm{p}(x,y)}{\mathrm{p}(x)\mathrm{p}(y)}\mathrm{d}x\mathrm{d}y \tag{2}$$

In Equation (2), p(x, y) is the joint probability density function of x and y; p(x) and p(y) are the marginal probability density function of variable x and y, respectively; MI[x, y] is mutual information of x and y.

Although MI has many advantages in assessing how closely two variables are associated, such as generality to all function types, it is hard to obtain the joint probability density function of two variables. Furthermore, MI fails in normalization.

3.1. Definition of MIC

The basic idea of MIC is to encapsulate the joint probability density function of variables x and y by dividing grids on the scatterplot of these two variables in every possible way, then computing MIs corresponding to every kind of dividing grid and normalizing the biggest MI. There are three steps in computing the MIC between two variables:

- Explore all grids up to a maximal grid resolution (dependent on the size of the sample) and estimate the joint probability density function of vibration limits and factors;
- Compute mutual information in every possible condition and find the biggest one;
- Normalize the biggest mutual information by considering the number of grid cells.

$$\mathrm{MIC}[x;y] = \max_{|X||Y|<B} \frac{\max\limits_{\text{allcells}}\left(\sum\limits_{x,y} P(x,y)\log_2 \frac{P(x,y)}{\sum\limits_{x}P(x,y)\sum\limits_{y}P(x,y)}\right)}{\log_2[\min(|X|,|Y|)]} \tag{3}$$

In Equation (3), $P(x, y)$ is the estimated joint probability density function by dividing grids on the scatterplot of variable x and y; $\sum\limits_{x} P(x,y)$ and $\sum\limits_{y} P(x,y)$ are the estimated marginal probability density function corresponding to variable x and y, respectively; $|X|$

and $|Y|$ are the number of segments by which the x-axis and y-axis are divided; $B = n^{0.6}$, n is the number of data.

Since the joint probability density function in the MIC is estimated by dividing the scatterplot into boxes and computing the frequencies, the MIC is not applicable when the scale of data is not big enough [9], which is another reason it was impossible for earlier researchers to utilize the MIC.

Although previous research showed that the MIC was better because it possessed qualities such as being normalized and general to all kinds of functional relationships, it is necessary to prove them by comparing the MIC with traditional correlation analysis tools. Taking the Pearson and Spearman correlation coefficients as examples, Table 3 shows the difference between these three coefficients when quantifying the correlation between perception limits (in the form of 0.5 s MTVV) and crest factors (the ratio of peak and root-mean-square values of acceleration). Vector x stands for crest factors and vector y stands for perception limits.

Table 3. Correlation coefficients between crest factors and perception limits.

	Forms of Relationship				
	$y \sim x$	$y \sim \ln(x)$	$y \sim e^x$	$y \sim x^2 - 2x$	$y \sim x^{-1}$
Pearson	0.3621	0.3466	−0.0063	0.3581	−0.3172
Spearman	0.7273	0.7273	0.7273	0.7273	−0.7273
MIC	0.4851	0.4851	0.4851	0.4851	0.4851

Table 3 shows that the MIC changes little if the relationship between the crest factors and perception limits are different in forms, while the Pearson and Spearman correlation coefficients change a lot.

Compared with traditional correlation analysis tools, the MIC was proven to possess many advantages [9]:

- Generality: The MIC between two groups of data is determined by the relevance of them, not the type of functional relationship;
- Equitability: The MIC gives similar scores to equally noisy relationships of different types;
- Normalization: The MIC ranges from 0 to 1 when the correlation between two groups of data increases.

Yet the MIC suits only a large data set, which is not a problem in this research. Hence the MIC performs better when the relationship is unknown.

3.2. MICs between Factors and Vibration Limits

There are many kinds of factors that have an influence on vibration serviceability. By summarizing earlier research [7,27,28], 13 representative factors were employed to explore the relationship between relevant factors and vibration limits, which were gender, age, body height, body weight, body mass index (BMI), number of stories, crest factor of vibration (short for CF), site, window condition, acts of volunteers, gesture, reason of vibration, usualness of occurrence.

Two kinds of vibration limits were discussed in this paper: perception limits (corresponding to the semantic label of 'weakly perceived') and comfort limits (corresponding to the semantic label of 'slightly uncomfortable'). Figures 10 and 11 show the MICs between factors and perception/comfort limits, respectively.

Researchers believed that vibration serviceability is a multivariable system [7]. However, this conclusion has not been proved before. The relatively low correlations shown in Figures 10 and 11 prove explicitly and quantitatively that vibration serviceability is influenced by many factors and no single factor would determine vibration limits completely.

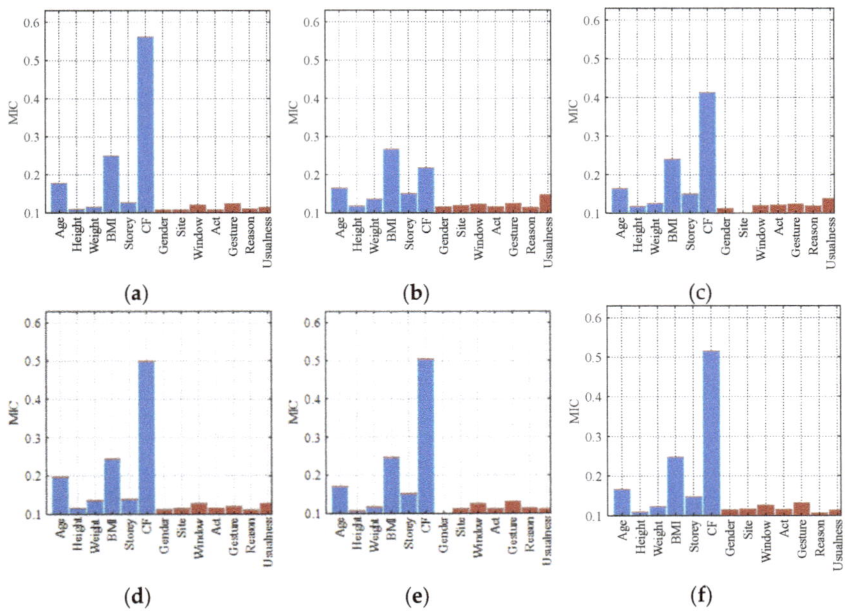

Figure 10. MICs between 13 factors and perception limits (Indices in (**a**–**f**) are peak/rms/VDV/rmq/1 s MTVV/0.5 s MTVV, respectively).

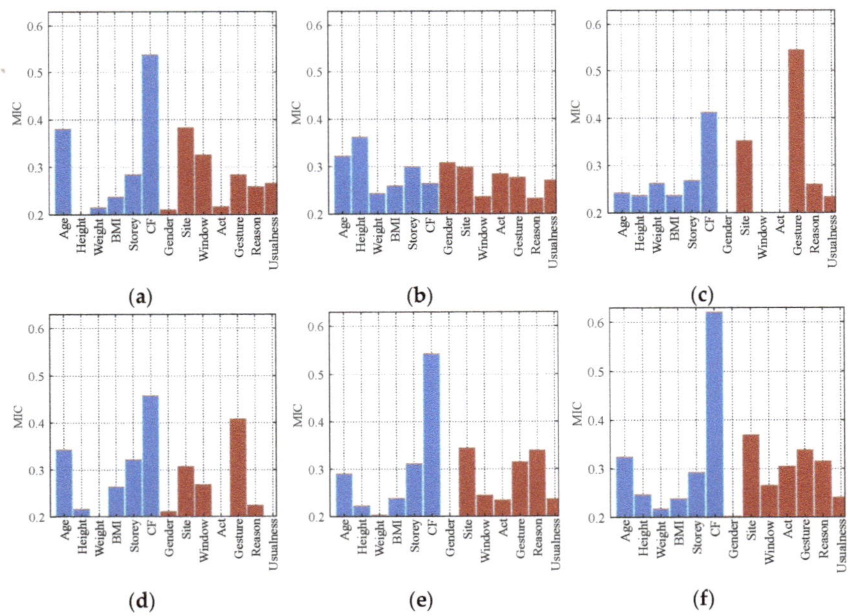

Figure 11. MICs between 13 factors and comfort limits (Indices in (**a**–**f**) are peak/rms/VDV/rmq/1 s MTVV/0.5 s MTVV, respectively).

It can be learned from Figures 10 and 11 that the crest factor has a much bigger MIC than other factors, which means the crest factor is a key factor of vibration serviceability.

The correlation between crest factor and comfort limits (using MTVV as an indicator) is very strong for the MIC exceeding 0.6.

BMI has the biggest MIC among the biological factors when perception limits are concerned. Since BMI is a comprehensive and widely used indicator of human health and earlier research [21] showed that human health has a great influence on vibration serviceability, BMI is chosen as a key biological factor.

Compared with other vibration indicators, the MIC of the same factor is much lower when rms is used as a vibration indicator. Since the rms method is considered unsuitable to estimate the influence of vibrations with high crest factors (higher than six) [21,29] on human comfort and Figure 12 shows that almost all data in this research (more than 99%) have crest factors higher than six, the results of this study are reasonable and rms is not recommended as an indicator in field research on vibration serviceability.

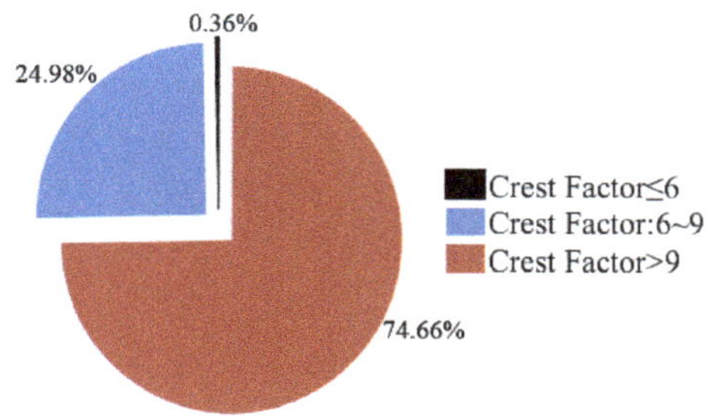

Figure 12. Proportion of data with different crest factors.

4. Curve Fitting of Key Factors and Vibration Limits

By comparing the MICs between different factors and limits, BMI and crest factor were chosen as key factors to fit functions with vibration limits. A 1 s MTVV and VDV were chosen as indicators of vibration magnitudes because of bigger MICs than other indicators.

Since there is more than one data corresponding to the same value of BMI or crest factor, the mean value of vibration magnitude in these data was chosen to fit functions with relevant factors. The Lilliefors test (LF test) and normal probability plot (Figure 13) were used to check the normality of residuals. In case residuals follow a normal distribution, the result of the LF test should be 0 and the data should distribute around the line between the first quartile and third quartile of the data in a normal probability plot. Finally, fitted functions between vibration limits and BMI/crest factor, respectively, were suggested with a 95% confidence interval (95% CI).

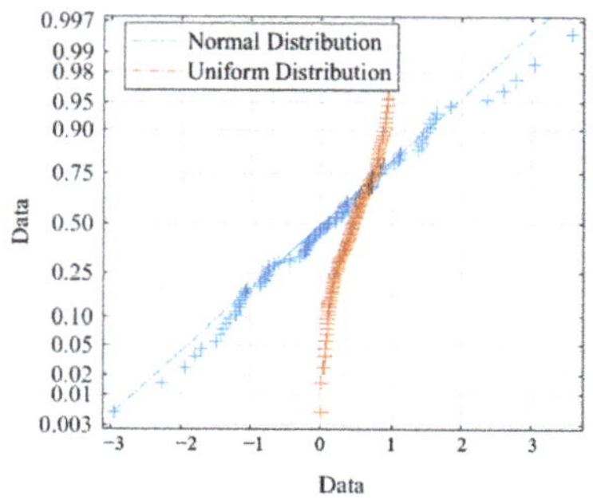

Figure 13. Normal probability plot of different distributions.

4.1. *Fitting Function of BMI and Vibration Limits*

Earlier research and standards found that human health had an important influence on vibration serviceability, and healthier people were more sensitive to vibration [21]. Since

BMI is a widely used indicator to assess human health [30], it is reasonable to assume that people with a certain range of BMI are more sensitive to vibration than others, which means extreme value exists in the "Vibration limits-BMI" curves. The WHO [31] suggests that healthy people's BMIs range from 18.5 to 24.9. Hence, this paper used quadratic polynomial to fit the relationship between BMIs and vibration limits Equation (4), with the breakpoint falling in the range of 18.5–24.9.

$$a_{\text{limit}} = a \cdot (\text{BMI} - b)^2 + c \tag{4}$$

In Equation (4), a_{limit} is vibration limit, there are two kinds of limits: perception limits and comfort limits; two kinds of indicators were used in each kind of limits: 1 s MTVV and VDV; $a/b/c$ are coefficients of fitting function, b refers to the BMI with which people have the maximal or minimal value of vibration limits.

Figures 14 and 15 show the fitting curves between female/male BMI and perception limits, respectively. Figures 16 and 17 show the fitting curves between female/male BMI and male comfort limits, respectively.

Coefficients of fitting between female/male BMI and perception/comfort limits are listed in Tables 4 and 5 when VDV or MTVV is used as an indicator of vibration magnitude, respectively.

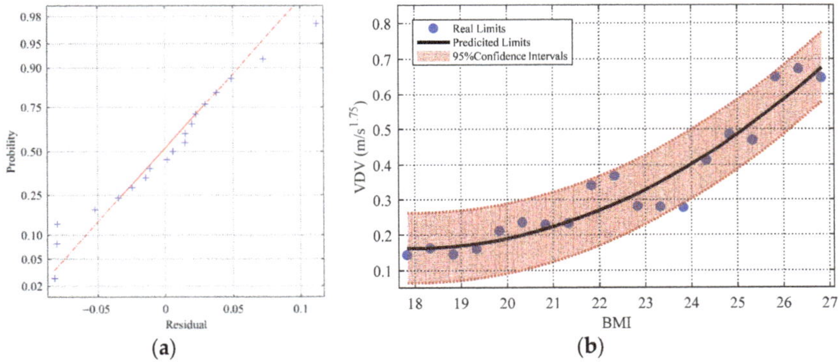

Figure 14. Normal probability plot (**a**) and fitting curves (**b**) (Female BMI and perception limits in VDV).

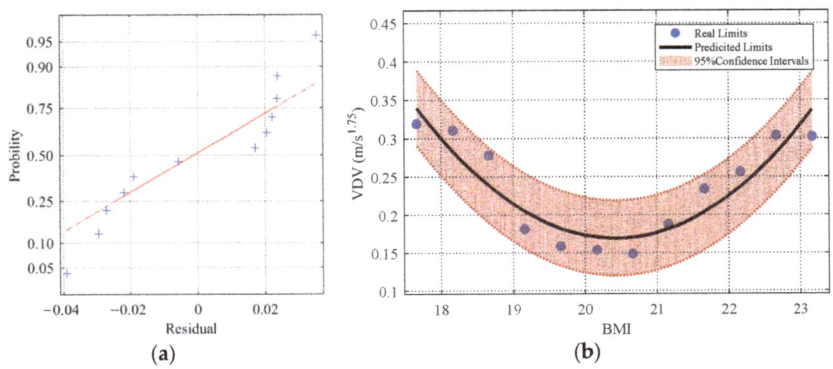

Figure 15. Normal probability plot (**a**) and fitting curves (**b**) (Male BMI and perception limits in VDV).

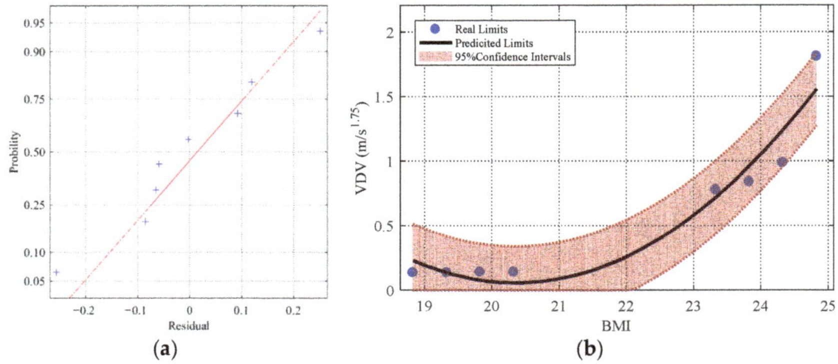

Figure 16. Normal probability plot (**a**) and fitting curves (**b**) (Female BMI and comfort limits in VDV).

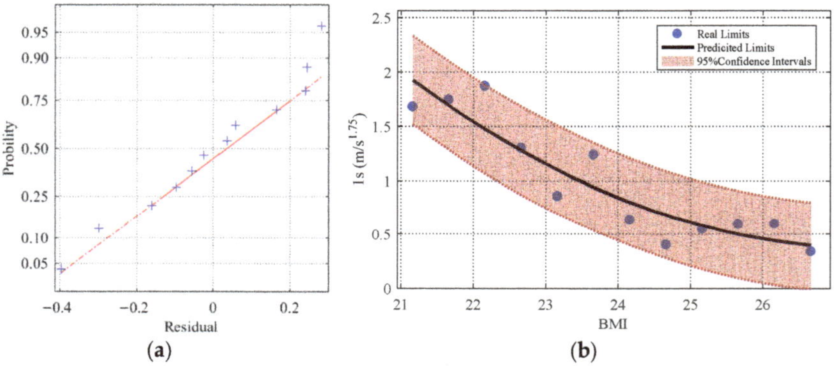

Figure 17. Normal probability plot (**a**) and fitting curves (**b**) (Male BMI and comfort limits in VDV).

Table 4. Coefficients of fitting between BMI and vibration limits (VDV, m/s$^{1.75}$).

Gender	Limits	a	b	c	R^2	t-Test	LF Test	95% CI ($\pm 1.96\sigma$)
Female	Perception	0.0066	17.98	0.1629	0.9121	0	0	0.0987
	Comfort	0.0747	20.35	0.0571	0.9344	0	0	0.2834
Male	Perception	0.0224	20.42	0.1692	0.8456	0	0	0.0494
	Comfort	0.0384	27.54	0.3655	0.8488	0	0	0.4062

Table 5. Coefficients of fitting between BMI and vibration limits (1 s MTVV, m/s^2)

Gender	Limits	a	b	c	R^2	t-Test	LF Test	95% CI ($\pm 1.96\sigma$)
Female	Perception	0.0019	17.05	0.0618	0.9072	0	0	0.0354
	Comfort	0.0368	20.93	0.0559	0.8016	0	0	0.1889
Male	Perception	0.0077	20.25	0.0732	0.8678	0	0	0.0354
	Comfort	0.0174	26.69	0.2104	0.8146	0	0	0.1273

The results show that these curves reflect the tendency of vibration limits to change with BMI well. Residuals between actual vibration limits and fitted values follow a normal distribution.

Women with a BMI of 17.0–18.0 possess lower perception limits than other women, which means they are more sensitive to vibration. The extreme value of the male's 'Perception limits—BMI' curves fall in a BMI of 20.2–20.4, hence the BMI of the most sensitive male is a little higher than the females. The results agree well with previous conclusions of authoritative research [31]. The range of female (BMI of 18–27) perception limits fall in 0.163–0.700 $m/s^{1.75}$ (VDV)/ 0.064–0.25 m/s^2 (1 s MTVV), with 95% confidence intervals of $\pm$0.099 $m/s^{1.75}$ (VDV) and $\pm$0.035 m/s^2 (1 s MTVV), respectively. The range of male (BMI of 17–24) perception limits falls in 0.169–0.456 $m/s^{1.75}$ (VDV)/0.073–0.181 m/s^2 (1 s MTVV), with 95% confidence intervals of $\pm$0.049 $m/s^{1.75}$ (VDV) and $\pm$0.035 m/s^2 (1 s MTVV), respectively.

As for comfort limits, the BMI of women who possess the lowest limits is also lower than men. The range of female (BMI of 18–25) comfort limits fall in 0.057–1.672 $m/s^{1.75}$ (VDV)/0.056–0.665 m/s^2 (1 s MTVV), with 95% confidence intervals of $\pm$0.283 $m/s^{1.75}$ (VDV) and $\pm$0.189 m/s^2 (1 s MTVV), respectively. The range of male (BMI of 21–27) perception limits fall in 0.377–2.008 $m/s^{1.75}$ (VDV)/0.210–0.774 m/s^2 (1 s MTVV), with 95% confidence intervals of $\pm$0.406 $m/s^{1.75}$ (VDV) and $\pm$0.127 m/s^2 (1 s MTVV), respectively.

It should be emphasized that extreme values fall in nearly the same interval when different indicators are used, which also proves the reasonability of the results. As a reason of extreme value and relatively small sample corresponding to some BMIs, lower bounds of 95% CI corresponding to these BMI are lower than 0. Hence, the lower bounds are altered to 0 when they are previously lower than 0.

4.2. Fitting Function of Crest Factor and Vibration Limits

Previous research [8] shows that vibration limit tends to increase when crest factor grows bigger, yet some other research shows that the upbound of limits growing with crest factor is finite. However, there is no more precise research on the relationship between the crest factor and vibration limits. By considering the trend of data collected and the existence of finite upbound, various models (such as the Logistic model, Weibull model and polynomial model) were used to fit the relationship between the crest factor (CF) and vibration limits. At last, Richards model Equation (5) was chosen to obtain the fit function between the crest factor (CF) and vibration limits. The coefficients were computed and listed in Table 6.

$$y = \alpha(1 + \exp(\beta - \gamma x))^{-\frac{1}{\delta}} \tag{5}$$

Table 6. Coefficients of fitting between CF and vibration limits.

Indice	Limits	α	β	γ	δ	R^2	*t*-Test	LF Test	95% CI ($\pm 1.96\sigma$)
sVDV ($m/s^{1.75}$)	Perception	0.6265	547.4	33.91	344.4	0.9871	0	0	0.0374
	Comfort	1.830	−6.914	0.2458	0.0007	0.6834	0	0	0.4281
1 s MTVV (m/s^2)	Perception	0.3090	30.31	2.001	15.50	0.9865	0	0	0.0184
	Comfort	0.7558	320.8	43.75	209.4	0.7843	0	0	0.1882

In Equation (5), x, y are two variables to be fitted; $\alpha/\beta/\gamma/\delta$ are parameters. α is the limit when x approaches infinite.

Figures 18 and 19 represent the result of the normal distribution test and the curve-fitting between the crest factors (CF) and perception limits.

Both perception and comfort limits fit well with crest factors (CF) by using the Richards models, and the residuals between the actual vibration limits and fitted values follow a normal distribution.

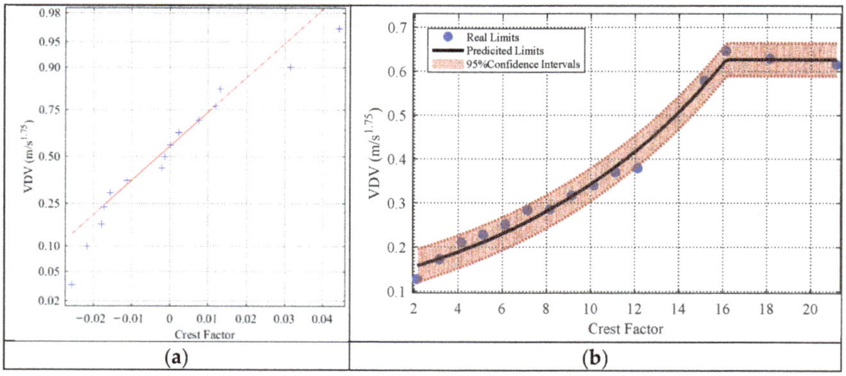

Figure 18. Normal probability plot (**a**) and fitting curves (**b**) (CF and perception limits in VDV).

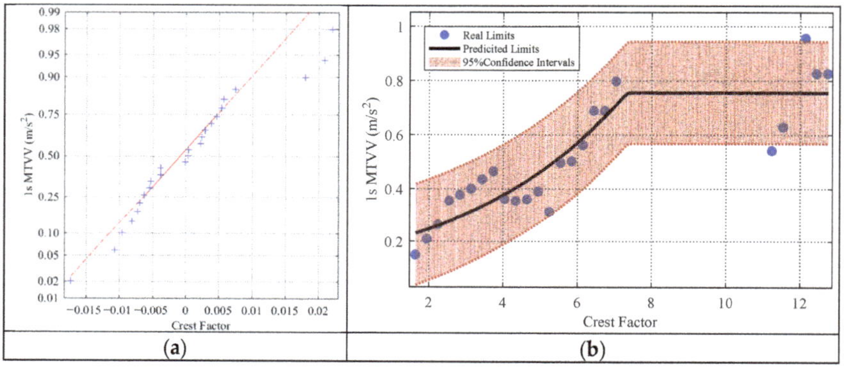

Figure 19. Normal probability plot (**a**) and fitting curves (**b**) (CF and perception limits in 1 s MTVV).

Since α is the limit when the crest factor approaches infinite and the fitting functions are all monotonously increasing, the upper bound of vibration limits is α. Previous research [3] and criteria [21,23] showed that vibration limits become bigger when the crest factor of vibration ascends, which is compatible with the conclusion of this research. For sinusoidal vibration, the perception limit is 0.062 m/s^2 (1 s MTVV) or 0.128 $m/s^{1.75}$ (VDV). The range of perception limits falls in 0.348–0.626 $m/s^{1.75}$ (VDV)/ 0.119–0.309 m/s^2 (1 s MTVV), with 95% confidence intervals of $\pm$0.037 $m/s^{1.75}$ (VDV) and $\pm$0.018 m/s^2 (1 s MTVV), respectively. When peak or rms is used as a vibration indicator, the perception limit is 0.062 m/s^2 (rms) or 0.087 m/s^2 (peak), respectively, which also fits well with the conclusions of previous research and criteria [22,32–35]. The values of comfort limits vary from 0.681 $m/s^{1.75}$ to 1.830 $m/s^{1.75}$ (indice as VDV) when the crest factor changes, which accords well with the criteria [32].

4.3. Probability Prediction of Vibration Serviceability

As a reason for the residuals between fitted values and actual limits following normal distribution, the probability of vibration serviceability can be predicted when the BMI of people or the crest factor is known. In order to achieve the prediction, the mean value (μ) and standard deviation (σ) of the vibration limits are necessary.

Once the value of BMI (distinguished by gender/crest factor) is known, the mean value of vibration limit (μ) can be predicted by using Equation (4)/Equation (5) and the coefficients in Tables 4–6. The standard deviation (σ) of vibration limits corresponding to the given BMI/CF can also be found in Tables 4–6. Then it is easy to obtain the probability density function of the vibration limits (perception/comfort) corresponding to the certain BMI/crest factor in Equation (6).

$$f(v) = \frac{1}{\sqrt{2\pi}\sigma} \exp\left(-\frac{(v-\mu)^2}{2\sigma^2}\right) \tag{6}$$

In case the magnitude of vibration is already known as v_0, P is the possibility of people perceiving this vibration or feeling uncomfortable due to this vibration in Equation (7).

$$P = \int_0^{v_0} \frac{1}{\sqrt{2\pi}\sigma} \exp\left(-\frac{(v-\mu)^2}{2\sigma^2}\right) \mathrm{d}v \tag{7}$$

5. Conclusions and Discussion

5.1. Conclusions

Novel research on quantifying the correlation between vibration limits and relevant factors is proposed based on big data and a new mathematical tool. There are several important improvements. First of all, by introducing the MIC to estimate correlations between factors and limits, it is the first time that the correlations are quantitatively compared and proves that vibration serviceability is a multivariable system. Secondly, this study finds that crest factor and BMI are key factors to vibration serviceability. Hence, researchers and designers should consider the crest factor of vibration and BMI more seriously than other factors. Finally, the Richards model and quadratic polynomial are used to fit the relationships between the limits and the crest factor/BMI, respectively, and the results are compatible with the authority conclusion. For the reason that the residuals between actual vibration limits and fitted values follow a normal distribution, a method based on probability is proposed to predict vibration serviceability when the value of the crest factor/BMI is known.

Since the database covers a sufficient range of relevant factors and the scale is much larger than previous research, along with the fact that the correlation analysis tool (MIC) is reliable and the results fit well with previous research, the authors consider that the conclusions are reasonable and recommend the functional relationship between the vibration limits and crest factor/BMI. Corresponding confidence intervals are also suggested for vibration serviceability performance design.

5.2. Discussion

Although the scale of the investigation is much larger than earlier research and the conclusions are reasonable, some improvements could be made.

Due to combinatorial explosion, it is impossible to obtain vibration limits corresponding to the combination of different relevant factors. Yet larger scales will draw a more precise conclusion. Extreme data and small samples in some groups result in the 95% confidence intervals corresponding to these groups reaching 0. Collecting more data might have corrected this problem.

More psychological principles could be used to make the questionnaire and curve-fitting models more reasonable. Other factors (such as noise and temperature) should be taken into consideration when studying correlations between the relevant factors and vibration limits.

Author Contributions: Conceptualization, L.C. and J.C.; methodology, L.C. and J.C.; software, J.C.; validation, L.C. and J.C.; formal analysis, L.C.; investigation, L.C. and J.C.; resources, J.C.; data curation, J.C.; writing—original draft preparation, L.C.; writing—review and editing, J.C.; visualization, L.C.; supervision, J.C.; project administration, L.C. and J.C.; funding acquisition, J.C. All authors have read and agreed to the published version of the manuscript.

Funding: This research was funded by the National Science Foundation of China, grant number 52178151.

Informed Consent Statement: Not applicable.

Conflicts of Interest: The authors declare no conflict of interest.

References

1. Reiher, H.; Meister, F.J. Human sensitivity to vibration. *Forsch. Auf Dem Geb. Des Ing.* **1931**, *2*, 381–386. [CrossRef]
2. Kanda, J.; Tamura, Y.; Fujii, K. Probabilistic evaluation of human perception threshold of horizontal vibration of buildings (0.125 Hz to 6.0 Hz). In Proceedings of the Structures Congress XII, Atlanta, GA, USA, 24–28 April 1994.
3. Zhou, Z.; Griffin, M.J. Response of the seated human body to whole-body vertical vibration: Discomfort caused by mechanical shocks. *Ergonomics* **2017**, *60*, 347–357. [CrossRef] [PubMed]
4. Jones, A.J.; Saunders, D.J. Equal comfort contours for whole body vertical, pulsed sinusoidal vibration. *J. Sound Vib.* **1972**, *23*, 1–14. [CrossRef]
5. Parsons, K.C.; Griffin, M.J. Whole-body vibration perception thresholds. *J. Sound Vib.* **1988**, *121*, 237–258. [CrossRef]
6. Liu, H.; Lian, Z.; Gong, Z. Thermal comfort, vibration, and noise in Chinese ship cabin environment in winter time. *Build. Environ.* **2018**, *135*, 104–111. [CrossRef]
7. Griffin, M.J. *Handbook of Human Vibration;* Harcourt Brace Jovanovich: San Diego, CA, USA, 1990.
8. Kowalska-Koczwara, A.; Stypuła, K. Influence of crest factor on evaluation of human perception of traffic vibration. *J. Meas. Eng.* **2018**, *6*, 250–255. [CrossRef]
9. Reshef, D.N.; Reshef, Y.A.; Finucane, H.K.; Grossman, S.R.; McVean, G.; Turnbaugh, P.J.; Lander, E.S.; Mitzenmacher, M.; Sabeti, P.C. Detecting novel associations in large data sets. *Science* **2011**, *334*, 1518–1524. [CrossRef] [PubMed]
10. Moore, D.S.; Notz, W.I.; Notz, W. *Statistics: Concepts and Controversies;* Macmillan: London, UK, 2017.
11. Chen, J.; Tan, H.; Pan, Z. Experimental validation of smartphones for measuring human-induced loads. *Smart Struct. Syst.* **2016**, *18*, 625–642. [CrossRef]
12. Grossi, M. A sensor-centric survey on the development of smartphone measurement and sensing systems. *Measurement* **2019**, *135*, 572–592. [CrossRef]
13. Cao, L.; Chen, J. Online Investigation of Vibration Serviceability Limitations using Smartphones. *Measurement* **2020**, *162*, 107850. [CrossRef]
14. Xia, H. *Traffic Induced Environmental Vibrations and Controls;* Nova Publ: Beijing, China, 2013.
15. Chen, P.W.; Robertson, L.E. Human perception thresholds of horizontal motion. *J. Struct. Div.* **1972**, *92*, 1681–1695. [CrossRef]
16. Morioka, M.; Griffin, M.J. Magnitude-dependence of equivalent comfort contours for fore-and-aft, lateral, and vertical vibration at the foot for seated persons. *J. Sound Vib.* **2010**, *329*, 2939–2952. [CrossRef]
17. Burton, M.D.; Kwok, K.; Abdelrazaq, A. Wind-Induced Motion of Tall Buildings: Designing for Occupant Comfort. *Int. J. High-Rise Build.* **2015**, *4*, 1–8.
18. Burton, M.D.; Kwok, K.; Hitchcock, P.A. Occupant comfort criteria for wind-excited buildings: Based on motion duration. In Proceedings of the 12th International Conference on Wind Engineering, Cairns, Australia, 1–6 July 2007.
19. Goto, T. Effects of Visual Views on the Perception Threshold of Building Motion. *Jpn. Informal Group Hum. Response Vib.* **1996**, *13–14*, 1–15.
20. Kawana, S.; Tamura, Y.; Matsui, M. Study on vibration perception by visual sensation considering probability of seeing. *Int. J. High-Rise Build.* **2012**, *1*, 283–300.
21. *ISO 2631-1:1997 (E);* Mechanical Vibration and Shock: Evaluation of Human Exposure to Whole-Body Vibration. Part 1: General Requirements. ISO: Geneva, Switzerland, 1997.
22. *VDI 2057-1;* Human Exposure to Mechanical Vibrations Whole-Body Vibration. VDI: Berlin, Germany, 2002.
23. *ISO 10137;* Bases for Design of Structures—Serviceability of Buildings and Walkways against Vibrations. ISO: Geneva, Switzerland, 2007.
24. Miwa, T. Evaluation methods for vibration effect. Part 1 Measurements of threshold and equal sensation contours of whole body for vertical and horizontal vibrations. *Ind. Health* **1967**, *5*, 183–205. [CrossRef]
25. AISC. *AISC DG11 Floor Vibrations: Due to Human Activity*, 2nd ed.; AISC: Chicago, IL, USA, 2003.
26. *BS 6841;* Guide to Measurement and Evaluation of Human Exposure to Whole-Body Mechanical Vibration and Repeated Shock. BSI: London, UK, 1987.
27. Lamb, S.; Kwok, K. The fundamental human response to wind-induced building motion. *J. Wind. Eng. Ind. Aerodyn.* **2017**, *165*, 79–85. [CrossRef]
28. Živanović, S.; Pavic, A.; Reynolds, P. Vibration serviceability of footbridges under human-induced excitation: A literature review. *J. Sound Vib.* **2005**, *279*, 1–74. [CrossRef]
29. Arnold, J.J.; Griffin, M.J. Discomfort caused by multiple frequency fore-and-aft, lateral or vertical whole-body vibration. In Proceedings of the 52nd UK Human Resopnse to Vibration Conference and Workshop, Shrivenham, UK, 5–6 September 2017.
30. Nuttall, F.Q. BMI, and health: A critical review. *Nutr. Today* **2015**, *50*, 117. [CrossRef] [PubMed]
31. WHO. *Mean Body Mass Index (BMI): Situation and Trends;* WHO: Geneva, Switzerland, 2013.
32. *BS 6472;* Guide to Evaluation of Human Exposure to Vibration in Buildings. BSI: London, UK, 2008.

33. Burton, M.D. *Effects of Low Frequency Wind-Induced Building Motion on Occupant Comfort. Doctor Degree*; The Hong Kong University of Science and Technology: Hong Kong, China, 2006.
34. Kanda, J.; Tamura, Y.; Fujii, K. Probabilistic criteria for human perception of low-frequency horizontal motions. In Proceedings of the Symposium/Workshop on Serviceability of Buildings, Ottawa, ON, Canada, 16–18 May 1988.
35. *RFC2-CT-2007-00033*; HiVoSS (Human Induced Vibrations of Steel Structures): Vibration Design of Floors Guideline. HiVoSS: Luxembourg, 2008.

buildings

Article

Train-Induced Vibration Predictions Based on Data-Driven Cascaded State-Space Model

Ziyu Tao [1,2], Zihao Hu [1], Ganming Wu [1], Conghui Huang [1], Chao Zou [1,*] and Zhiyun Ying [3]

1 School of Civil and Transportation Engineering, Guangdong University of Technology, Guangzhou 510006, China; 201710101340@mail.scut.edu.cn (Z.T.); 3120002922@mail2.gdut.edu.cn (Z.H.); 3120002941@mail2.gdut.edu.cn (G.W.); 3120002923@mail2.gdut.edu.cn (C.H.)
2 School of Civil Engineering and Transportation, South China University of Technology, Guangzhou 510641, China
3 School of Economics and Management, North China Electric Power University, Baoding 071066, China; 201806000326@ncepu.edu.cn
* Correspondence: chao.zou@gdut.edu.cn

Abstract: Over-track buildings above metro depots have become common in megacities due to urban land shortages. The transmission of vibrations into the over-track buildings during routine train operations has the potential to adversely impact the occupants in terms of perceptible vibration and noise. There is a need to quantify the potential impacts before construction for planning and design purposes. Train-induced vibration measurements were carried out on a six-story over-track building at the Luogang metro depot in Guangzhou, China, which is located adjacent to the tracks. The measurements were used to develop a data-driven cascaded state-space model, which can be applied to planned over-track buildings located in track areas to predict and assess whether train-induced vibrations would adversely affect the buildings' future occupants. Vibration levels in the platform of the building's columns were used as inputs to the models, thereby avoiding the complexity of modeling the transfer behavior of the platform. The predicted vibration levels corresponded with measurements in the existing building. This comparison validated the use of the model for future residential buildings where the predictions indicate that the impacts on its occupants will be within the applicable criteria.

Keywords: building vibration predictions; human comfort; train-induced vibration measurements; metro depot; over-track buildings; state-space model

Citation: Tao, Z.; Hu, Z.; Wu, G.; Huang, C.; Zou, C.; Ying, Z. Train-Induced Vibration Predictions Based on Data-Driven Cascaded State-Space Model. *Buildings* **2022**, *12*, 114. https://doi.org/10.3390/buildings12020114

Academic Editors: Jun Chen and Haoqi Wang

Received: 20 December 2021
Accepted: 19 January 2022
Published: 25 January 2022

1. Introduction

Urban rail transit, with the advantages of large capacity and lower energy consumption, is an efficient solution to road traffic congestion and environmental pollution in metropolitan cities. By the end of 2020, 45 cities in mainland China had put urban rail transit systems into service with a total length of 7969.7 km [1]. Current design standards stipulate that metro depots are necessary logistical support centers to the metro system [2]. Depots function as a metro system's base for cleaning, storing, testing, and maintaining trains, covering a large land area. In order to balance the problem of urban land shortage and the construction of depots and to provide financial support to the sustainable operations of urban rail transit systems, exploiting air space over metro depots with over-track buildings is becoming a modern development trend [3]. Further, the distances between depots and nearby residential buildings are also decreasing. Given these facts, the side impact of train-induced perceptible vibrations and noise on human comfort and the performance of vibration-sensitive equipment becomes a social concern [4].

To control vibration impacts, it is of economic importance to predict vibrations before constructing over-track buildings in order to determine the need for vibration mitigation measures. Prediction models can normally be classified into three types: empirical, numerical, and analytical.

For characterizing train-induced vibrations, field measurement is the most direct and efficient method [5–16]. General vibration transmission characteristics at metro depots and within over-track buildings could be drawn. For instance, Xia et al. [9] investigated train-induced vibration transmissions in a six-story masonry building adjacent to the Beijing-Guangzhou railway line. The study showed train-induced vibrations increased with train speed and attenuated with increased distance from the track. However, train-induced vibration is a complex phenomenon, so field measurements were site-specific. In [6], it was concluded that train-induced ground vibrations were affected by various factors, including soil profiles, train type, train operation speed, and track unevenness.

Empirical models are developed based on past designs and construction experiences and need a great number of field measurement studies. They can account for factors such as the distance between the receiver and the tracks, train speeds, and geometric and material attenuation [17]. The shortcomings of empirical models lie mostly in predicting accuracy and applicability.

Numerical models always need to integrate three sub-models: the train-track, soil, and building structure models, representing the vibration source, vibration transmission path, and receiver, respectively. Vibrating loads can be calculated through a train-track model that accounts for the quasi-static loads from train axle loads and dynamic loads from track irregularities, wheel wear, and/or eccentric wheel mountings [18–20]. The finite element method (FEM) and boundary element method (BEM) are the most frequently adopted methods to establish a 2D or 3D soil model. The 2D or 3D building structure model is generally developed based on the FEM because of the geometric complexity of buildings. Train-induced vibration levels within a building depend not only on the natural frequency of the structure but also on the soil properties [21–24]. This results from the problem of coupling or dynamic interaction between soil and building foundations. The disadvantages of numerical models are their formidable computational cost and the uncertainty of numerical model parameters. The computational efficiency was improved by extensive simplification and optimization. More efficient numerical models have been proposed, such as 2.5D finite element models [25,26] and finite difference models [27].

Analytical models (i.e., the white-box model) are physics-based models developed from first principles and based on strict theoretical deduction with a closed-form and high prediction accuracy. Simplified 1D or 2D impedance models for predicting the axial vibration propagation within the buildings' support structures belong to analytical models and proved efficient [28–30]. However, analytical closed-form solutions do not exist for most practical engineering problems with complex 3D geometries. In addition, analytical models normally introduce many unrealistic hypotheses to simplify the solutions.

Due to construction policies, over-track buildings above metro depots contain two stages. As seen in Figure 1, the footprint of first stage over-track buildings is away from the tracks, and these buildings are generally built synchronously with the platform. The buildings are used as office space or apartments for people working in the metro depot. The second stage over-track buildings are built after the construction of first stage over-track buildings, further exploiting the air space above metro depots. They have various usages such as residences, offices, schools, clinics, stores, and restaurants.

For simulating a dynamic system, another method called the system identification technique exists, which generally derives data-driven models from experimental data (i.e., the black-box model). In order to predict train-induced vibrations within future second stage over-track buildings and assess vibration impacts prior to construction, this paper develops and validates the cascaded state-space model based on the measurements from a first stage over-track building at Luogang metro depot in Guangzhou, China.

State-space models of sub-structural systems are efficient and have great applicability to similar structures. By applying vibration measurements from the platform to the cascaded state-space model as inputs, vibration responses in future second stage over-track buildings with similar structures can be predicted and assessed. It is of economic importance to predict train-induced vibration responses prior to construction to assess whether vibration mitigation measures are needed.

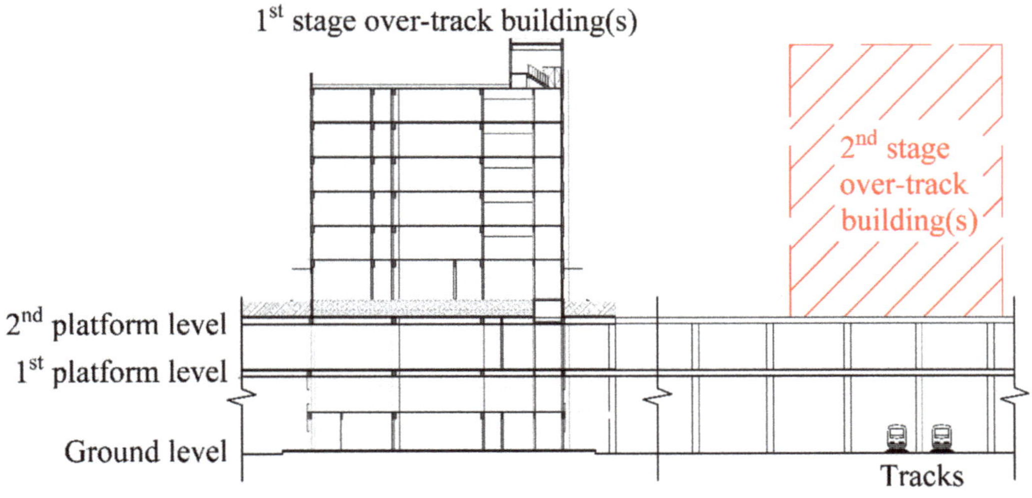

Figure 1. Two-stage construction strategy.

This research is comprised of four parts. Firstly, a cascaded state-space model for predicting vibrations within the building is proposed. Its modeling procedure and realization methods are presented in Section 2. Secondly, field measurements to develop and validate the cascaded state-space model have been carried out, and the setup and results are displayed in Section 3. Thirdly, a data-driven cascaded state-space model for the first stage over-track building is obtained based on the measurements. The predicted and measured vibrations are compared to verify the model's accuracy and applicability. The cascaded state-space model is also applied to the future second stage over-track building to assess train-induced vibration impacts, as shown in Section 4. Finally, natural frequencies and damping ratios related to a typical floor of the first stage over-track building are identified to gain insights into structural dynamic characteristics, as exhibited in Section 5. The organization diagram of this paper is shown in Figure 2.

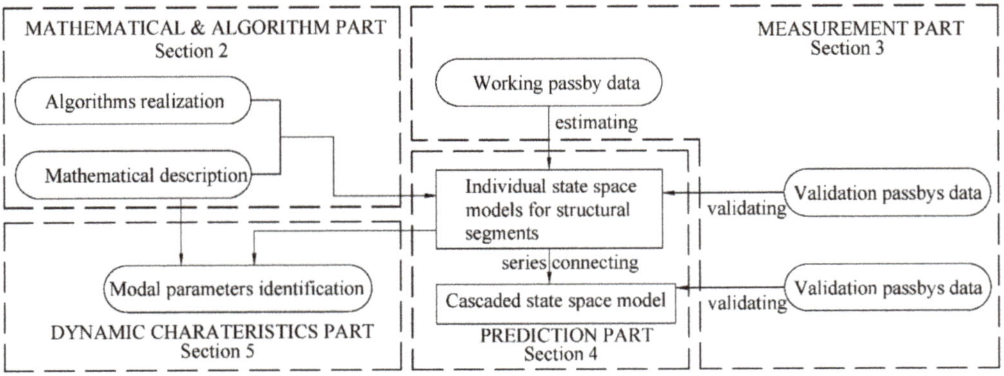

Figure 2. Paper organization diagram.

2. Prediction Method Based on State-Space Model

Unlike structural dynamic problems where time-variant forcing functions are directly applied to the system DOFs as inputs, train-induced structural vibration is affected by the problem of foundation excitation, which is widely studied in civil engineering [31]. Structural foundation excitation, induced by vibrations in the surrounding soil through soil-foundation dynamic interaction, can be caused by various sources, such as earthquakes, vehicular loads, and construction machines operated near a building. The corresponding system inputs to the problem of foundation excitation are the motions of the structure boundary. Each column or load-bearing wall segment between different floors can be considered a supported–excited structural component, through which vibrations in the foundation or lower floor will be transmitted to the upper floor [28].

2.1. Modeling Procedure

A data-driven cascaded state-space model was developed and validated based on an experimental study. As illustrated in Figure 3a, the ith column segment and the $(i+1)^{\text{th}}$ floor structures constitute a single-input and single-output (SISO) sub-structural system. A state-space model with the system order n_i (see Equation (1)) can be used to describe the sub-structural system and its state variables, and coefficient matrices can be estimated from the measured input and output data using algorithms based on the subspace methods mentioned in Section 2.2. Such sub-structural systems of the different floors are assumed to be independent of each other, and the dynamic coupling between them is ignored. Successive sub-structural systems can be series connected through their communal joint vibrations (see Figure 3b). The predicted output of the lower i^{th} system is taken as the "measured" input of the upper $(i+1)^{\text{th}}$ system.

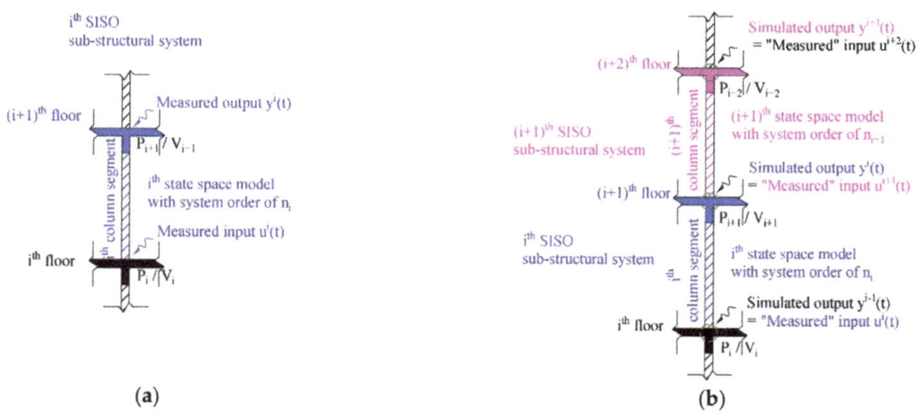

Figure 3. Schematic diagram of individual and cascaded structural systems: (**a**) Individual SISO system; (**b**) Cascaded structural system.

A state-space model is typically used to simulate the dynamic behavior of a state-determined system that can be physically modeled with one-port elements such as masses, springs, and dampers. It can be expressed in either continuous-time or discrete-time form by using differential or difference equations, respectively. Theoretically, the system state is completely characterized by the state variables. The number of internal state variables of a system is physically equal to the number of independent energy storage elements in the system. This number is also defined as the system order. At any time (t), the value of each state variable represents the energy of the corresponding energy storage element. The time derivatives of state variables specify the changing rate of the system energy.

Considering the context of this study is small-amplitude structural vibrations induced by running metro trains, each individual sub-structural system in Figure 3a can be considered a linear time-invariant (LTI) system with a single input and single output. An nth-order discrete-time SISO state-space model for a combined deterministic–stochastic LTI system is mathematically described in Equation (1):

$$\begin{cases} \vec{x}(k+1) = [A]\vec{x}(k) + [B]\vec{u}(k) + \vec{w}(k) \\ \vec{y}(k) = [C]\vec{x}(k) + [D]\vec{u}(k) + \vec{v}(k) \end{cases} \quad (1)$$

where $\vec{u}(k)$ is the input vibration at time step k, $\vec{y}(k)$ is the output vibration at time step k, and $\vec{x}(k)$ is the n-dimensional discrete state vector at time step k. $[A]$ is the dynamical system matrix describing the dynamic characteristics of the sub-structural system. $[B]$ is the input matrix representing a linear transformation through which the deterministic inputs influence the next system state. $[C]$ is the output matrix implying how the system's internal state is transferred to the measured output vibrations $\vec{y}(k)$. D is a matrix consisting of feedthrough elements. $\vec{w}(k)$ and $\vec{v}(k)$ are the process noise and measurement noise at time step k, respectively. They are both assumed to be stochastic and uncorrelated Gaussian zero-mean white noise processes, whose covariance matrices can be defined by Equation (2):

$$E\left[\begin{pmatrix} \vec{w}(k) \\ \vec{v}(k) \end{pmatrix}\left(\vec{w}(q)^T, \vec{v}(q)^T\right)\right] = \begin{bmatrix} [Q] & [S] \\ [S^T] & [R] \end{bmatrix}\delta_{kq} \quad (2)$$

where $[Q]$, $[R]$, and $[S]$ are covariance matrices of the noise vectors $\vec{w}(k)$ and $\vec{v}(k)$.

Since this model considers the unknown disturbances coming from two additional stochastic processes, process and measurement noise, functioning as a data-driven model compared with a determined state-space model is more realistic. The task of simulating an LTI system through the combined deterministic–stochastic system identification method mentioned in Section 2.2 can be summarized as making the optimal estimation of matrices $[A]$, $[B]$, $[C]$, and $[D]$ when given measured input and output vectors with typically infinite measuring time [32,33]. The flow chart depicting the development of individual state-space models is shown in Figure 4. The realization of sub-structural system identification through subspace method-based algorithms is described in Section 2.2.

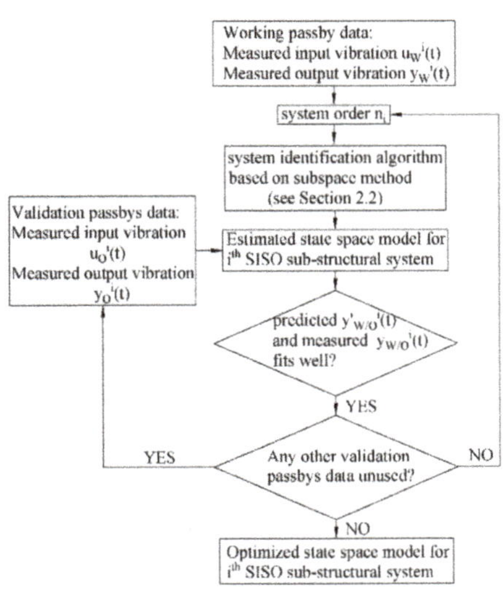

Figure 4. Flow chart of developing individual state-space model.

2.2. Sub-Structural System Identification

Developing a data-driven state-space model is also referred to as subspace state-space system identification (4SID), which are generally classified into two main types: realization-based and direct 4SID methods [34].

Realization-based methods extract state-space models by using the extended observability matrix [35], which is estimated directly from the Markov parameters (MPs). MPs indicate estimations of system impulse responses, and their reliable estimation is the backbone of realization-based 4SID methods.

Direct 4SID methods also are referred to as subspace methods, data-driven subspace identification, or subspace system identifications [36]. Unlike realization-based methods, direct 4SID methods estimate state-space models directly from an arbitrary set of input and output data (working data) without estimating system impulse response functions and then use other sets of input and output data (validation data) to verify the estimated state-space model. The measured system output $\left\{\vec{y}\right\}$ generally can be divided into three components: the forced vibration response, free vibration response, and colored noise response, where the colored noise response $\left\{\vec{y}^{S}\right\}$ is the result of the unmeasured colored noise disturbance. From this point of view, system output $\left\{\vec{y}\right\}$ can be correspondingly decomposed into three orthogonal matrices $[H]\left\{\vec{u}\right\}$, $[O]\left\{\vec{x}\right\}$, and $\left\{\vec{y}^{S}\right\}$, which are parallel to the span of the system input $\left\{\vec{u}\right\}$, the span of the joint null space of the input and the colored noise output $\left\{\left\{\vec{u}\right\}^{\perp}, \left\{\vec{y}^{S}\right\}^{\perp}\right\}$, and the span of the colored noise output $\left\{\vec{y}^{S}\right\}$, respectively. The matrix $[O]\left\{\vec{x}\right\}$ is the product of the system extended observability matrix $[O]$ and state sequence $\left\{\vec{x}\right\}$. Direct 4SID methods estimate the matrix $[O]\left\{\vec{x}\right\}$ along the span of $\left\{\left\{\vec{u}\right\}^{\perp}, \left\{\vec{y}^{S}\right\}^{\perp}\right\}$ using subspace operations based on measured data since free vibration responses of the system are only related to the system dynamic characteristics. Then, the state-space matrices $[A]$, $[B]$, $[C]$, and $[D]$ of the system can be further extracted from the estimated matrix $[O]\left\{\vec{x}\right\}$.

Plenty of numerical algorithms based on direct 4SID methods are well established, such as multivariable output-error state-space (MOESP) and the numerical algorithm for subspace state-space system identification methods (N4SID) [33,37,38]. The key to each algorithm is to project the measured system output $\left\{\vec{y}\right\}$ to the span of $\left\{\left\{\vec{u}\right\}^{\perp}, \left\{\vec{y}^{S}\right\}^{\perp}\right\}$. MOESP uses orthogonal projection, while N4SID uses oblique projection. Even though MOESP and N4SID algorithms use different numerical procedures to obtain state-space matrices $[A]$, $[B]$, $[C]$, and $[D]$, the research demonstrated that the accuracy of these two algorithms was comparable [39]. Detailed algorithms were not discussed, and the N4SID algorithm was adopted in this research with the help of MATLAB.

3. Train-Induced Vibration Measurements

3.1. Measurement Program

3.1.1. Location

The measurements were conducted at the Luogang depot (Figure 5) in Guangzhou, China. The Luogang depot is used for parking, testing, cleaning, and maintaining the Line 6 metro trains of Guangzhou Metro, covering an area of 0.354 km^2.

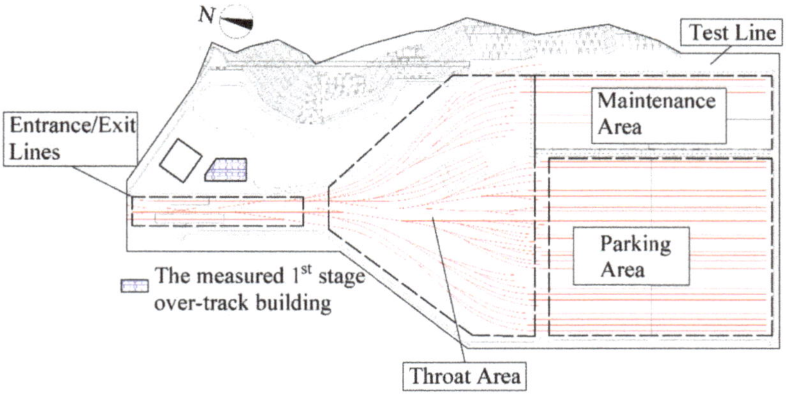

Figure 5. Plan view of the Luogang depot.

3.1.2. Setup

The tested first stage over-track building, located beside the entrance/exit tracks (Figure 6), is a six-story residential building with a concrete frame structure. The instrumentations were wireless units with built-in accelerometers and a JM3873 Data Acquisition System [14]. The sampling frequency was 512 Hz, which was enough for researching railway-induced vibration responses in buildings up to 200 Hz [40].

Figure 6. Entrance/exit tracks and first platform structure.

Two columns (Figure 7a) of the first stage over-track building were measured from floor to floor. They are 9 m apart and of the same construction materials (see Table 1), the same sectional dimensions (0.8 m × 0.8 m), and the same distance to the nearest entrance/exit line (33.5 m, Figure 7b).

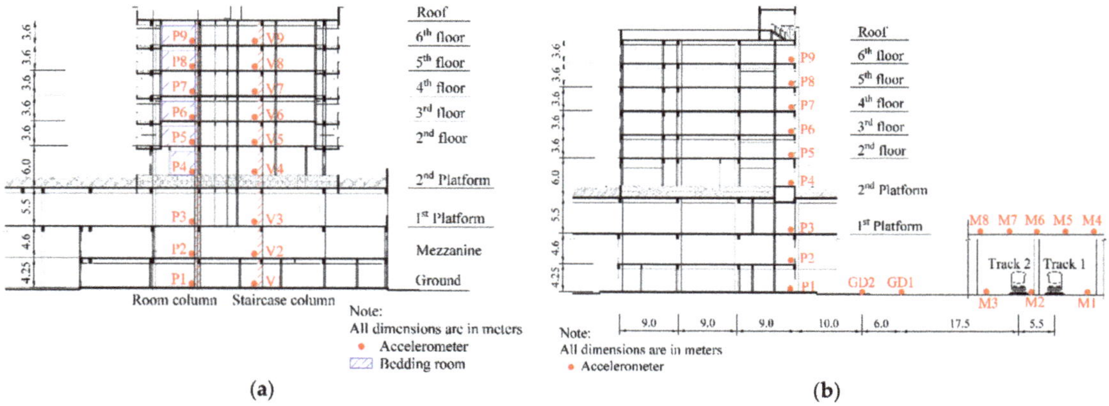

Figure 7. Measurement setup: (**a**) Building profile along the tracks; (**b**) Building profile perpendicular to the tracks.

Table 1. Structural parameters of the two measured columns.

Level	Length (m)	Cross Area (m^2)	Young's Modulus (GPa)	Density (kg/m^3)	Floor Thickness (m)
Ground–Mezzanine	4.45		36.0	2440	–
Mezzanine–1st platform	4.4		35.5	2430	0.12
1st platform–2nd platform	5.5		34.5	2420	0.15
2nd platform–2nd floor	5.95	0.64	33.5	2410	0.25
2nd floor–3rd floor	3.6		32.5	2400	0.12
Upper floors	3.6		31.5	2390	0.12

The column including measured points P1 to P9 was called the room column. The data from 11 train passbys were obtained: 4 passbys on Track 1 and 7 passbys on Track 2. The column including measured points V1 to V9 was named the staircase column. The data from 10 train passbys were recorded, 5 passbys on Track 1 and 5 passbys on Track 2. The main difference between the room column and the staircase column is that the latter connects to a staircase.

3.2. *Vertical Vibration Levels on Different Floors in the First Stage Over-Track Building*

3.2.1. Room Column

Figure 8 shows the vertical vibration transmission along the room column from the ground into the building. The averaged acceleration levels from four passbys on Track 1 and seven passbys on Track 2 were calculated and shown, respectively. Due to an instrumentation problem on the third floor, obtaining the P6 failed. As shown in Figure 7, Track 2 is 5.5 m closer to the first stage over-track building than Track 1. These two tracks are of the same design and construction but heading to different parking lines.

The building vibration levels caused by train passbys on Track 1 and Track 2 are comparable. The dominant frequency range of train-induced vibration is 8–80 Hz. Vibration transmission patterns along the room column caused by passbys on Track 1 and 2 are consistent. The vertical vibration amplified as the elevation increased from the ground to the first platform, which may have resulted from the structure's stiffness changing. For upper building floors, the vibration reduction is minimal. Vibrations on the sixth floor were also amplified because of wave reflection from the roof.

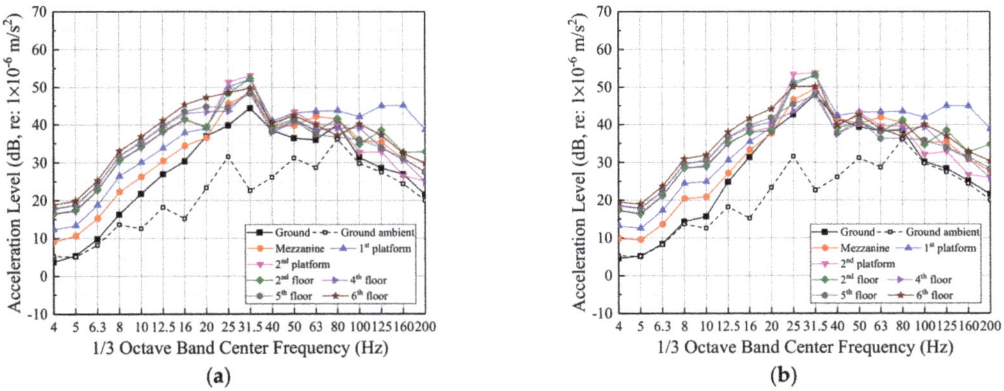

Figure 8. Vertical vibration transmission along the working column: (**a**) Track 1; (**b**) Track 2.

3.2.2. Staircase Column

Figure 9 displays the vertical vibration transmission along the staircase column. The averaged acceleration level from five passbys on Track 1 and five passbys on Track 2 were calculated and shown, respectively. The dominant vibration frequencies are also 8–80 Hz.

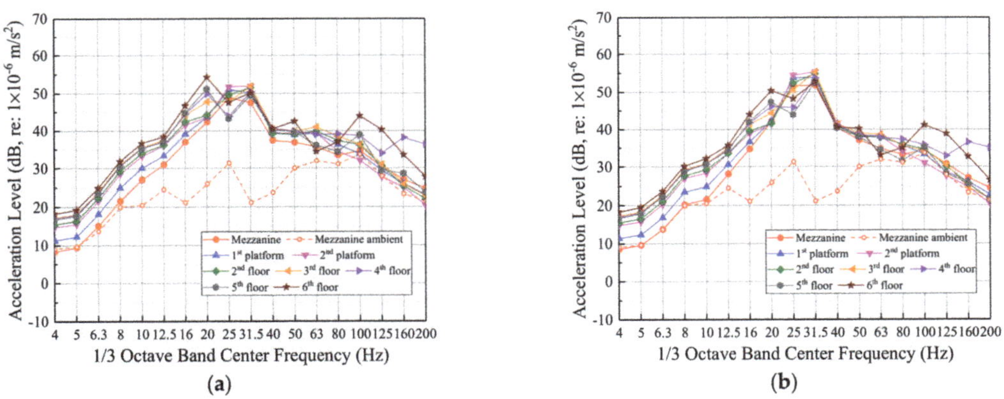

Figure 9. Vertical vibration transmission along the validation column: (**a**) Track 1; (**b**) Track 2.

Comparing Figures 8 and 9, the general vibration transmission rule for the two columns is the same. Little vibration reduction was shown as the floor level increased, and the vibration was amplified at the top floor.

3.3. Ground Vibration Levels at Different Distances from the Track

According to FTA guidelines [41], the propagation of a basic curve for rapid transit or light rail vehicles is:

$$L_v = 90.17 - 1.06 \log D - 2.32 \log D^2 - 0.87 \log D^3 \tag{3}$$

where D is the distance to the track centerline, measured in m; L_v is the velocity level, measured in dB. The speed adjustment factor is $20 \log \frac{speed}{reference\ speed}$; however, while the train speed used in this study is 15 km/h, the reference speed used by the FTA is 80 km/h. The FTA propagation curve with the added speed adjustment factor is shown in Figure 10b. Figure 10a shows the measured velocity levels at different distances from the Track 1 centerline. P1 was set on the ground floor within the building.

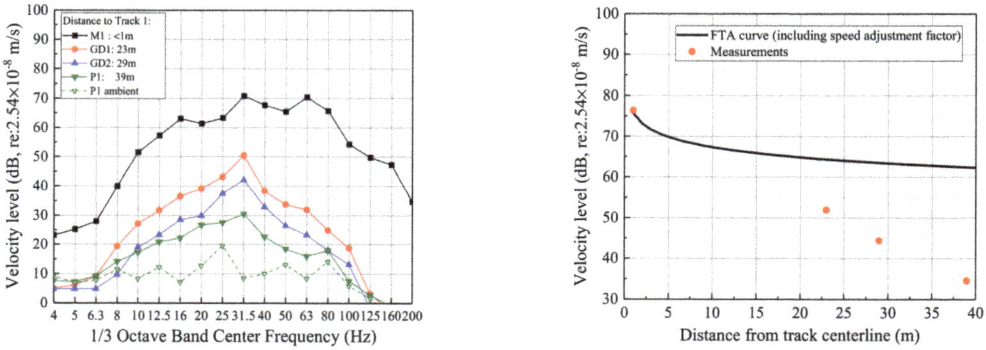

(a) (b)

Figure 10. Vibration variation with distance comparison between measurements and FTA: (**a**) Measured vibration variation with distance; (**b**) Comparison between measurements and FTA.

Although near-track measurements fit the FTA prediction well in Figure 10, far-field measurements are 12 dB, 19 dB, and 28 dB lower than the curve, respectively. The measurement at P1 is still 15 dB lower than the FTA curve even when adding an adjustment factor of -13 dB to account for coupling loss.

3.4. Different Vibration Transmissions from Ground to First Platfrom

Figure 11 shows the different vibration transmissions from the ground to the first platform between two setups. One is from M1 to M4, and the other is from P1 to P3. The distances between the Track 1 centerline and M1/P1 are 1 m and 39 m, respectively. P1 and P3 are under the first stage over-track building.

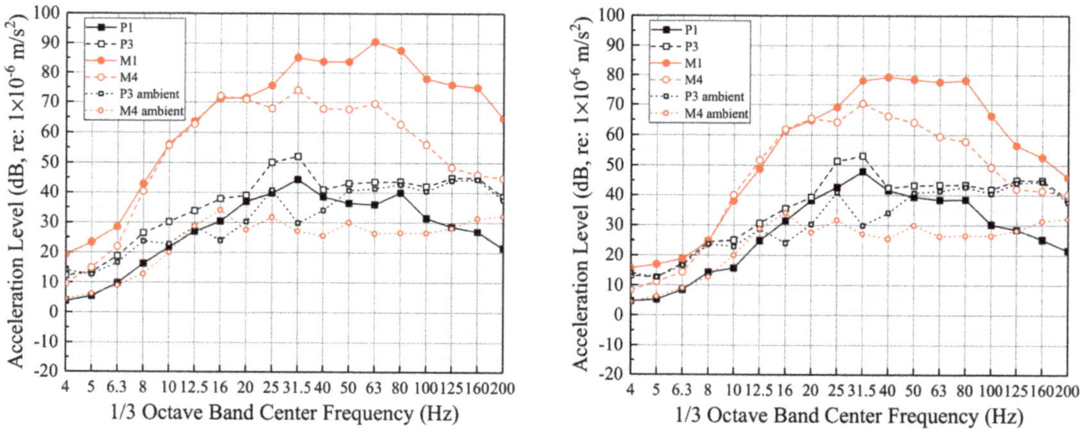

(a) (b)

Figure 11. Different vibration transmissions from ground to first platform: (**a**) Track 1; (**b**) Track 2.

From Figure 11, the dominant frequencies of train-induced vibration at different locations on the first platform are different. For M4, which is just above tracks, the vibration could transmit directly into the platform through the ground columns. Its dominant frequency range is 4–200 Hz. The vibration amplitudes for P3, which is further away from the tracks horizontally, reduced more than 15 dB, and the dominant frequencies narrowed to 10–50 Hz.

When vibrations are transmitted from the ground at M1 to the first platform at M4, the vibrations are reduced by 5–25 dB at 25–200 Hz. Vibration components below 20 Hz were barely reduced. However, for vibrations transmitting from P1 to P3, the vibration is amplified by 6–8 dB. The different transmission pattern is related to different structural configurations.

4. Train-Induced Vibration Simulations and Predictions

The cascaded state-space model has been developed and validated and functions as a case study for the first stage over-track building. In order to save space, only Track 2 passbys are discussed in this section.

4.1. Model Versatility for Typical Floors

For the first stage over-track building, the third and sixth floors are standardized design. These floors are typical floors with the same construction materials, spans, column/floor structure dimensions, etc. Figures 12 and 13 show the simulations and predictions of train-induced vibrations on the third floor over time and in a one-third octave band spectra, respectively.

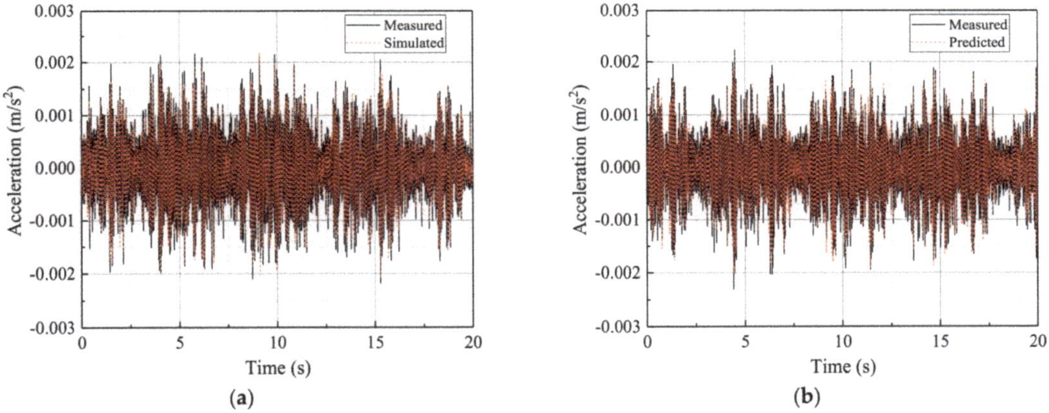

Figure 12. Third floor model: time comparison: (**a**) Working passby; (**b**) Validation passby.

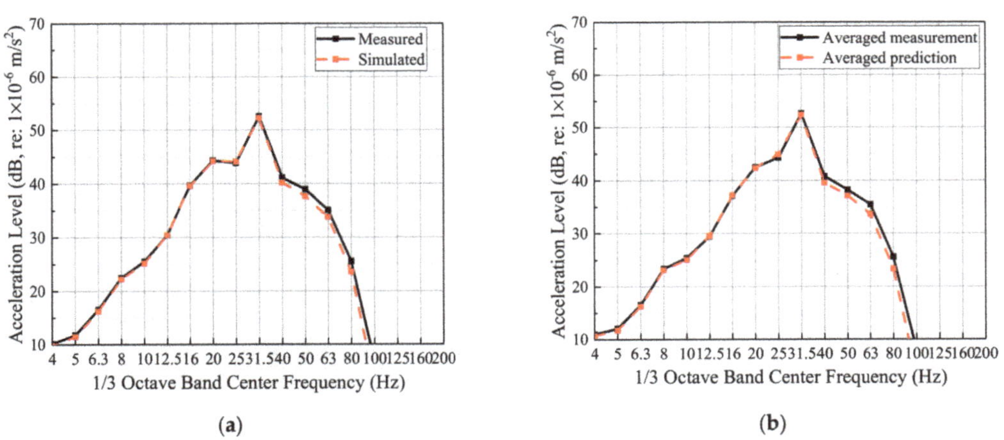

Figure 13. Third floor model: 1/3 octave band spectra comparison: (**a**) Working passby; (**b**) Validation passby.

As seen in Figures 8 and 9, the dominant frequency range of train-induced vibration lies in the 8–80 Hz range. Thus, all measurements use a band-pass filter of 8–80 Hz for this section. The system order of the third floor state-space model was selected to be 29. Although the state-space model is a time–domain method, Figure 13 demonstrates that the vibration's dominant frequency components are captured accurately, either for working passby simulation or for validation passby predictions.

In order to demonstrate the model's versatility for typical floors, the third floor state-space model was applied to the fourth and fifth floors, using measurements at V7 and V8 as inputs, respectively. Figure 14 shows the comparison between the averaged measurement and prediction. In Figure 14a, the third floor state-space model is also effective for the fourth floor. However, when applying the third floor model to the fifth floor, the prediction accuracy was reduced though still acceptable. This is because the sixth floor is the top floor, and vibrations at V9 will normally amplify [13], which is consistent with the data in Figure 14b: the measurement was amplified and greater than predicted.

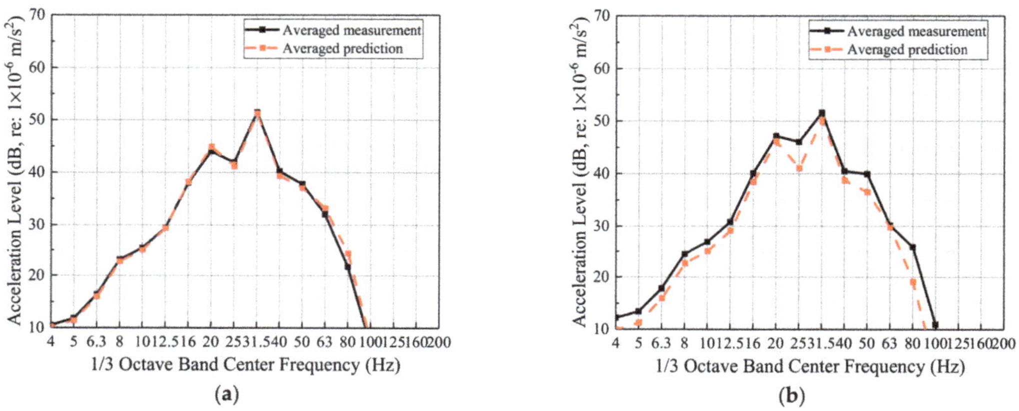

Figure 14. Demonstration of model versatility for typical floors: (**a**) Applying sixth floor model to seventh floor; (**b**) applying sixth floor model to eighth floor.

4.2. Cascaded State-Space Model Validation

According to the building drawings and analysis in Section 4.1, a cascaded state-space model can be generated by a series connecting two individual state-space models representing lower floors and typical floors, respectively. The predictions of one floor are used as inputs for the floor above.

The measurement pairs V4/V5 (first floor) and V6/V7 (third floor) are used to construct the lower floor and typical floor state-space models. Through comparison and selection, their system order was set at 26 and 29, respectively. Figure 15 shows the comparison of floor-to-floor vibration transmission between 8–80 Hz band-pass filtered measurements and cascaded state-space model predictions. This demonstrates that the cascaded state-space model is useful for predicting train-induced vibrations within a building.

4.3. Vibration Predictions for the Future Second Stage Over-Track Building

From measurements in this research and previous research [14], it was found that the vibration levels of the first and second platforms are comparable. Given this, the M4 measurements are used as the second platform input of the cascaded state-space model to predict vibration responses within the future second stage over-track building, assuming it is constructed similarly to the first stage over-track building. Figure 16 shows the floor-to-floor vibration transmission predicted by the cascaded state-space model. In order to compare the predictions with the Chinese standard limit [40] (a nighttime limit of 67 dB), the acceleration levels in Figure 16 were frequency weighted. As predicted, the train-induced maximum frequency-weighted acceleration level is quite close to the Chinese nighttime limit. Since vibrations in the center of the floor are higher than near the column bases [14], train-induced vibrations in the center of the floor of the future second stage over-track building have the potential to exceed the limit.

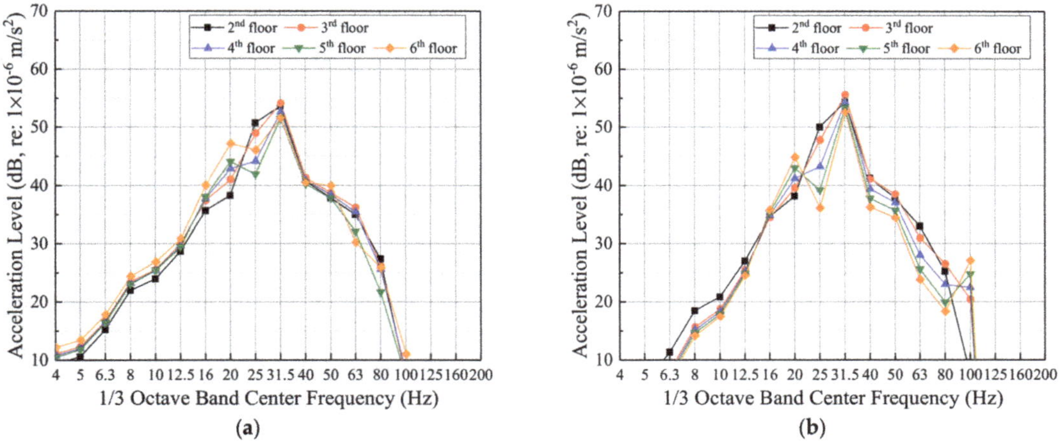

Figure 15. Comparison of filtered measurements and cascaded state-space model predictions: (**a**) 8–80 Hz band-pass filtered measurements; (**b**) cascaded state-space model predictions.

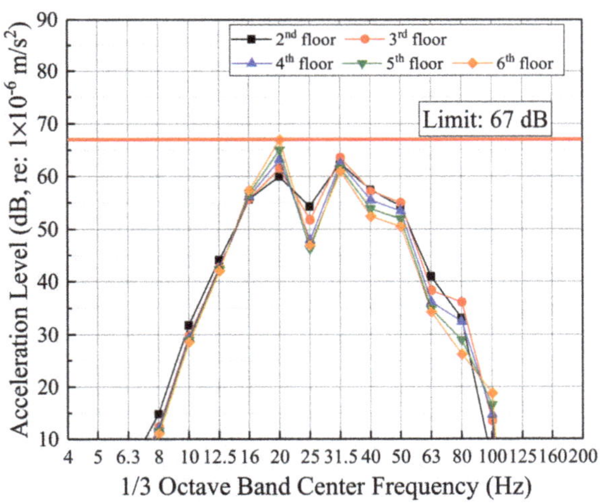

Figure 16. Comparison of over-track building predictions and Chinese standard.

5. Modal Parameters Identification

The natural frequencies can be extracted from the estimated state-space matrices, $[A]$ and $[C]$ of the system, using Equation (4):

$$\omega_i = \frac{1}{\Delta h}\sqrt{\ln \lambda_i \cdot \ln \lambda_i^*} \tag{4}$$

where λ_i are the eigenvalues of matrix $[A]$. λ_i^* represents the conjugate complex number of λ_i, and Δh is the sampling time interval. ω_i are the identified natural frequencies of the system measured in rad/s. This procedure is normally called modal parameter identification and uses a stabilization diagram for assistance.

In this section, the fourth floor room and staircase column's natural frequencies were identified to understand the dynamic characteristic differences caused by structural appurtenances such as staircases. The natural frequency stabilization diagrams for the fourth floor column segments are shown in Figure 17. The system order ranges from 0 to 65 with an increment of 5. Structural modes with damping ratios of more than 3% were discarded.

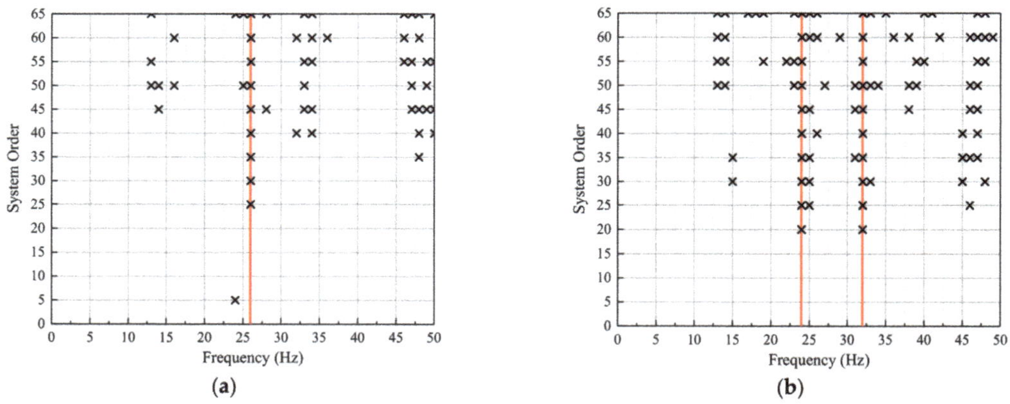

Figure 17. Natural frequency stabilization diagrams: (**a**) Room column; (**b**) Staircase column.

For the room column, modes consistently appear at 26 Hz after a system order of 25. For the staircase column, modes consistently appear at 24 Hz and 32 Hz after a system order of 20. This means the system order of a typical floor state-space model in Section 4 is reasonably set as 29. Comparing the staircase column to the room column, the staircase column has more resonance frequencies, which may make its vibration responses more complex.

6. Discussion

6.1. Vibration Transmission within the First Stage Over-Track Building

The dominant train-induced vibration frequencies in the first stage over-track building are 8–80 Hz. Vibrations transmitted along columns into the building amplify in floors under the first platform and the top floor of the building, which may be attributed to stiffness changes and wave reflections, respectively.

6.2. Comparison between Measurements and FTA Guidelines

Figure 18 shows the floor-to-floor variation of the overall velocity levels within the first stage over-track building. Vibration amplifications of 2–4 dB/floor are shown under the first platform. The vibration levels of the first and second platforms are comparable. Vibration level differences among building floors above the second platform (except the top floor) are small, especially for the staircase column measurements where vibration level differences are within 2 dB. Vibrations are amplified by around 2 dB on the top floor. The FTA guidelines for floor-to-floor changes in overall levels are −2 dB per floor for the first through fifth floors and −1 dB per floor for the fifth through tenth floors, which is significantly different from these measurements.

As seen in Figure 10b, comparing the measured ground vibration levels at different distances from the track centerline to the FTA propagation curve demonstrated that the FTA propagation curve effectively estimated vibration levels near the track but overestimated the vibration levels at 23–39 m away from the track's centerline. The FTA propagation curve underestimates the vibration transmission loss when the distance to the track's centerline increases.

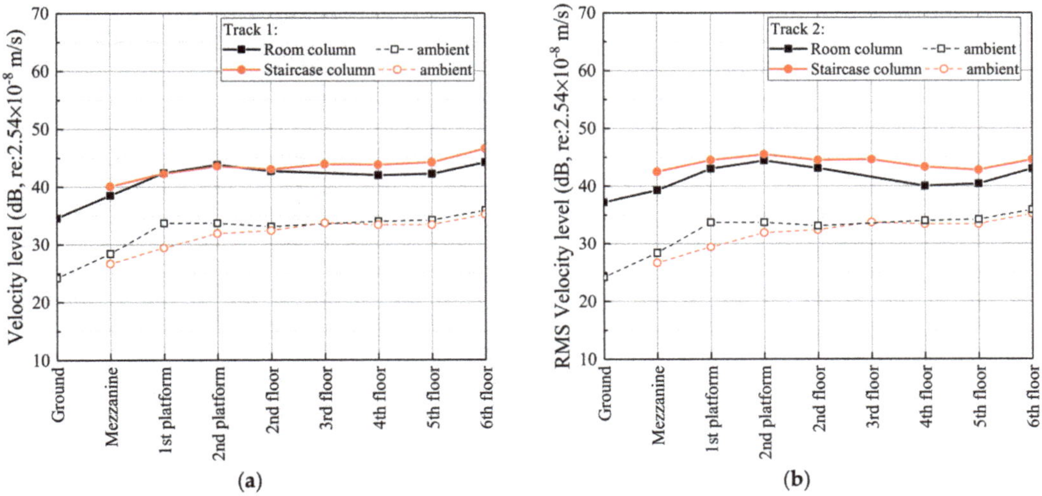

Figure 18. Floor-to-floor variation of overall velocity levels: (**a**) Track 1; (**b**) Track 2.

Measurements in this research enriched the database regarding train-induced vibrations at metro depots with over-track buildings. Train-induced vibration predictions based on measurements in this study are more accurate than the FTA's empirical guidelines, which, therefore, offers a significant reference for similar research at metro depots.

6.3. Selection of State-Space Model System Order

The system order is the only parameter when estimating a state-space model from input/output measurements. Selecting a proper system order requires plenty of trials. In general, the more complex the system, the greater the system order. The natural frequency stabilization diagram is a helpful tool if the selected system order is stabilized.

6.4. Significance of the Cascaded State-Space Model

Over-track buildings above a metro depot can be classified into two types according to their construction period and the relationship between their footprints and the tracks. The first stage over-track building is further away from the tracks and usually built synchronously with the platform structure. The second stage over-track building is just above the tracks and usually requires transfer girders at the platform level to transfer building loads.

A cascaded state-space model for the first stage over-track building was developed and validated using measured vibrations from the second platform as input. When applying this model to the future second stage over-track building, it is assumed the second stage building is constructed similarly to the first stage building. Using the measured vibrations of the platform as an input, the vibration predictions for the future second stage building above the input level can be obtained and assessed. Predicting vibrations before the construction of over-track buildings is of economic significance to determine whether vibration mitigation measures should be designed and considered in advance. The measurements and predictions in this paper can be referred to when developing future metro depots with over-track buildings made of similar concrete-framed structures. The results enriched the database for train-induced vibrations in concrete-framed over-track buildings at metro depots. Future applications and validations of over-track buildings of other structural types, such as shear wall supported buildings, need further field measurements.

7. Conclusions

This paper proposed a data-driven cascaded state-space model for predicting train-induced vibrations within over-track buildings at metro depts. The method has been successfully applied and validated based on field measurements from the Luogang metro depot in Guangzhou, China. The insights from this study will be useful for understanding vibration transmission within a first stage over-track building with concrete-frame columns:

(1) The data-driven cascaded state-space model predicts train-induced structural vibration responses with reasonable accuracy. It provides a practical method to assess train-induced vibration impacts prior to construction when designing similar buildings at metro depots in the future.
(2) The system order of the estimated state-space model is related to the structural system complexity. In general, the more complex the system, the greater the system order.
(3) Considering over-track buildings' different supporting methods, it is advisable to use the measured vibration levels from the second platform as the inputs to the cascaded state-space model, which avoids the added complexity of modeling the transfer behavior of the platform and expands the applicability of the model.
(4) Vibration levels within the first stage over-track building were amplified by 2–4 dB/floor from the first platform and barely reduced from one floor to the floor above. The FTA guidelines overestimate the vibration transmission loss within buildings.
(5) The FTA propagation curve for rapid transit and light rail vehicles effectively estimates vibration levels near the track. However, it underestimates the vibration transmission loss when the distance to the track's centerline increases.

Author Contributions: Conceptualization, Z.T. and C.Z.; methodology, Z.T.; software, Z.H.; validation, Z.T.; formal analysis, G.W.; investigation, C.H.; resources, Z.Y.; data curation, Z.H.; writing—original draft preparation, Z.T.; writing—review and editing, C.Z.; visualization, Z.T.; supervision, C.Z.; project administration, C.Z.; funding acquisition, C.Z. All authors have read and agreed to the published version of the manuscript.

Funding: This research was funded by the National Natural Science Foundation of China, grant number 51908139, and the Guangdong Basic and Applied Basic Research Foundation, grant number 2021A1515012605. The APC was funded by the National Natural Science Foundation of China.

Institutional Review Board Statement: Not applicable.

Data Availability Statement: New data were created or analyzed in this research. Data will be shared upon request and consideration of the authors.

Conflicts of Interest: The authors declare no conflict of interest.

References

1. Lin, D.; Nelson, J.D.; Beecroft, M.; Cui, J. An overview of recent developments in China's metro systems. *Tunn. Undergr. Space Technol.* **2021**, *111*, 103783. [CrossRef]
2. Tao, Z.Y.; Zou, C.; Wang, Y.M.; Wu, J. Vibration transmissions and predictions within low-rise buildings above throat area in the metro depot. *J. Vib. Control* **2021**, 10775463211057644. [CrossRef]
3. Tao, Z.Y.; Wang, Y.M.; Zou, C.; Li, Q.; Luo, Y. Assessment of ventilation noise impact from metro depot with over-track platform structure on workers and nearby inhabitants. *Environ. Sci. Pollut. Res.* **2019**, *26*, 9203–9218. [CrossRef]
4. Connolly, D.P.; Marecki, G.P.; Kouroussis, G.; Thalassinakis, I.; Woodward, P.K. The growth of railway ground vibration problems—A review. *Sci. Total Environ.* **2016**, *568*, 1276–1282. [CrossRef] [PubMed]
5. Ma, M.; Jiang, B.; Gao, J.; Liu, W. Experimental study on attenuation zone of soil-periodic piles system. *Earthq. Eng. Struct. Dyn.* **2019**, *126*, 105738. [CrossRef]
6. Kouroussis, G.; Conti, C.; Verlinden, O. Experimental study of ground vibrations induced by Brussels IC/IR trains in their neighborhood. *Mech. Ind.* **2013**, *14*, 99–105. [CrossRef]
7. Connolly, D.P.; Kouroussis, G.; Woodward, P.K.; Costa, P.A.; Verlinden, O.; Forde, M.C. Field testing and analysis of high speed rail vibrations. *Soil Dyn. Earthq. Eng.* **2014**, *67*, 102–118. [CrossRef]
8. Connolly, D.P.; Costa, P.A.; Kouroussis, G.; Galvin, P.; Woodward, P.K.; Laghrouche, O. Large scale international testing of railway ground vibrations across Europe. *Soil Dyn. Earthq. Eng.* **2015**, *71*, 1–12. [CrossRef]

9. Xia, H.; Chen, J.; Wei, P.; Xia, C.; De Roeck, G.; Degrande, G. Experimental investigation of railway train-induced vibrations of surrounding ground and a nearby multi-story building. *Earthq. Eng. Eng. Vib.* **2009**, *8*, 137–148. [CrossRef]
10. Sanayei, M.; Maurya, P.; Moore, J.A. Measurement of building foundation and ground-borne vibrations due to surface trains and subways. *Eng. Struct.* **2013**, *53*, 102–111. [CrossRef]
11. Sanayei, M.; Moore, J.A.; Brett, C.R. Measurement and prediction of train-induced vibrations in a full-scale building. *Eng. Struct.* **2014**, *77*, 119–128. [CrossRef]
12. Zou, C.; Wang, Y.M.; Wang, P.; Guo, J.X. Measurement of ground and nearby building vibration and noise induced by trains in a metro depot. *Sci. Total Environ.* **2015**, *536*, 761–773. [CrossRef] [PubMed]
13. Zou, C.; Wang, Y.M.; Moore, J.A.; Sanayei, M. Train-induced field vibration measurements of ground and over-track buildings. *Sci. Total Environ.* **2017**, *575*, 1339–1351. [CrossRef]
14. Tao, Z.Y.; Wang, Y.M.; Sanayei, M.; Moore, J.A.; Zou, C. Experimental study of train-induced vibration in over-track buildings in a metro depot. *Eng. Struct.* **2019**, *198*, 109473. [CrossRef]
15. Cao, Z.L.; Guo, T.; Zhang, Z.Q.; Li, A.Q. Measurement and analysis of vibrations in a residential building constructed on an elevated metro depot. *Measurement* **2018**, *125*, 394–405. [CrossRef]
16. He, W.; Zou, C.; Pang, Y.; Wang, X. Environmental noise and vibration characteristics of rubber-spring floating slab track. *Environ. Sci. Pollut. Res.* **2021**, *28*, 13671–13689. [CrossRef]
17. Verbraken, H.; Lombaert, G.; Degrande, G. Verification of an empirical prediction method for railway induced vibrations by means of numerical simulations. *J. Sound Vib.* **2011**, *330*, 1692–1703. [CrossRef]
18. Zhai, W.M.; Wang, K.Y.; Cai, C.B. Fundamentals of vehicle–track coupled dynamics. *Veh. Syst. Dyn.* **2009**, *47*, 1349–1376. [CrossRef]
19. Yang, J.J.; Zhu, S.Y.; Zhai, W.M.; Kouroussis, G.; Wang, Y.; Wang, K.Y.; Lan, K.; Xu, F.Z. Prediction and mitigation of train-induced vibrations of large-scale building constructed on subway tunnel. *Sci. Total Environ.* **2019**, *668*, 485–499. [CrossRef]
20. Jin, H.; Tian, Q.; Li, Z.; Wang, Z. Ability of vibration control using rubberized concrete for tunnel invert-filling. *Constr. Build. Mater.* **2022**, *317*, 125932. [CrossRef]
21. Lopes, P.; Costa, P.A.; Ferraz, M.; Calçada, R.; Cardoso, A.S. Numerical modeling of vibrations induced by railway traffic in tunnels: From the source to the nearby buildings. *Soil Dyn. Earthq. Eng.* **2014**, *61–62*, 269–285. [CrossRef]
22. Xia, H.; Zhang, N.; Cao, Y.M. Experimental study of train-induced vibrations of environments and buildings. *J. Sound Vib.* **2005**, *280*, 1017–1029. [CrossRef]
23. Kuo, K.A.; Papadopoulos, M.; Lombaert, G.; Degrande, G. The coupling loss of a building subject to railway induced vibrations: Numerical modelling and experimental measurements. *J. Sound Vib.* **2019**, *442*, 459–481. [CrossRef]
24. Xu, L.; Ma, M.; Cao, R.; Tan, X.; Liang, R. Effect of longitudinally varying characteristics of soil on metro train-induced ground vibrations based on wave propagation analysis. *Soil Dyn. Earthq. Eng.* **2022**, *152*, 107020. [CrossRef]
25. François, S.; Schevenels, M.; Galvín, P.; Lombaert, G.; Degrande, G. A 2.5 D coupled FE–BE methodology for the dynamic interaction between longitudinally invariant structures and a layered halfspace. *Comput. Methods Appl. Mech. Eng.* **2010**, *199*, 1536–1548. [CrossRef]
26. Liang, X.; Yang, Y.B.; Ge, P.; Hung, H.H.; Wu, Y. On computation of soil vibrations due to moving train loads by 2.5 D approach. *Soil Dyn. Earthq. Eng.* **2017**, *101*, 204–208. [CrossRef]
27. Katou, M.; Matsuoka, T.; Yoshioka, O.; Sanada, Y.; Miyoshi, T. Numerical simulation study of ground vibrations using forces from wheels of a running high-speed train. *J. Sound Vib.* **2008**, *318*, 830–849. [CrossRef]
28. Sanayei, M.; Zhao, N.; Maurya, P.; Moore, J.A.; Zapfe, J.A.; Hines, E.M. Prediction and mitigation of building floor vibrations using a blocking floor. *J. Struct. Eng.* **2012**, *138*, 1181–1192. [CrossRef]
29. Zou, C.; Moore, J.A.; Sanayei, M.; Wang, Y.M.; Tao, Z.Y. Efficient impedance model for the estimation of train-induced vibrations in over-track buildings. *J. Vib. Control* **2020**, *27*, 924–942. [CrossRef]
30. Zou, C.; Moore, J.A.; Sanayei, M.; Wang, Y.M. Impedance model for estimating train-induced building vibrations. *Eng. Struct.* **2018**, *172*, 739–750. [CrossRef]
31. Richart, F.E.; Hall, J.R.; Wood, R.D. *Vibrations of Soils and Foundations*; Prentice Hall: Englewood Cliffs, NJ, USA, 1970.
32. Kim, J.; Lynch, J.P. Subspace system identification of support excited structures—Part I: Theory and black-box system identification. *Earthq. Eng. Struct. Dyn.* **2012**, *41*, 2235–2251. [CrossRef]
33. Verhaegen, M. Identification of the deterministic part of MIMO state-space models given in innovations form from input-output data. *Automatica* **1994**, *30*, 61–74. [CrossRef]
34. Viberg, M. Subspace-based methods for the identification of linear time-invariant systems. *Automatica* **1995**, *31*, 1835–1851. [CrossRef]
35. Ho, B.L.; Kalman, R.E. Effective construction of linear state-variable models from input-output functions. *Regelungtechnik* **1965**, *12*, 545–548.
36. Peeters, B.; Ventura, C.E. Comparative study of modal analysis techniques for bridge dynamic characteristics. *Mech. Syst. Signal Process.* **2003**, *17*, 965–988. [CrossRef]
37. Van Overschee, P.; De Moor, B. N4SID: Subspace algorithms for the identification of combined deterministic-stochastic systems. *Automatica* **1994**, *30*, 75–93. [CrossRef]

38. Van Overschee, P.; De Moor, B. Subspace algorithm for the stochastic identification problem. *Automatica* **1993**, *29*, 649–660. [CrossRef]
39. Viberg, M.; Ottersten, B.; Wahlberg, B.; Ljung, L. Performance of Subspace-Based System Identification Methods. In Proceedings of the 12th Federation of Automatic Control Symposium System Identification (IFAC SYSID), Sydney, Australia, 18–23 July 1993.
40. *JGJ/T 170-2009*; Standard for Limit and Measuring Method of Building Vibration and Secondary Noise Caused by Urban Rail Transit. Ministry of Housing and Urban-Rural Development of the People's Republic of China: Beijing, China, 2009.
41. Federal Transit Administration. *Transit Noise and Vibration Impact Assessment Manual*; U.S. Department of Transportation: Washington, DC, USA, 2018.

buildings

MDPI

Article

A Case Study on Structural Serviceability Subjected to Railway-Induced Vibrations at TOD Developed Metro Depot

Yingying Liao [1,2], Peijie Zhang [2], Qiong Wu [3] and Hougui Zhang [1,3,*]

1 State Key Laboratory of Mechanical Behavior and System Safety of Traffic Engineering Structures, Shijiazhuang Tiedao University, Shijiazhuang 050043, China; sjzlyy820@163.com
2 School of Civil Engineering, Shijiazhuang Tiedao University, Shijiazhuang 050043, China; 1202001104@student.stdu.edu.cn
3 Institute of Urban Safety and Environment Science, Beijing Academy of Science and Technology, Beijing 100054, China; 18600053885@163.com
* Correspondence: zhanghougui@bmilp.com

Abstract: As a sustainable mode of metro-development strategy, transit-oriented development (TOD) is rapidly growing to finance the transport infrastructure investment. The main negative consequence of constructing residential buildings directly over metro depots is railway-induced vibration, that may affect structural serviceability. The residents may feel uncomfortable, as the metro trains start running very early in the morning and finish daily operations very late at night. In order to evaluate the level of human comfort subject to the special situation, a case study was provided in this paper. Directed by the academic review, there were four common comfort evaluation methods, with difference indexes to describe the influence of vibrations. Therefore, a measurement campaign was conducted and both acceleration and velocity sensors were simultaneously installed at the same measurement points, to reduce the influence of the conversion accuracy. The results show that there are certain differences between the evaluation methods in assessing the vibration comfort, but considering the most adverse effects together, the over-track building at this particular TOD-developed depot can ensure that 90% of the occupants would not be highly annoyed by the vibrations. The main negative effect on human comfort at the TOD depot is that the high-level vibrations would cause interruptions in sleep. Among them, the vibrations in this case would affect the rest of 17% of the occupants in the bedrooms on the seventh floor, and make it difficult for 9% of the occupants to fall asleep. Therefore, the evaluation index was suggested to consider more factors related to sleep difficulties and awake threshold values.

Keywords: structural serviceability; human comfort; metro railway vibrations; TOD development; sleep disturbance

Citation: Liao, Y.; Zhang, P.; Wu, Q.; Zhang, H. A Case Study on Structural Serviceability Subjected to Railway-Induced Vibrations at TOD Developed Metro Depot. *Buildings* **2022**, *12*, 1070. https://doi.org/10.3390/buildings12081070

Academic Editors: Jun Chen and Haoqi Wang

Received: 28 June 2022
Accepted: 19 July 2022
Published: 22 July 2022

1. Introduction

The development of urban rail transit systems affects, to a certain extent, the development level of cities. According to the information released by the China Urban Rail Transit Association [1], by the end of 2021, an urban rail transit system had been built and put into operation in a total of 50 cities in China, with a total length of 9192.62 km, including 7253.73 km of subways, accounting for 78.9% of the total rail transit system length. A metro depot is a basic ancillary facility of a metro system, that is used for the storage, cleaning, maintenance, and performance test of subway trains, and which usually covers a large land area [2]. With the large-scale construction and development of a subway system, a metro depot with a low building density and large floor space has not been an economical use of urban land [3]. In recent years, under the guidance of the transit-oriented development (TOD) model [4,5], many cities have started to develop over-track buildings at metro depots, which not only improved the urban land utilization, but also compensated for the deficits of the construction and operation of a metro.

The emergence of the over-track buildings has allowed the traditional roof of the metro depot to be replaced by a large reinforced concrete platform. The platform is supported by columns in the middle of the train tracks and divided horizontally, according to construction joints, to accommodate thermal expansion and create a modular aseismic structure [3]. The buildings located on these platforms are designed as multistoried or high-rise buildings and are intended for residential and commercial activities, such as residences, schools, stores, and restaurants. Due to the comprehensive advantages of high land-use efficiency, the short walking distance to subway stations, and high return on investment, these buildings are increasingly favored by designers and investors [6]. The vibration characteristics of a metro depot are related to many factors, such as speed, track type, sleeper, ballast, roadbed, building foundation, and structure [7,8]. However, the vibrations of the train power system and track structure and the dynamic interaction of the wheel-rail and the wheel-rail's unevenness are the main sources of vibrations in a track structure [9]. As a result of slow running, the vibrations generated by trains have been reduced, providing the possibility of building the over-track buildings above the metro depots. Unlike ordinary railroads, subways operate intensively at night and in the early morning. However, during the operation of a metro depot, the vibrations are transmitted without soil attenuation, thus having a large influence on the over-track buildings. The vibrations caused by extreme events, such as earthquakes, can generate large amplitudes and damage the safety of buildings; in contrast, train-induced building vibrations bring smaller amplitudes and do not affect building safety, but can cause discomfort to the occupants [10]. In recent studies, field measurements of the effects of train-induced vibrations caused by train operation on over-track buildings have been conducted. The results have shown that train-induced vibrations can be transmitted directly through the columns to the platform, and subsequently to the over-track buildings, thus possibly causing annoyance to the occupants [3,11–13].

Railway vibrations will have an impact on the comfort of the human body, and the subject has gradually attracted people's attention in recent years. As a result, the assessment of railway vibrations has become more common. [14]. The Chinese-issued vibration standards have often been used by developers as a control target. However, many of the projects that comply with vibration standards are still receiving a large number of complaints from occupants. Moreover, many foreign studies have shown that the occupants of buildings disturbed by vibrations cannot live freely, although the vibrations do not exceed the vibration standards [15]. The reason for such a situation is that the quality of living is not related to vibration indicators, but it is directly related to the comfort of the occupants. In China, research on the effect of vibrations on human comfort has been insufficient; also, there have been fewer application cases of the results of the vibration comfort research to the over-track building design in foreign countries. Aiming to improve the living quality and reduce the complaints of occupants, this paper focuses on the main factors of discomfort caused by the structural vibrations of the over-track buildings and provides a case reference for studying the effect of vibrations on human comfort in the over-track buildings.

The vibrations and noise generated by a railroad during its movement can affect the comfort of the occupants in the surrounding buildings, and this effect must be considered in the development of new lines or the reconstruction of the existing lines. Compared to noise, vibrations are often overlooked. However, due to an increase in public awareness and the success of noise mitigation measures, vibrations have become an increasingly important issue [16]. Human responses to railroad-induced vibrations include sleep disturbance, annoyance, and non-vibration factors.

In recent years, many studies on the effects of vibrations on human comfort were conducted, and significant results were achieved. The methods used to study the effects of vibrations on sleep include both objective and subjective measurements of sleep disturbance. The objective measurements of sleep disorders were mainly conducted by polysomnography (PSG), while subjective measurements of sleep disorders were usually performed through questionnaires. Research has shown that vibrations can adversely affect the sleep quality of occupants. Arnsberg et al. [17] simulated the vibrations of heavy traffic and found that the vibrations can cause changes in sleep architecture and a reduction in rapid eye movement sleep. It has also been found that increased vibrations can increase the probability of waking up during the night and early morning [18], that is, it was demonstrated that the occupants can distinguish between train-induced vibrations and noise, and as the vibration amplitude increased, occupants' heart rate amplitudes and sleep disturbances increased, and sleep quality decreased [19]. It has been known that the vibrations caused by freight trains can increase the heart rate of people who are sleeping and may affect the cardiovascular function of occupants near the railroad [20]. The number of trains passing through a metro depot and the amplitude of induced vibrations have a negative effect on the sleep macrostructure, that is, a large number of trains and high vibration conditions increase the occurrence of sleep depth changes in the occupants, interrupt the continuity of slow-wave sleep, and increase the number of night-time awakenings [21]. The effects of traffic-induced vibrations on sleep are summarized in Table 1.

Table 1. Summary of the effects of traffic-induced vibrations on sleep [16].

Effect		Significant Findings
Biological changes	Change in cardiovascular activity	Increase in heart rate [19,20]
	Change in sleep structure	Reduction in REM sleep [17] Greater number of sleep stage shifts [22] Shorter period between falling asleep and first awakening [22] Shorter maximum length of uninterrupted time spent in slow wave sleep [22]
	EEG awakening	Increase in probability of EEG awakening [22]
Sleep quality	Waking in the night/too early	Increase of reported awakenings/waking too early (Figure E.2 in [18])
	Difficulty in getting back to sleep	Greater difficulty in getting back to sleep once awoken for higher amplitudes of vibration [22]
	Self-reported sleep disturbance from vibrations	Increase in proportion of people reporting sleep disturbances (Figure E.2 in [18] and Figure E.3 in [23]) Self-reported sleep disturbances related to vibration amplitude [22] Decrease in self-reported sleep quality [22]
	Self-reported sleep disturbances from noise	Vibration related to increase in proportion of people reporting sleep disturbances from noise [24]
	Decreased restoration	Decrease in self-reported restoration [22]

Guski et al. [25] identified that annoyance is associated with disturbance, aggravation, dissatisfaction, concern, bother, displeasure, harassment, irritation, nuisance, vexation, exasperation, discomfort, uneasiness, distress, and hate. According to the EU FP7 project, CargoVibes, annoyance is a concept that has been widely used to evaluate the negative effect of environmental stressors on a population. It is a broad concept that describes the negative effects of vibrations on the environment from three aspects: activity disturbance, emotional responses, and attitudinal responses to the source of the annoyance. The effect of vibrations on people's annoyance is usually examined through questionnaires and field tests. It should be noted that the subjective responses of people to vibrations are significantly influenced by individual differences; therefore, the reference significance of individual responses to vibrations is not high, and only the proportion of people's responses to vibrations obtained by statistical laws on a large number of samples is valuable for studying the relationship between the annoyance rate and the vibration intensity. The EU FP7 project, CargoVibes, collected data from social vibration surveys conducted in seven countries. The collected data included 4490 samples. By analyzing these data, the curves of people's annoyance caused by railroad vibrations were plotted. These experimental data will be used in this study to evaluate the annoyance rate due to train-induced vibrations.

In addition, non-exposure factors cause vibrations to have an impact on human comfort. Table 2 summarizes the existing studies on the effect of non-exposure factors on people's annoyance. For the over-track buildings, the vibrations are generated at the arrival and departure times of metro trains, which are mostly arranged in the evening and early morning hours. Since occupants can commonly see a metro depot through the windows, the vibrations' source can be considered visible. The presence of the non-exposure factors will further increase the impact of vibrations on the occupants' comfort.

Table 2. Summary of the effects of the situation, attitudinal, and socio-demographic factors on annoyance [16].

Factor		Significant Findings
Time of day	Evening	Annoyance greater during the evening than during the day at the same level of vibration exposure [26]
	Night	Annoyance is greater during the night than during the day and evening at the same level of vibration exposure [26]
Situational	Situational	Annoyance greater if the source is visible [27,28]
	Time spent at home	Annoyance greater for people who spend less than 10 h per day at home [27]
	Type of area	Annoyance greater for people living in rural areas [27]
Attitudinal	Concern of damage	Annoyance greater for those concerned that vibrations are damaging their property or belongings [26,28]
	Expectation regarding future vibrations	Annoyance greater for those expecting vibrations to get worse in the future [27]
	Necessity of source	Annoyance greater for those considering the source unnecessary [28]
	Noise sensitivity	Annoyance from vibrations greater for those considering themselves as noise sensitive [28]
Sociodemographic	Age	Annoyance greater for those in the middle age group in [26], no significant effect in [28]

In studies on vibration comfort, it is usually necessary to collect a large amount of actual measurement data and conduct a large number of questionnaire surveys, which is time-consuming and costly. Recently, some countries and institutions have fitted vibration–response curves and used them in studies on railroad line-induced vibrations. However, there have been fewer application cases of the research results for studying the vibration effect on over-track buildings.

This study provides the field test data of the over-track buildings in China and uses the exposure–response curve to analyze the vibrations impact on occupants' comfort in over-track buildings. The results presented in this study can help to develop strategies for providing better occupant comfort under train-induced vibrations and possible vibration reduction measures, which can help to improve the quality of living.

2. Description of Metro Depot and Over-Track Building and Vibration Measurement

The over-track buildings considered in this study are located above the operation depot of a metro depot. The over-track buildings consist of ten 11-storey residential buildings with a kindergarten and a shopping center. The number of available households in the over-track buildings is 613, and the total construction area is 111,311.31 m^2 with a frame structure system.

The plan view of the metro depot is shown in Figure 1. The north side of the metro depot is the maintenance depot, responsible for the daily maintenance of trains. On the south side of the metro depot is the testing line for the high-speed testing and performance evaluation of trains to ensure safe operation. The throat area is located on the west side of the operation depot, connecting the train entry and exit lines. The over-track building under the test is located above the operating depot, and the test tracks are 14–18. Table 3 shows the train and track parameters in the operating metro depot. We measured the combined stiffness of the type I separate fastener system (see Figure 2). The average measurement results show that the combined static stiffness of the fastener is 40.35 KN/mm, and the calculated combined dynamic stiffness is 38.9~43.7 KN/mm.

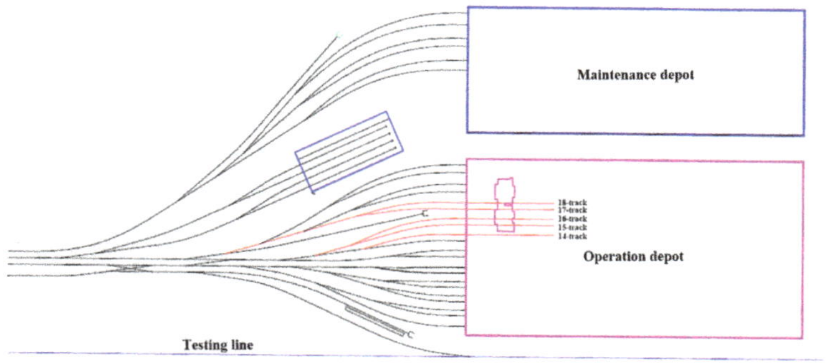

Figure 1. The plan view of the presented depot.

Table 3. Train and track parameters in the operating metro depot.

Rail Types	Rail Weight	Train Speed	Fastener
Long sleeper embedded ballast less track	60 kg/m	The warehouse door-10 km/h The middle of the operating depot-5 km/h	Type I separate fastener

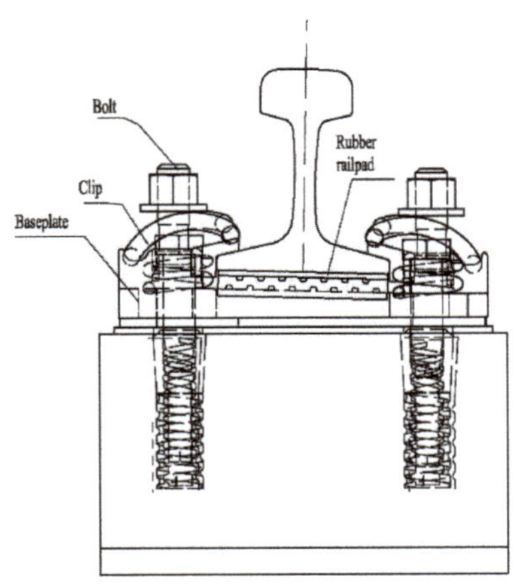

Figure 2. Type I separate fastener.

Since the speed of the train is the highest at the entrance/exit, the train will generate more vibrations in the building near the entrance/exit when the train is running, which will bring higher annoyance to the occupants. Therefore, the 11-story building in Figure 1 was selected as the test. Figure 3 shows the sectional view of the over-track building under the test, which has 11 floors; the test floors included the third, fifth, seventh, ninth, and 11th floors. The test floors are all affected by the vibrations of trains on tracks 14–18 under the platform.

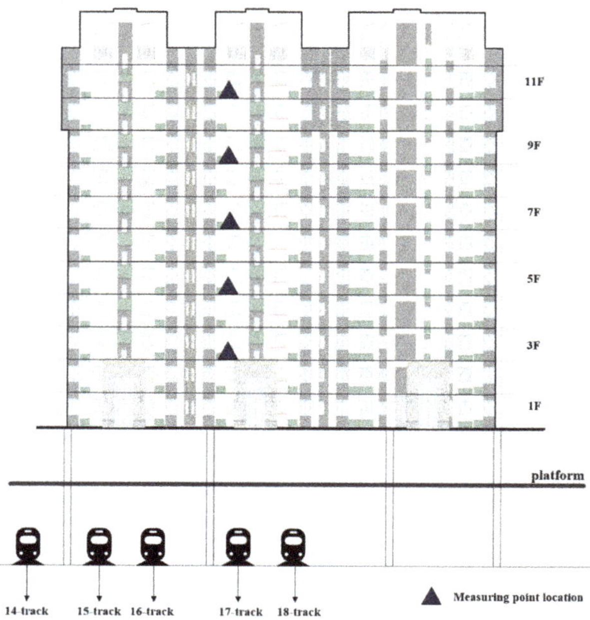

Figure 3. The sectional view of the over-track building.

To consider the vibration comfort in the rooms with different usage functions, field experiments were conducted to measure the amplitude in the drawing room and bedroom of each test floor, as shown in Figure 4. When a train passes, the vertical vibrations in a building are significantly greater than the horizontal vibrations [11]. Considering that the impact of vibration intensity and human comfort are directly negatively correlated, the vibration measurement and analysis considered only the vertical vibrations in this study. The test conditions were divided into two groups of working conditions: normal operation and scheduled shunting. The normal operation conditions included four peak-hour periods: 10:30–11:30 p.m.; 11:30–12:30 p.m.; 4:30–5:30 a.m.; and 5:30–6:30 a.m. Scheduled shunting was performed during the daytime to allow trains to pass through the test tracks as scheduled. Figure 5 shows the operational depot used in the field test.

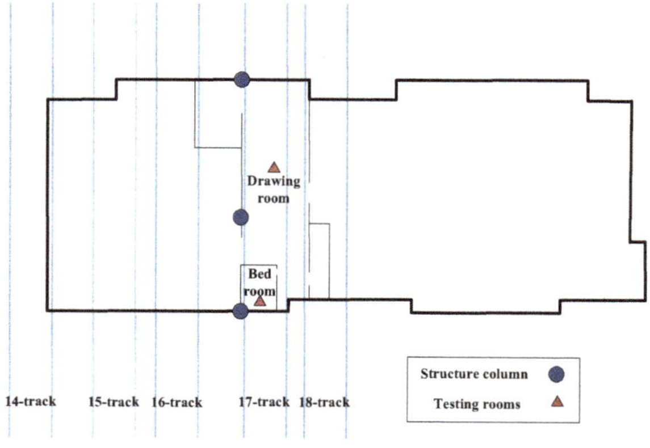

Figure 4. Measuring point location plan.

(**a**) (**b**)

Figure 5. Field testing in the operating metro depot. (**a**) Test train; (**b**) Test tracks.

The test train in this research is a B-type metro train with six cars and has a length of 118 m. In the case of no-load, the mass of the train is 202 t, and the axle load is less than 14 t. Figure 6 shows the instruments used in the measurement, which included the INV3062C1 data acquisition and the signal processing systems (China Orient Institute of Noise and Vibration, Beijing, China) used to collect the data on eight channels simultaneously. Since the acceleration and velocity were used as evaluation indexes of the vibration effect on human comfort, accelerometers and velocimeters were used for the measurements. The B&K8344 accelerometer with a sensitivity of 5 mV/g and the pickup 941B were installed at the target location. The pickup 941B contained the velocity and acceleration gears, and the velocity gear was used in the test. The B&K8344 accelerometer, the pickup 941B, and the acquisition system were calibrated before the test. A sampling rate of 2048 Hz at Nyquist frequency provided a meaningful level of spectrum below 1024 Hz. This sampling rate provided a large enough range to include the dominant frequencies for analysis.

(**a**) (**b**) (**c**)

Figure 6. (**a**) INV3062C1 data acquisition; (**b**) Pickup 914B; (**c**) B&K8344 accelerometer.

3. Vibration Response Result Analysis

The collected data were analyzed, using the DASP analysis system and MATLAB software, to obtain the vibration response in the considered area.

3.1. Track Effect on Vibration Response

We selected the third-floor drawing room for analysis. The vibration response of the measured area of the building was evaluated when the train passed over the different tracks. As mentioned above, the test tracks included five tracks, 14–18.

The vibration responses of the accelerometer and velocity sensor in the time domain for the selected drawing room for a train passing over track 16 are shown in Figure 7. As shown in Figure 7, in the time–domain diagram, there was a clear spindle-shaped waveform at both ends, because the unified track was divided into two sections with a clear speed limit between them, that is, the speed was at first fast and then slow on the entry of track 16, but it was at first slow and then fast on the exit of track 16. The velocity and acceleration of the test building exhibited the same trend of vibration variation in the time domain. When the train speed is higher, the vibration generated in the building is also higher.

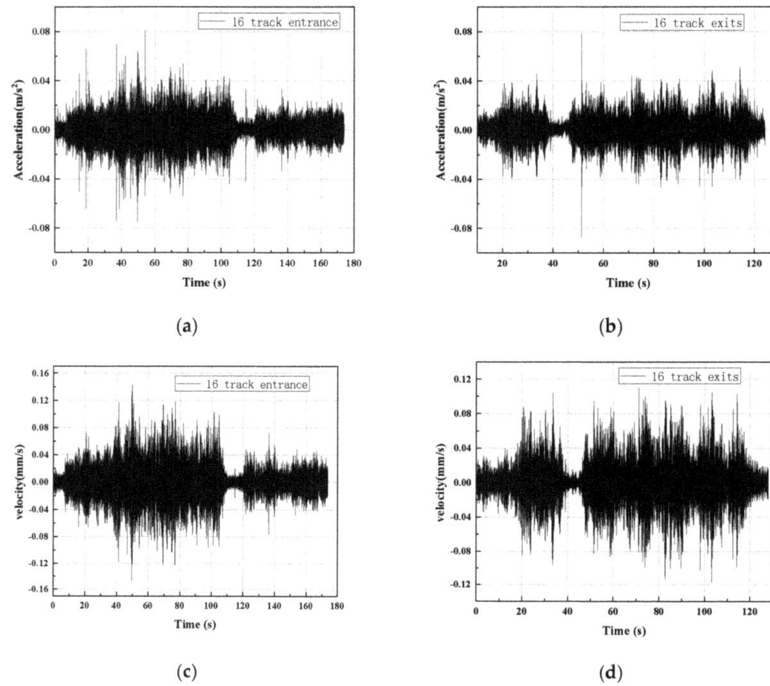

Figure 7. (**a**) Vibration acceleration response of the third-floor drawing room at the entry of track 16; (**b**) vibration acceleration response of the third-floor drawing room at the exits of track 16; (**c**) vibration velocity response of the third-floor drawing room at the entry of track 16; (**d**) vibration velocity response of the third-floor drawing room at the exits of track 16.

Figure 8 shows how the frequency spectrum and one-third octave spectra of the vertical vibrations of the drawing-room acceleration and velocity varied among the tracks; the reference velocity was 2.54×10^{-8} m/s, and the reference acceleration was 1×10^{-6} m/s^2.

The results show that for the different tracks, the vibrations transmitted to the building showed a peak of 30 Hz at both acceleration and velocity, with the main frequencies mostly between 20 Hz and 60 Hz. In the one-third octave spectrum, the velocity and acceleration followed the same trend in frequency, having a peak at 31.5 Hz. The peak frequency was related to the resonance of the vibration amplification in the corresponding frequency band with the vertical vibrations in a specific room. The ambient vibrations had a lower acceleration and velocity than the over-travel vibrations in the frequency domain, but the peak frequency was the same as that of the over-travel vibrations. When the train passed over different tracks, the vibrations in the building showed the trend that the vibrations of the vertically downward track of the tested room was larger than the non-vertically downward track vibrations, that is, the vibrations on tracks 14 and 15 were smaller than the vibrations on tracks 16–18.

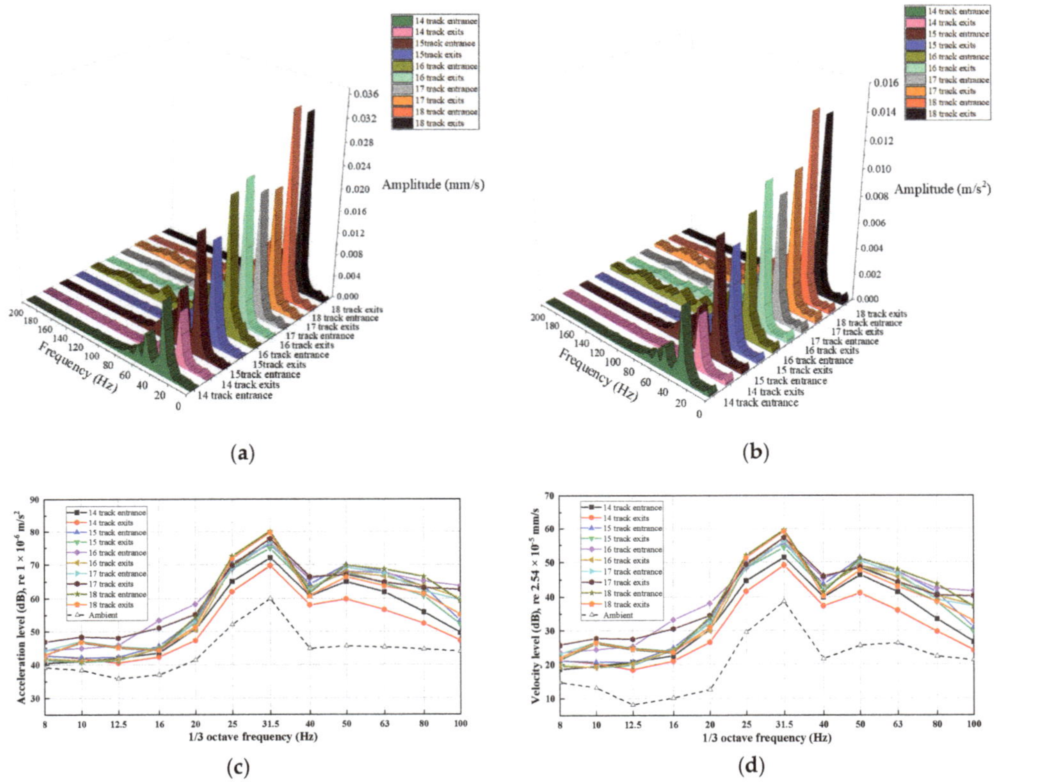

Figure 8. (**a**) Results of the building vibration acceleration caused by a train passing over different tracks in the frequency domain; (**b**) results of the building vibration velocity caused by a train passing over different tracks in the frequency domain; (**c**) results of the building vibration acceleration over one-third octave caused by a train passing over different tracks; (**d**) results of building vibration velocity over one-third octave caused by a train passing over different tracks.

3.2. Floor Effect on Vibration Response

This test was conducted uniformly on the same floors as the previous test (i.e., third, fifth, seventh, ninth, and 11th floors). The first floor was temporarily unavailable for testing due to construction problems, and the 11th floor was the top floor of the building.

The analysis was conducted for the entrance of track 15 and the exit of track 16 to determine the vibration responses at the entrance and exit of the operation depot on different floors. The acceleration was analyzed in the time and frequency domains in the bedroom on each of the test floors, to illustrate the transmission pattern of vibrations between floors.

The time–domain analysis results of the vibration response from the third to the 11th floors showed that the vibration intensity first decreased, then increased, and finally decreased; the vibration intensity was the largest on the upper-middle floor. The vibration attenuation between the floors showed a zigzag trend. Therefore, when assessing the vibration impact on a building, the lowest floor should not be selected for measurement and evaluation. The frequency-domain analysis results of the vibration response on the different floors showed that the main frequency band of vibrations was 25–80 Hz. The peak frequency of the third, fifth, ninth, and 11th floors was 63 Hz, while the peak frequency of the seventh floor was 50 Hz. The results in Figure 8 show that the vibration peak frequency of the drawing room was 31.5 Hz; meanwhile, the vibration peak frequencies of the bedroom were 63 Hz and 50 Hz, as shown in Figures 9 and 10. In the subsequent analysis of the vibration effect on occupant comfort in the over-track buildings, the third and seventh floors were considered.

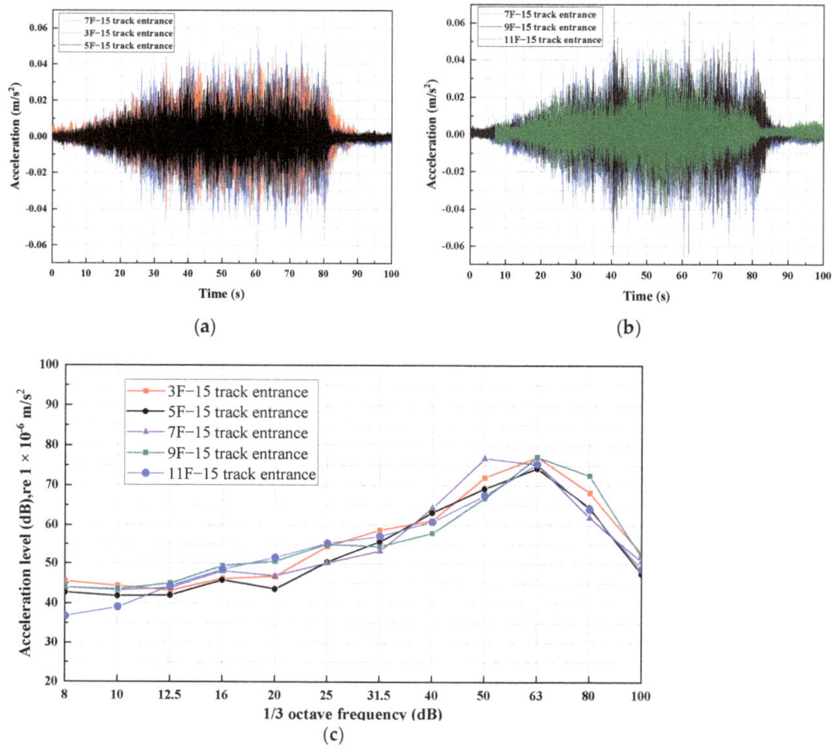

Figure 9. Analysis results of the vibration effect on human comfort on different floors in the time and frequency domains when the train enters track 15. (**a**) Time domain diagrams of measurement points on the 3rd, 5th and 7th floors; (**b**) Time domain diagrams of measurement points on the 7th, 9th and 11th floors; (c) One-third octave frequency diagrams of measurement points on the 3rd, 5th, 7th, 9th, and 11th floors.

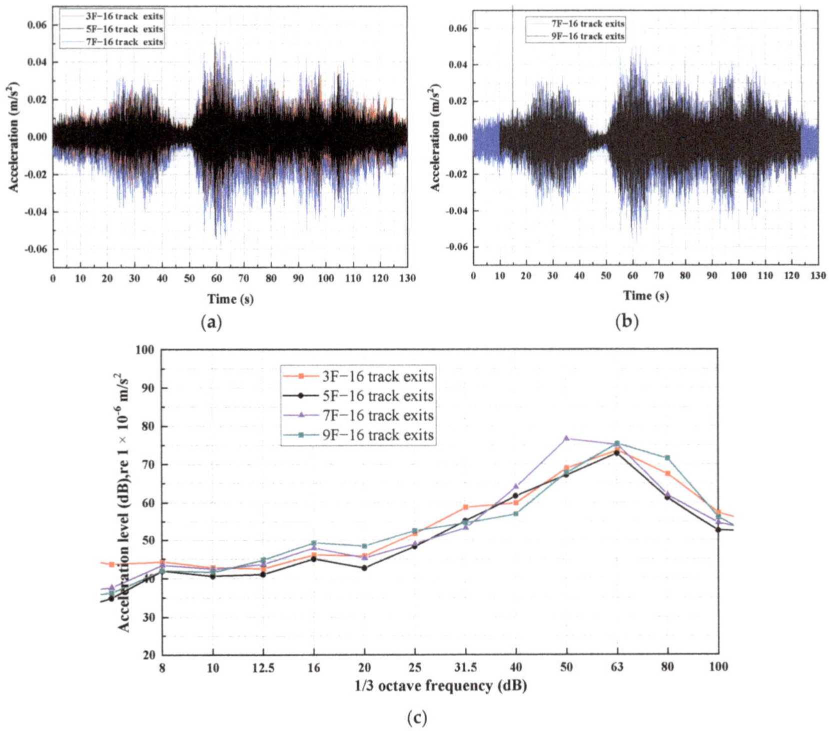

Figure 10. Analysis results of the vibration effect on human comfort on different floors in the time and frequency domains when the train exits track 16. (**a**) Time domain diagrams of measurement points on the 3rd, 5th and 7th floors; (**b**) Time domain diagrams of measurement points on the 7th and 9th floors; (**c**) One-third octave frequency diagrams of measurement points on the 3rd, 5th, 7th, and 9th floors.

4. Evaluation of Vibration Impact on Human Comfort

When evaluating the impact of vibrations on human comfort, it is usually referred to as the vibration standard evaluation. However, the vibration limit values defined by the standards around the world do not ensure that 100% of the population is not disturbed, the evaluation of vibration results cannot reflect the continuity of human subjective feelings, and the vibration standards cannot provide an evaluation of large complex systems. The analysis based on the annoyance rate could be beneficial to the evaluation of quantitative vibration comfort. In an environment with vibrations, the percentage of people who experience annoyance at the same vibration intensity in the total number of people in the environment represents the structural vibration annoyance rate. The annoyance rate caused by the vibrations of the over-track building is studied by using the two evaluation systems of acceleration and velocity, so as to summarize the influence of the vibrations of the over-track building on human comfort.

4.1. Human Comfort Analysis with Vibration Velocity as Evaluation Index

The effects of vibrations on human comfort have been extensively investigated and analyzed in many countries. The research results of developed countries (e.g., USA, Norway) on human comfort were selected in this study to evaluate human comfort in over-track buildings. It should be noted that different countries use different evaluation indicators and weighting methods, for example, the USA and Norway have adopted velocity indicators.

4.1.1. Annoyance Rate Analysis Based on Vibration Dosage-Response Curve Released by USA

The FTA (Federal Transit Administration) and FRA (Federal Railroad Administration) guidelines state that the human body responds to an average vibration amplitude. Because the net average of a vibration signal is zero, the root mean square (rms) amplitude is used to describe the "smoothed" vibration amplitude. Decibel notation is commonly used for vibration assessment (L_v dB).

According to the FTA and FRA guidelines [29,30], the peak velocity level of the building structure, L_v, is calculated as follows:

$$L_v = 20\log_{10}\left[\frac{v}{v_{\text{ref}}}\right] \quad (1)$$

where v is the rms velocity.

The reference velocity v_{ref} is calculated as follows:

$$v_{\text{ref}} = 2.54 \times 10^{-8}\ \text{m/s} \quad (2)$$

The vibration data of the drawing room and bedroom on the third and seventh floors were analyzed according to the above formula, and the results are shown in Table 4. The Transportation Cooperative Research Program (TCRP), sponsored by the FTA in cooperation with Transit Development Corporation and administered by the Transportation Research Board of the National Academy of Sciences, has studied the annoyance rate due to vibrations, and the annoyance rate curve is shown in Figure 11, where 72 dB is the limit for frequent events in residential areas.

For a normal distribution, the probability that a train would exceed the mean +2σ level is 5%, so this level would correspond to the "loudest" trains that passed by the site, and, hence, is the more appropriate measure if it is believed that people are more likely to be disturbed by the loudest trains in the fleet, rather than the fleet-average train. The results obtained, using the method based on the mean plus two times the standard deviation in the American Social Vibration Surveys-Annoyance, are presented in Table 5.

The results indicated that the vibrations in the seventh-floor drawing room would have a 9% probability of making the residents feel highly annoyed, and a 17.5% probability of making them feel moderately or highly annoyed; the vibrations in the seventh-floor bedroom would have a 6.5% probability of making residents feel highly annoyed, and a 11.5% probability of making them feel moderately or highly annoyed. Further, the vibrations in the third-floor drawing room would have a 6.2% probability of making residents feel highly annoyed and a 11.2% probability of making them feel moderately or highly annoyed; the vibrations in the third-floor bedroom would have a 3.5% probability of making residents feel highly annoyed and a 6.5% probability of making them feel moderately or highly annoyed. Figure 12 compares the train-induced vibration levels in the over-track building with the vibration limits categorized under frequent events in the US surveys. Only the bedroom vibrations on the third floor did not exceed the limits, with the maximum vibration level occurring in the living room on the seventh floor, which exceeded the limit by 6.93 dB.

Table 4. Evaluation of vibration levels at the measurement point according to the US vibration standards.

Track	Drawing Room-3F L_V (dB)	Bed Room-3F L_V (dB)	Drawing Room-7F L_V (dB)	Bed Room-7F L_V (dB)
14-track entrance	57.81	58.29	60.02	61.03
14-track exits	57.89	55.10	59.35	59.16
15-track entrance	67.01	62.92	65.20	66.62
15-track exits	64.22	63.27	63.20	65.10
16-track entrance	66.60	64.04	77.17	70.68
16-track exits	66.42	65.10	77.26	67.27
17-track entrance	66.90	63.32	65.71	70.77
17-track exits	70.34	64.34	67.57	69.60
18-track entrance	68.61	62.14	67.21	68.08
18-track exits	66.78	59.09	69.66	67.88

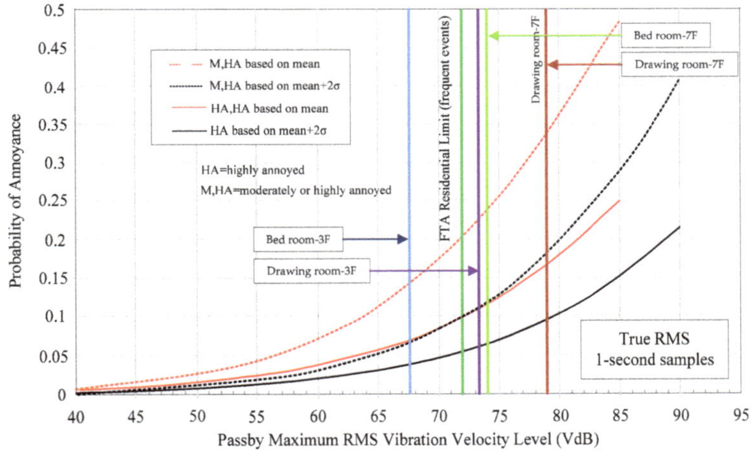

Figure 11. Results of the American Social Vibration Surveys-Annoyance [30].

Table 5. Annoyance assessment results obtained by the (Mean + 2σ) method.

Floors and Rooms	Velocity Level Mean + 2σ (dB)	Probability of Highly Annoyed	Probability of Moderately or Highly Annoyed
Drawing room-3F	73.24	0.062	0.112
Bed room-3F	67.85	0.035	0.065
Drawing room-7F	78.93	0.090	0.175
Bed room-7F	73.99	0.065	0.115

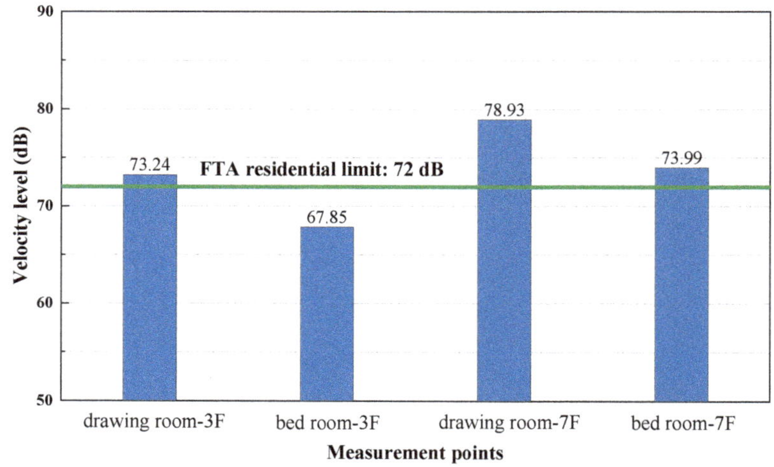

Figure 12. Comparison of vibration of the over-track building and the US standard.

4.1.2. Annoyance Rate Analysis Based on Vibration Exposure–Response Curve Released by Norway

The Norwegian standard (NS 8176:2005) suggests using the statistical maximum weighted acceleration or velocity level ($a_{w,95}$ or $v_{w,95}$) when assessing the vibration effect on human comfort [31]. These indicators are calculated from the 1-s rms averages of w_m-weighted acceleration or velocity signals. In this study, the velocity index was selected for evaluation.

The $v_{w,95}$ descriptor was calculated as follows:

$$v_{w,95} = \overline{v_{w,\max}} + 1.8\sigma_v \tag{3}$$

where $\overline{v_{w,\max}}$ is the average value of the maximum weighted speed of all trains passing during the evaluation period; and σ_v is the standard deviation of the maximum weighted speed of all of the trains passing during the evaluation period.

The $\overline{v_{w,\max}}$ and σ_v values are, respectively, calculated as follows:

$$\overline{v_{w,\max}} = \frac{\sum_{j=1}^{N} v_{w,\max,j}}{N} \tag{4}$$

$$\sigma = \sqrt{\frac{1}{N-1}\sum_{j=1}^{N}\left(v_{w,\max,j} - \overline{v_{w,\max}}\right)^2} \tag{5}$$

where $v_{\mathrm{w,max}}$ is the maximum 1-s average weighted speed of a single train passing during the evaluation period; and N is the total number of trains passing during the evaluation period.

The value of $v_{\mathrm{w,max}}$ can be calculated using various standard arithmetic methods, but this study adopts the w_m-frequency weighting method. The exposure effect curves for the rest and daily activity periods were evaluated according to the curves defined by the standard. As shown in Figure 13, there were between 10 and 15 reports of disturbances during the rest and sleep periods, respectively, and $v_{w,95}$ was about 0.1 mm/s. This test was conducted for the trains running in and out of the depot on tracks 14–18, for a total of 10 trains, to simulate the normal operation of the train. Finally, a standard assessment was performed for all of the 10 trains.

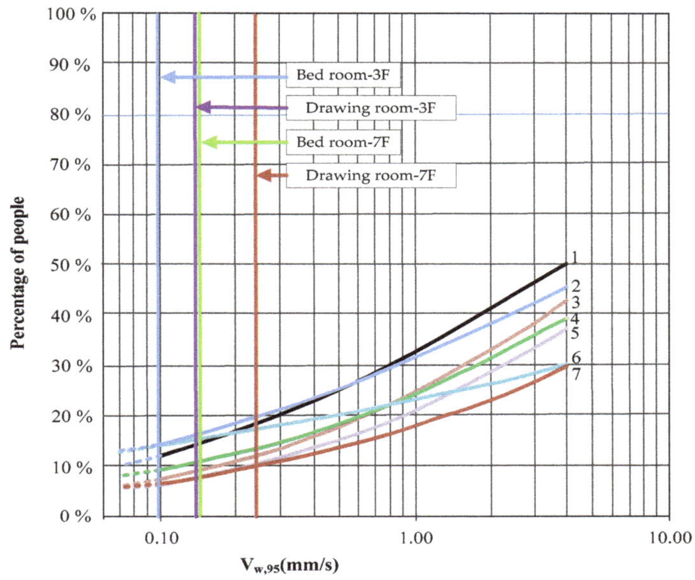

Figure 13. Percentage of people who were disturbed during rest or during daily activities by the vibrations in dwellings. The results are plotted against the calculated statistical maximum value for the weighted velocity, $v_{w,95}$, expressed in mm/s [17].

Where key:

1. Disturbance when watching TV or listening to the radio;
2. Disturbance during the rest period;
3. Waking up too early;
4. Waking up during the night;
5. Disturbance during telephone usage;
6. Disturbance during conversations;
7. Difficulty in falling asleep.

Table 6 shows the results of $v_{\mathrm{w,max}}$ and $v_{w,95}$ on the third floor when the train passed through tracks 14–18. The $v_{w,95}$ value in the drawing room was 0.14; considering the function of this room, the vibrations affected 14% of people watching TV or engaging in other sources of entertainment, 8% of people talking on the phone, and 15% of people talking to each other. The $v_{w,95}$ value in the bedroom was 0.09; considering the function of this room, the vibrations affected 12% of people who were resting, and 7% of people reported difficulty in sleeping.

Table 7 shows the results of $v_{\mathrm{w,max}}$ and $v_{w,95}$ on the seventh floor when the train passed through tracks 14–18. The $v_{w,95}$ value in the drawing room was 0.23; considering the function of this room, the vibrations affected 19% of people watching TV or engaging in other sources of entertainment, 11% of people talking on the phone, and 18% of people talking to each other. The $v_{w,95}$ value in the bedroom was 0.15; considering the function of this room, the vibrations affected 17% of people who were resting, and 9% of people experienced certain difficulties in sleeping. Figure 14 compares the train-induced vibration levels in the over-track building with the Norwegian standard vibration limits, where vibrations above Class B causes a certain level of vibration disturbance to residents; the living room and bedroom on the seventh floor exceed the limits of Class B.

Table 6. Evaluation indicators for the third-floor rooms.

Table.	Drawing Room $v_{w,max}$ (mm/s)	Bed Room $v_{w,max}$ (mm/s)	Drawing Room $v_{w,max}$ (mm/s)	Bed Room $v_{w,max}$ (mm/s)
14-track entrance	0.0385	0.03777	0.14	0.09
14- track exits	0.0319	0.03178		
15-track entrance	0.0777	0.064694		
15- track exits	0.06817	0.057754		
16-track entrance	0.10331	0.087049		
16- track exits	0.0839	0.065383		
17-track entrance	0.0873	0.067918		
17- track exits	0.1266	0.079552		
18-track entrance	0.11804	0.056475		
18- track exits	0.11611	0.040929		

Table 7. Evaluation indicators for the seventh-floor rooms.

Track	Drawing Room $v_{w,max}$ (mm/s)	Bed Room $v_{w,max}$ (mm/s)	Drawing Room $v_{w,95}$ (mm/s)	Bed Room $v_{w,95}$ (mm/s)
14-track entrance	0.0461	0.0404	0.23	0.15
14-track exits	0.0411	0.0374		
15-track entrance	0.0862	0.0831		
15-track exits	0.0760	0.0695		
16-track entrance	0.3093	0.1378		
16-track exits	0.1076	0.0929		
17-track entrance	0.0939	0.1263		
17-track exits	0.0939	0.1326		
18-track entrance	0.1013	0.0907		
18-track exits	0.1095	0.0940		

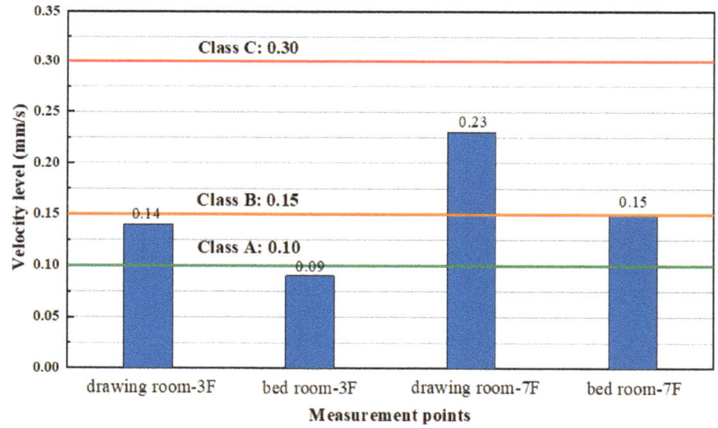

Figure 14. Comparison of vibration of the over-track building and Norway standard. Class A, normally not be expected to notice vibration; Class B, can be expected to be disturbed by vibration to some extent; Class C, about 15% of the affected persons in the dwellings can be expected to be disturbed by vibration.

4.2. Human Comfort Analysis with Vibration Acceleration as Evaluation Index

4.2.1. Annoyance Rate Analysis Based on Vibration Exposure–Response Curve Released by Europe

The EU FP7 project, CargoVibes, conducted a meta-analysis of a total of 4490 existing samples of surveys to investigate the effect of vibrations on the population, providing the exposure–response relationship curves for Germany, Norway, Japan, the United States, Canada, the United Kingdom, Sweden, the Netherlands, and Poland [16]. The curves are plotted for three different vibration exposure descriptors to provide a reasonable estimate of annoyance from the vibration exposure measurement, according to the major standards. Figure 15 shows the annoyance rate plotted for the *rms* indicator. The impact of the vibrations generated by train operation on the comfort of occupants in the building was evaluated considering the peak-hour operation at night and early morning and a single-train operation for daytime shunting conditions. The three vibration exposure descriptors are as follows:

(1) $V_{dir,max}$: Maximum w_k-weighted fast exponentially filtered rms velocity over the entire evaluation period;
(2) *rms*: w_k-weighted rms acceleration over the entire evaluation period;
(3) *VDV*: w_k-weighted vibration intensity over the entire evaluation period.

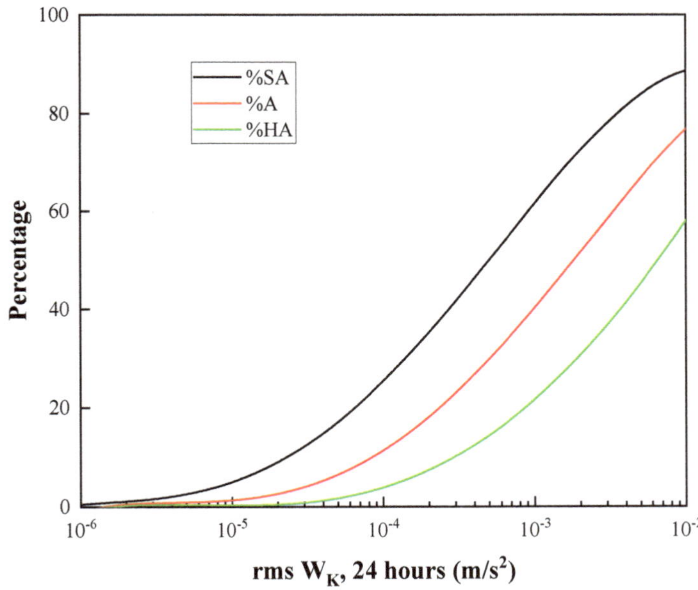

Figure 15. Annoyance rate for the *rms* (w_k-weighted root-mean-square acceleration over the entire evaluation period) indicator.

For the human comfort evaluation in the over-track building using *rms*, the annoyance rate is calculated as follows:

$$\%SA_{rms} = -1.806X^4 - 3.198X^3 + 11.812X^2 + 35.059X + 25.390 \quad (6)$$

$$\%A_{rms} = -1.648X^4 - 0.013X^3 + 13.826X^2 + 22.510X + 11.380 \quad (7)$$

$$\%HA_{rms} = -0.527X^4 + 2.089X^3 + 9.850X^2 + 10.785X + 3.910 \quad (8)$$

where SA_{rms} denotes slight annoyance; A_{rms} denotes annoyance; HA_{rms} denotes severe annoyance; and X is related to the rms value of the acceleration.

The X descriptor is calculated as follows:

$$X = \frac{\log_{10}(\text{rms}) + 4}{1.1564} \quad (9)$$

It is important to note that the above equations must not be used when the value of rms is out of range of $(0.001 \times 10^{-3}, 10 \times 10^{-3})$ m/s^2.

(1) The annoyance rate during peak hours of train operation

Next, the annoyance rates were analyzed in the rooms on the third and seventh floors, and the results showed that the third-floor drawing room had a high annoyance rate during the peak hours, with the highest HA of 8.27%, which severely affected the comfort of the residents. The bedroom on the third floor also had a high annoyance rate during the peak hours, with the highest HA of 8.49%, which occurred between 5:30 a.m. and 6:30 a.m. The results are shown in Table 8. However, this period comprises the residents' sleep time; thus, any annoyance in this period can affect the residents' sleep.

Table 8. Vibration annoyance rate of the third-floor test room during peak hours of train operation.

Single Peak Hour	Test Room-3F	%HA	%A	%SA
10:30 p.m. to 11:30 p.m.	Drawing room	4.24	12.07	26.46
11:30 p.m. to 12:30 p.m.	Drawing room	6.30	16.09	32.36
4:30 a.m. to 5:30 a.m.	Drawing room	5.18	13.95	29.28
5:30 a.m. to 6:30 a.m.	Drawing room	8.27	19.69	37.31
10:30 p.m. to 11:30 p.m.	Bed room	8.71	20.47	38.34
10:30 p.m. to 12:30 p.m.	Bed room	7.91	19.05	36.45
4:30 a.m. to 5:30 a.m.	Bed room	7.96	19.14	36.57
5:30 a.m. to 6:30 a.m.	Bed room	8.49	20.08	37.82

The annoyance rate measured in the bedroom on the seventh floor was smaller compared to that on the third floor; thus, did not have a significant impact on the comfort of the residents, as shown in Table 9.

Table 9. Vibration annoyance rate of the seventh-floor test room during peak hours of train operation.

Single Peak Hour	Test Room-7F	%HA	%A	%SA
10:30 p.m. to 11:30 p.m.	Bed room	0.59	3.13	10.28
10:30 p.m. to 12:30 p.m.	Bed room	0.69	3.49	11.13
4:30 a.m. to 5:30 a.m.	Bed room	0.58	3.09	10.18
5:30 a.m. to 6:30 a.m.	Bed room	0.84	3.98	12.22

The analysis results of the peak hours showed that the annoyance rate during peak hours was higher than that during other periods of the day. Since the trains stop operating at night and depart again early in the morning, and the late evening and early morning are sleeping times, in these times, the vibration annoyance rate in the bedroom has a high impact on people's comfort, from the point of view of the room's function.

(2) The annoyance rate in a single-train operation

For a more detailed assessment of the impact of the different tracks on building vibrations and thus human comfort, calculations were performed for all of the tracks, and the results are presented in Tables 10 and 11.

The calculation was performed with a 5% annoyance rate limit to ensure that the comfort of 95% of people is relatively guaranteed. In the drawing room on the third floor, the annoyance rate exceeded the limit for all of the tracks. The train ran 10 times a night, and the vibrations generated in 8 of 10 cases caused discomfort to the residents. It should be noted that even the vibrations from only one operation caused discomfort to the residents in the bedrooms on both the third and the seventh floors.

Table 10. Vibration annoyance rate of the third-floor test room during a single-train operation.

A Single Car	Test Room-3F	%HA	%A	%SA
14-track entrance	Drawing room	4.70	12.99	27.85
14-track exits	Drawing room	2.80	8.99	21.55
15-track entrance	Drawing room	6.98	17.37	34.15
15-track exits	Drawing room	5.00	13.59	28.75
16-track entrance	Drawing room	5.26	14.10	29.49
16-track exits	Drawing room	6.56	16.59	33.06
17-track entrance	Drawing room	6.03	15.58	31.64
17-track exits	Drawing room	5.91	15.36	31.33
18-track entrance	Drawing room	8.31	19.77	37.41
18-track exits	Drawing room	7.59	18.48	35.68
14-track entrance	Bed room	4.88	13.36	28.41
14-track exits	Bed room	2.58	8.48	20.69
15-track entrance	Bed room	5.24	14.07	29.45
15-track exits	Bed room	3.92	11.39	25.41
16-track entrance	Bed room	3.68	10.90	24.64
16-track exits	Bed room	3.72	10.98	24.76
17-track entrance	Bed room	3.44	10.39	23.83
17-track exits	Bed room	3.21	9.90	23.04
18-track entrance	Bed room	4.36	12.31	26.82
18-track exits	Bed room	3.33	10.15	23.45

Table 11. Vibration annoyance rate of the seventh-floor test room during a single-train operation.

A Single Car	Test Room-7F	%HA	%A	%SA
14-track entrance	Bed room	2.38	8.02	19.90
14-track exits	Bed room	1.20	5.04	14.41
15-track entrance	Bed room	3.23	9.93	23.09
15-track exits	Bed room	2.44	8.16	20.15
16-track entrance	Bed room	5.92	15.37	31.34
16-track exits	Bed room	3.60	10.72	24.36
17-track entrance	Bed room	4.15	11.87	26.16
17-track exits	Bed room	4.72	13.03	27.91
18-track entrance	Bed room	4.36	12.31	26.83
18-track exits	Bed room	2.58	8.49	20.70

4.2.2. Annoyance Rate Analysis Using the Set-Value Statistical Method and Psychological Annoyance Rate

Due to the vagueness and randomness of the subjective vibration response judgment, the set-value statistical method and psychological annoyance rate were used to evaluate human comfort. The structural vibration annoyance rate calculation method, introduced by Tu et al. [32] and Song [33], was adopted.

For the case of discrete distribution, the annoyance rate is calculated as follows:

$$A(a_{wi}) = \frac{\sum\limits_{j=1}^{m} v_j n_{ij}}{\sum\limits_{j=1}^{m} n_{ij}} = \sum_{j=1}^{m} v_j p(i,j) \tag{10}$$

where $A(a_{wi})$ is the annoyance rate at the ith vibration intensity a_{wi}; n_{ij} is the number of people with the jth subjective response at the ith vibration intensity; v_j is the conceptual affiliation of the "unacceptable" category of the jth subjective response; $\sum\limits_{j=1}^{m} n_{ij}$ is the total number of people reporting "unacceptable" vibration intensity; and $p(i, j)$ reflects the variability in the annoyance degree among people.

The value of v_j is calculated as follows:

$$v_j = \frac{j-1}{m-1} \tag{11}$$

where m is the number of levels of the subjective response of people, and usually, m is set to 5 or 11; when m = 5, the levels of the subjective response are: "no vibration"; "light vibration"; "moderate vibration"; "strong vibration"; and "unbearable."

For the case of continuous distribution, since there is a variability in the human perception of vibrations, the variability in people's responses can be described by a log-normal distribution [34].

Considering the distribution characteristics of ambiguity and randomness, the annoyance rate under continuous distribution at the vibration acceleration x is given by:

$$A(x) = \int_{u_{\min}}^{\infty} \frac{1}{\sqrt{2\pi} u \sigma_{\ln}} \exp\left(\frac{-\left(\ln(u/x) - 0.5\sigma_{\ln}^2\right)^2}{2\sigma_{\ln}^2} \right) v(u) du \tag{12}$$

where x is the expected value of u; and $v(u)$ is the vibration intensity fuzzy affiliation function.

Further, the value of $\sigma_{\ln}$ is calculated as follows:

$$\sigma_{\ln} = \sqrt{\ln(1+\delta^2)} \tag{13}$$

and $v(u)$ is calculated as follows:

$$v(u) = \begin{cases} 0 & u < u_{\min}, \\ a\ln(u) + b & u_{\min} \leq u \leq u_{\max} \\ 1 & u > u_{\max}, \end{cases} , \tag{14}$$

where δ is the coefficient of variation of u, and it is usually set to 0.3 [33,35]; $u_{\min}$ is the upper limit of the vibration intensity that a human defines as "not felt"; $u_{\max}$ is the lower limit of the vibration intensity that a human being senses as "unbearable". Based on the experiments, the values of $u_{\min}$ and $u_{\max}$ are 0.05 m/s^2 and 1.5 m/s^2, respectively; a and b are coefficients to be determined, and they are calculated as follows:

$$\begin{cases} a\ln(u_{\min}) + b = 0, \\ a\ln(u_{\max}) + b = 1, \end{cases} \tag{15}$$

The annoyance rate can be regarded as a resistance R in the reliability analysis, and its distribution function can be expressed by a log-normal distribution function. Therefore, approximating the annoyance rate curve before the analysis of the annoyance rate could be a good solution to obtain objective results, and the calculation result approximation does not cause significant deviations [33].

The function $A(x)$ is expressed as follows:

$$A(x) \approx CDF_{\log norm}(x, \mu_{\ln x}, \sigma_{\ln x}) \tag{16}$$

where, in the vertical direction of the train-induced vibrations, $\mu_{\ln x}$ has a value of −4.247, and $\sigma_{\ln x}$ equals 0.473.

By using the calibration method, an engineering acceptable design level of vibration comfort corresponding to an allowable annoyance rate of 7% is obtained [33]. The vibration data of the third and seventh floors of the over-track buildings were analyzed and evaluated, using Equation (16).

For a more detailed assessment of the impact of the different tracks on the building vibrations, calculations were performed for all of the tracks, as shown in Tables 12 and 13. In the drawing room on the third floor, the annoyance rate for all of the other tracks, except for track 14, exceeded the limit. The train ran 10 times a night, and the vibrations generated in 7 of 10 cases caused discomfort to the residents. For occupants living in the bedrooms of the third and seventh floors, the annoyance rate caused by vibrations did not exceed the 7% limit; thus, the vibrations did not have a noticeable effect on the comfort of the residents in these bedrooms.

Table 12. Vibration annoyance rate of the third-floor test room during a single-train operation based on annoyance rate model.

A Single Car	Test Room-3F	Annoyance Rate
14-track entrance	Drawing room	0.086%
14-track exits	Drawing room	0.14%
15-track entrance	Drawing room	15.89%
15-track exits	Drawing room	6.16%
16-track entrance	Drawing room	19.82%
16-track exits	Drawing room	17.41%
17-track entrance	Drawing room	21.75%
17-track exits	Drawing room	51.71%
18-track entrance	Drawing room	37.2%
18-track exits	Drawing room	21.19%
14-track entrance	Bed room	0.00059%
14-track exits	Bed room	0.0003%
15-track entrance	Bed room	0.16%
15-track exits	Bed room	0.088%
16-track entrance	Bed room	0.29%
16-track exits	Bed room	0.35%
17-track entrance	Bed room	0.11%
17-track exits	Bed room	0.2%
18-track entrance	Bed room	0.072%
18-track exits	Bed room	0.00018%

Table 13. Vibration annoyance rate of the seventh-floor test room during a single-train operation based on annoyance rate model.

A Single Car	Test Room-7F	Annoyance Rate
14-track entrance	Bed room	0.001%
14-track exits	Bed room	0.000%
15-track entrance	Bed room	0.0159%
15-track exits	Bed room	0.0089%
16-track entrance	Bed room	0.1479%
16-track exits	Bed room	0.0207%
17-track entrance	Bed room	0.1713%
17-track exits	Bed room	0.0898%
18-track entrance	Bed room	0.0508%
18-track exits	Bed room	0.0498%

5. Findings and Discussion

Based on the above measurement campaign and human comfort evaluation, it was clear that the neighborhood residents were indeed suffering from the railway-induced vibrations under certain circumstances. However, as the human comfort would be related to the vibration level and period of period, the serviceability and particularity of the TOD developed depot still need to be fully discussed.

5.1. The Particularity of Railway Vibration at Tod Developed Depot

Firstly, a metro depot is a facility where trains are regularly parked for maintenance, testing, and storage. Therefore, there were rush hours when the metro trains started going out very early in the morning and coming back very late at night. The main uncomfortable influence due to railway vibrations would be that causing residents difficulty in sleeping at night or waking early in the morning.

Secondly, based on the measurement results, the highest vibrations happened when the train was running though the track immediately under the particular bedroom. Generally, during the rush hours, the resident may have to suffer one or two episodes of high-level vibration shock and four to six episodes of sensible vibrations.

Thirdly, the level of the vibrations was highly related to train speed, as limited train speed would obviously reduce the vibrations.

Besides that, the structure of the building plays an important role in the vibration transit. The measurement results showed that when the vibration is transmitted to the upper floors, the vibration intensity first decreases, then increases, and then again decreases; so, the vibrations are the largest on the upper-middle floors. Thus, when assessing the vibration impact, not only the lowest floors should be selected for measurement and evaluation, but also the upper-middle floors.

5.2. Human Comfort Analysis with Vibration Velocity as Evaluation Index

There are two types of human response index associated with vibration velocity in the available national standards. The US standards provide a single-figure exposure descriptor related to the energy equivalent *rms* velocity value and vibration velocity level, considering frequency weightings, which in the presented case study, means that the residents in the seventh floor drawing room will be most disturbed, with a 9% probability of being highly annoyed and a 17.5% probability of being moderately or highly annoyed. In the Norwegian standards, detailed human comfort responses to vibration velocity are provided, including having sleeping and living activities disturbed, which considers a frequency weighting index w_m in the range of 0.5 to 160 Hz. In the presented case study, the most serious impact occurs in the bedroom on the seventh floor; that particular vibration affected 17% of the people who were resting and 9% of people who had difficulties in sleeping. It is suggested that the sleep disturbed index should be more significant in the TOD developed depot, as the rush hours happen very late in the night and very early in the morning.

5.3. Analysis of Human Comfort Using Train Acceleration as Evaluation Index

To assess human comfort due to vibrations, the cumulative vibration acceleration values are more commonly used in some national standards as well. The acceleration indicators defined by the EU FP7 project, CargoVibes, and the annoyance rate based on the set-value statistical method and psychological annoyance rate are used for comfort assessment.

According to the EU FP7 project, CargoVibes, which weighted the root-mean-square acceleration for the entire evaluation period, using w_k frequency weighting considering 1–80 Hz., in the study, the bedroom on the third floor reached a high annoyance rate during the peak hours, with an HA of 8.71%, occurring between 10:30 PM and 11:30 PM, which is the sleep time of the residents, and therefore affecting their sleep. The bedroom on the seventh floor has a lower annoyance rate, which did not significantly affect the comfort of the residents. During the daytime scheduled shunting, the annoyance rate was calculated with a 5% limit to ensure that the comfort of 95% of the people concerned was relatively guaranteed. In the drawing room on the third floor, the annoyance rate exceeded the limit for all of the tracks. According to the train timetable, the trains run 10 times during the nighttime, and the vibrations generated by the trains caused discomfort to the residents in 8 out of the 10 cases. Only one vibration would cause discomfort to the residents of the third and seventh floor bedrooms.

According to the annoyance rate based on the set-value statistical method and psychological annoyance rate, which uses w_i frequency weighting considering 1–80 Hz., in the study, in the drawing room on the third floor, the annoyance rate for all of the tracks, except for track 14, exceeds the limit. The trains run 10 times at night, but in seven cases, the vibrations generated by the trains causes discomfort to the residents. In the bedrooms on the third and seventh floors, the vibration limit is not reached; thus, the vibrations do not have a noticeable effect on human comfort.

Evaluating the acceleration effects in the over-track building, using the EU FP7 project, CargoVibes, and the annoyance rate based on the set-value statistical method and psychological annoyance rate, the vibrations are less annoying for people in the bedroom during the daytime scheduled shunting. However, at 10:30 to 11:30 p.m., the people in the third-floor bedrooms suffer from a high level of annoyance of 8.71%, which merits attention.

6. Conclusions

The present paper provided a case study to describe the effects of the metro railway-induced vibrations on human comfort at a particular TOD developed depot. In this paper, a measurement campaign was conducted at an operated metro depot, where the residential buildings were directly located on the cover structure of the metro train storage. Considering that there was not a proper standard to describe the human comfort subject to railway vibrations, several evaluation indexes were employed to analyze the structural serviceability in this paper. Some interesting findings could be found from this particular project, as below:

(1) It was clear that the neighborhood residents were indeed suffering from the railway-induced vibrations under certain circumstances. However, the results indicated that 90% of the occupants were not highly annoyed by the train-induced vibrations;

(2) The vibration events that happened at the TOD depot related to many factors, such as the train speed, building structure, and the track location that the trains were running on. It is possible to reduce the vibration effect by using a particular solution;

(3) At the particular situation of the metro depot, the main negative effect on human comfort was that the high level vibrations regularly happened in the rush hours, very early in the morning and very late at night, which would cause an interruption in sleep. Therefore, the evaluation index should consider more factors related to sleep difficulties and the awake threshold value;

(4) Based on the review of the current available standards, there are differences in terms of the single-figure or comprehensive indexes' descriptors, frequency weightings, measurement methods, and the guidelines' values for detailed impact. However, the current descriptors were insufficient to assess the effect of the vibrations on human comfort in such a particular situation, as it is difficult to derive exposure–response relationships or threshold values for impact on sleep and other living activities. Future studies should therefore focus on self-reported sleep difficulties and the impact on activities undertaken when awake from the residents living in TOD developed metro depots.

Author Contributions: Investigation, Y.L., P.Z., Q.W. and H.Z.; methodology, Y.L. and P.Z.; software, P.Z.; project administration, H.Z. and Y.L.; supervision, H.Z. and Y.L.; writing-original draft, P.Z. and H.Z.; writing-review and editing, H.Z. and Y.L. All authors have read and agreed to the published version of the manuscript.

Funding: This research was funded by the Beijing Natural Science Foundation, grant number 8202040, and the National Natural Science Foundation of China, grant number 12072208, and the Opening Foundation of State Key Laboratory of Shijiazhuang Tiedao University, grant number KF2021-15, ZZ2021-13.

Institutional Review Board Statement: Not applicable.

Informed Consent Statement: Not applicable.

Data Availability Statement: The data presented in this research can be requested from the corresponding author or the first author.

Conflicts of Interest: The authors declare no conflict of interest.

References

1. Overview of Urban Rail Transit Lines in Mainland China in 2021. Available online: https://www.camet.org.cn/xxfb/9283 (accessed on 2 April 2022).
2. *GB 50157-2013*; Code for Design of Metro. Ministry of Housing and Urban-Rural Development of the People's Republic of China: Beijing, China, 2013.
3. Tao, Z.; Hu, Z.; Wu, G.; Huang, C.; Zou, C.; Ying, Z. Train-Induced Vibration Predictions Based on Data-Driven Cascaded State-Space Model. *Buildings* **2022**, *12*, 114. [CrossRef]
4. Newman, P.W.; Kenworthy, J.R. The land use—Transport connection: An overview. *Land Use Policy* **1996**, *13*, 1–22. [CrossRef]
5. Handy, S. Smart Growth and The Transportation-Land Use Connection: What Does the Research Tell Us? *Int. Reg. Sci. Rev.* **2005**, *28*, 146–167. [CrossRef]
6. He, W. Field measurement and evaluation of noise and vibrations induced by bridge cranes in full-scale over-track buildings. *Environ. Sci. Pollut. Res. Int.* **2019**, *26*, 36515–36537. [CrossRef]
7. Ma, M.; Li, M.; Qu, X.; Zhang, H. Effect of passing metro trains on uncertainty of vibration source intensity: Monitoring tests. *Measurement* **2022**, *193*, 110992. [CrossRef]
8. Xu, L.H.; Ma, M. Dynamic response of the multilayered half-space medium due to the spatially periodic harmonic moving load. *Soil Dyn. Earthq. Eng.* **2022**, *157*, 107246. [CrossRef]
9. Kouroussis, G.; Connolly, D.P.; Verlinden, O. Railway-induced ground vibrations-a review of vehicle effects. *Int. J. Rail Transp.* **2014**, *2*, 69–110. [CrossRef]
10. Ziyu, T.; James, A.M.; Masoud, S.; Yimin, W.; Chao, Z. Train-induced floor vibration and structure-borne noise predictions in a low-rise over-track building. *Eng. Struct.* **2022**, *255*, 113914.
11. Chao, Z.; Yimin, W.; Peng, W.; Jixing, G. Measurement of ground and nearby building vibration and noise induced by trains in a metro depot. *Sci. Total. Environ.* **2015**, *536*, 761–773.
12. Chao, Z.; Yimin, W.; James, A.M.; Masoud, S. Train-induced field vibration measurements of ground and over-track buildings. *Sci. Total. Environ.* **2017**, *575*, 1339–1351.
13. Zhiliang, C.; Tong, G.; Zhiqiang, Z.; Aiqun, L. Measurement and analysis of vibrations in a residential building constructed on an elevated metro depot. *Measurement* **2018**, *125*, 394–405.
14. Connolly, D.P.; Marecki, G.P.; Kouroussis, G.; Thalassinakis, I.; Woodward, P.K. The growth of railway ground vibration problems—A review. *Sci. Total Env.* **2016**, *568*, 1276–1282. [CrossRef] [PubMed]
15. Beben, D.; Maleska, T.; Bobra, P.; Duda, J.; Anigacz, W. Influence of Traffic-Induced Vibrations on Humans and Residential Building-A Case Study. *Environ. Res. Public Health* **2022**, *19*, 5441. [CrossRef] [PubMed]
16. Waddington, D.; Woodcock, J.; Smith, M.G.; Janssen, S.; Persson Waye, K. CargoVibes: Human response to vibration due to freight rail traffic. *Int. J. Rail Transp.* **2015**, *3*, 233–248. [CrossRef]
17. Arnberg, P.W.; Bennerhult, O.; Eberhardt, J.L. Sleep disturbances caused by vibrations from heavy road traffic. *J. Acoust. Soc. Am.* **1990**, *88*, 1486–1493. [CrossRef]
18. Klæboe, R.; Turunen-Rise, I.H.; Hårvik, L.; Madshus, C. Vibration in dwellings from road and rail traffic—Part II: Exposure–effect relationships based on ordinal logit and logistic regression models. *Appl. Acoust.* **2003**, *64*, 89–109. [CrossRef]
19. Michael, G.S.; Ilona, C.; Mikael, O.; Kerstin, P.W. On the influence of freight trains on humans: A laboratory investigation of the impact of nocturnal low frequency vibration and noise on sleep and heart rate. *PLoS ONE* **2013**, *8*, e55829.
20. Croy, I.; Smith, M.G.; Waye, K.P. Effects of train noise and vibration on human heart rate during sleep: An experimental study. *BMJ Open* **2013**, *3*, e002655. [CrossRef]
21. Smith, M.G.; Croy, I.; Hammar, O.; Persson, W.K. Vibration from freight trains fragments sleep: A polysomnographic study. *Sci. Rep.* **2016**, *6*, 24717. [CrossRef]
22. Waye, K.P.; Smith, M.G.; Croy, I.; Hammar, O.; Ögren, M. *D1.4 Final Report of the Results of the Sleep Disturbance Study*; CargoVibes: Delft, The Netherlands, 2013.
23. Waddington, D.C.; Woodcock, J.; Peris, E.; Condie, J.; Steele, A. Human response to vibration in residential environments. *J. Acoust. Soc. Am.* **2014**, *135*, 182–193. [CrossRef]
24. Öhrström, E.; Gidlöf Gunnarsson, A.; Ögren, M.; Jerson, T. Effects of railway noise and vibration in combination: Field and laboratory studies. In Proceedings of the 8th European Conference on Noise Control, Edinburgh, UK, 26–28 October 2009; pp. 1204–1213.
25. Guski, R.; Felscher-Suhr, U.; Schuemer, R. The concept of noise annoyance: How international experts see it. *J. Sound Vib.* **1999**, *223*, 513–527. [CrossRef]
26. Peris, E.; Woodcock, J.; Sica, G.; Moorhouse, A.T.; Waddington, D.C. Annoyance due to railway vibration at different times of the day. *J. Acoust. Soc. Am.* **2012**, *131*, L191–L196. [CrossRef]
27. Peris, E.; Woodcock, J.; Sica, G.; Sharp, C.; Moorhouse, A.T.; Waddington, D.C. Effect of situational, attitudinal and demographic factors on railway vibration annoyance in residential areas. *J. Acoust. Soc. Am.* **2014**, *135*, 194–204. [CrossRef] [PubMed]
28. Janssen, S.A.; Vos, H.; Koopman, A. *Deliverable D1.2-Exposure Response Relationships and Factors Influencing These Relationships*; TNO: Delft, The Netherlands, 2013.
29. USA Department of Transportation Federal Transit Administration. *Transit Noise and Vibration Impact Assessment*; USA Department of Transportation Federal Transit Administration: Washington, DC, USA, 2006.

30. USA Department of Transportation Federal Railroad Administration. *High-Speed Ground Transportation Noise and Vibration*; USA Department of Transportation Federal Railroad Administration: Washington, DC, USA, 2012.
31. *NS 8176(2005)*; Vibration and Shock—Measurement of Vibration in Buildings from Landbased Transport and Guidance to Evaluation of Effects on Human Beings. Standard Norge: Oslo, Norway, 2005.
32. Ruihe, T.; Danqun, F. Evaluation of urban residents' subjective responses to environmental vibration using fuzzy mathematical principles. *China Environ. Sci.* **1990**, *10*, 356–359.
33. Zhigang, S. A New Annoyance-based Vibration Comfort Design Theory on Engineering Structures. Ph.D. Thesis, Zhejiang University, Hangzhou, China, 2003.
34. Xia, H.; Cao, Y.; Guido, R.D.; Geert, D. Environmental problems of vibrations induced by railway traffic. *Front. Struct. Civ. Eng.* **2007**, *1*, 142–152. [CrossRef]
35. Griffin, M.J.; Whitham, E.M. Individual variability and its effect on subjective and biodynamic response to whole-body vibration. *J. Sound Vib.* **1978**, *58*, 239–250. [CrossRef]

buildings

MDPI

Article

Numerical and Statistical Evaluation of the Performance of Carbon Fiber-Reinforced Polymers as Tunnel Lining Reinforcement during Subway Operation

Ishola Valere Loic Chango and Jun Chen *

School of Civil Engineering, Tongji University, Shanghai 200092, China
* Correspondence: cejchen@tongji.edu.cn

Abstract: Ground vibrations during train operations have become a serious problem in recent years. Local residents often feel disturbed by the vibrations emanating from the railroad line. This inconvenience is particularly pronounced in loose areas traversed by subways. However, improving the mechanical properties of tunnels has been the subject of several studies. Among these works, the widely discussed fiber-reinforced polymer (FRP) is considered as a material that can be incorporated into the tunnel structure to increase stiffness, durability, and corrosion resistance. However, the function of FRP in the interaction between the soil and the tunnel during operation has scarcely been studied. In this study, the effectiveness of carbon fiber-reinforced polymers (CFRP) as reinforcement of tunnel lining on ground vibration is investigated. For this purpose, a nonlinear 3D finite element model was developed based on a subway section in Shanghai to simulate the dynamic behavior of the system. The moving subway load was modeled as a transient dynamic load via a DLOAD subroutine, in which the rail irregularities are taken into account. The numerical model was efficiently validated by field tests. Then, the efficiency of using CFRP as concrete reinforcement of the tunnel lining during the subway operation was investigated. In addition, a statistical analysis of the ground dynamic response depending on the CFRP bars properties is presented, evaluated, and discussed.

Keywords: ground vibration; tunnel; subway train; carbon fiber-reinforced polymer; 3D finite element model

Citation: Chango, I.V.L.; Chen, J. Numerical and Statistical Evaluation of the Performance of Carbon Fiber-Reinforced Polymers as Tunnel Lining Reinforcement during Subway Operation. *Buildings* **2022**, *12*, 1913. https://doi.org/10.3390/buildings12111913

Academic Editors: Jan Fořt and Elena Ferretti

Received: 2 October 2022
Accepted: 31 October 2022
Published: 7 November 2022

1. Introduction

In recent years, the increasing modernization of cities has led to a general awareness of the serious effects of traffic-induced vibrations on the environment and urban life [1]. In all major cities of the world, the demand for subway transportation is increasing, which brings a major environmental problem related to the vibrations caused by subway operations [2,3]. The effects of vibrations caused by subways mainly affect the surrounding buildings, which are at risk of collapse if they become obsolete. In addition to the risk of affecting sensitive equipment in industry and research, vibrations affect people's work and health [4]. Exposure of the human body to indoor vibrations is evaluated in a certain range, which corresponds to the same range of vibrations caused by the subway [5]. Since the ground vibration and structure-borne noise caused by trains can be disturbing to humans, many countermeasures have been developed to reduce the effects of railroad vibration. Various types of isolation, such as open and filled trenches, concrete walls or piles, and flexible gas cushions, have been used [6].

Recently, due to unforeseen geological conditions leading to aging and deterioration of the structure [7,8], a new tunnel design based on fiber application has become one of the most important and promising technologies [9,10]. FRP was first used in the automotive, marine, and aerospace industries as a lightweight, high-strength, and high-modulus material [11]. With the advancement of technology, FRPs have become an attractive alternative for reinforcing concrete structures due to their advantages [12]. The performance of FRP grids embedded in PCM-shotcrete was estimated by numerical analysis considering various factors, such as soil classes, degree of deterioration of the lining, and condition of the tunnel [10,13]. The experimental study conducted on a pre-damaged tunnel showed that the use of carbon fiber-reinforced polymers (CFRP) mesh as reinforcement improved the bearing capacity and stiffness of the tunnel [14]. In some countries, FRP as internal reinforcement, such as bars, has been developed to reinforce non-prestressed and prestressed concrete structures, and the overall level of research, demonstration, and commercialization has increased significantly [15–17]. Strengthening tunnel linings with FRP bars increases the strength and ductility of the structure and reduces the width of cracks [18]. Several methods for reliable reinforcement design have been developed to optimize the use of FRP reinforcement bars in tunnel linings [19–21]. A blast test on a tunnel vault showed that FRP steel bars can prevent the occurrence of cracks in concrete to some extent [22].

Several studies have looked at FRP-reinforced structures. Most of these studies are based on the behavior or ability of FRP to reinforce concrete and resist corrosion through laboratory experiments or numerical modeling. Other studies have focused on design methods to optimize FRP-reinforced concrete structures. However, there are very few, if any, studies that focus on the real objective: Investigating the role of FRP on the behavior of the entire system; namely, the actual impact of FRP as bunker reinforcement during an explosion or to improve the dynamic response of the soil during subway operation. Since FRP is a material that is likely to be used permanently in structures, it would be imperative to study in detail the actual effects of FRP in an operational structure. In this way, the function of FRP could be better evaluated to prevent damage.

Objectives

The main objective of this study is to evaluate the effectiveness of FRP bars as reinforcement of tunnel lining in the presence of ground vibrations during subway operation. Then, the main factors that may affect the performance of FRP bars in improving dynamic ground behavior are evaluated. For this purpose, carbon fiber-reinforced polymers (CFRP) were selected for tunnel concrete reinforcement due to their high corrosion and chemical resistance, low density, high fatigue strength, and high elastic modulus [23,24]. In this context, a nonlinear 3D FE model of a double tunnel system was created using the Abaqus program [25]. The wheel load of the subway train was designed as a transient dynamic load implemented by a user-defined Fortran subroutine. For a more accurate characterization of the dynamic behavior of the tunnel lining, the plasticity behavior during concrete damage was included in the concrete properties. The reliability of the FE model was confirmed by comparing the calculated results with the vibration accelerations measured on the tunnel wall and the ground surface of the Shanghai subway line.

Subsequently, the influence of the CFRP bar reinforcement on the ground vibration was investigated. Statistical analysis was also performed to investigate the relationship between the distribution of ground vibration and the influencing factors, such as the type of CFRP, the CFRP bar used, and the location of the calculation point, as well as the effects of the interaction between the factors. The measured values were then recorded according to an orthogonal factorial matrix of the full fractional size L27 (3^{13-10}). The orthogonal matrix L27 (3^{13-10}) is a full fractional factorial design in which there are 13 columns that can be used to assign test factors and their interaction. In this case, with three factors and three levels, the total number of tests to be performed is 3^3 = 27. Therefore, analysis of variance was applied to investigate the influence of each factor and its interaction on the ground vibrations.

2. The Three-Dimensional Finite Element Model of the Investigated Double Metro Tunnel

2.1. Numerical Model Description

2.1.1. Concepts of Rail Track Dynamic Analysis

The wheelset loads of a moving subway train moving over the tracks at different speeds are undeniably dynamic. Therefore, the contact stress and the position of the axle loads are time dependent and can be considered as the sum of the static load and a constantly changing load. Various factors, such as the unevenness of the rail surface, the speed of the subway train, the weight, and the suspension system can affect the constantly changing dynamic load. The oscillating motion of the subway train due to the suspension system of the cars causes the dynamic axle loads to fluctuate by their average amplitudes during subway operation.

According to the usual FE procedure [26,27], the main equation of a nonconservative dynamical system with material damping can be defined by Equation (1). This mathematical statement can be solved by the implicit or explicit integration method in Abaqus. In this study, the implicit direct integration method is adopted since it is more efficient at the frequencies observed in the simulation of systems subjected to a moving load [28].

$$[M]\{\ddot{u}\} + [C]\{\dot{u}\} + [K]\{u\} = F(t) \tag{1}$$

where $[M]$ represents the mass matrix; $[C]$ represents the damping matrix; $[K]$ represents the stiffness matrix; $\{\dot{u}\}$ represents the velocity vector; $\{\ddot{u}\}$ represents the acceleration related to nodes; $\{u\}$ represents the displacement vector; and $F(t)$ represents the external force vector related to the structural dynamic system.

2.1.2. Subway Structure Model Geometry and Element Mesh

Shanghai Metro Line 8 (China) was selected as a reference to analyze the dynamic behavior of a structure at the intersection of two trains. The 44.1 km Shanghai Metro Line 8 starts at Shiguang Road in Yangpu District and ends at Shendu Highway in Minhang. The line consists of various route profiles, including tunnels and bridges. According to the objective, this study focused on the Shanghai Hongkou District Youth Sports Center-Xinghua Community section, which is defined by a double tunnel with a burial depth of 9 to 15 m.

The three-dimensional FE model, implemented in Abaqus, is 260 m long, 120 m wide, and 50 m high. The tunnel was built with a layered lining of segmental walls. The track system consists of rails, track slabs, and the track bed. As shown in Figure 1, the two tunnels are located at a depth of 11.1 m and have a spacing of 12.4 m, with the outer diameter of the tunnel lining being 6.2 m in each case. The soil in which the tunnel is buried consists of several partial layers corresponding to the Shanghai soil. The Euler-Bernoulli beam element was used to model the rail. A bond connection was used to ensure the mutual contact between the layers under the track slab with the continuity of the deformation of the interface [29]. A Cartesian coordinate system was used in which the transverse direction of the rail is indicated by the X-axis, the vertical downward direction by the Y-axis, and the longitudinal direction (direction of travel of the subway train) by the Z-axis.

When analyzing FE, the consistency of the mesh size has some influence on the analysis result. Meshing should be conducted to obtain the most accurate results. Several researchers have investigated the appropriate mesh size for a dynamic model. In [30], it was suggested that 20 mm should be used as the mesh length in the flow direction and between 15 and 18 mm in the lateral direction in the loading region. In the present study, the meshing of the FE model was carried out to increase the accuracy of the model. A relatively fine mesh was used along the path of the wheels since the stresses and displacements were high. A dense mesh was used near the loading area, while a relatively coarse mesh was used outside the loading area (Figure 2). The depth of the 3D mesh in the model was chosen appropriately depending on the thickness of each layer to avoid errors and warnings related to the mesh during the analysis. To improve the convergence rate, a continuum 3D-reduced-integration element (C3D8R) with eight nodes was used in the finite element field, while a continuum 3D infinite integration element (CIN3D8R) with eight nodes was chosen to define infinite boundaries on each side of the model of the tunnel system (Figure 2) [31].

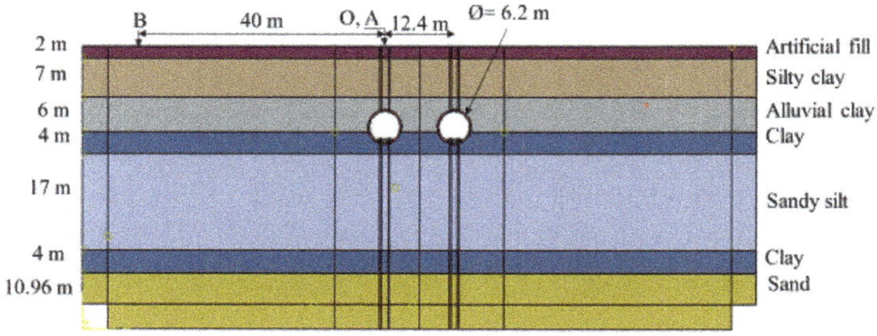

Figure 1. Tunnel description.

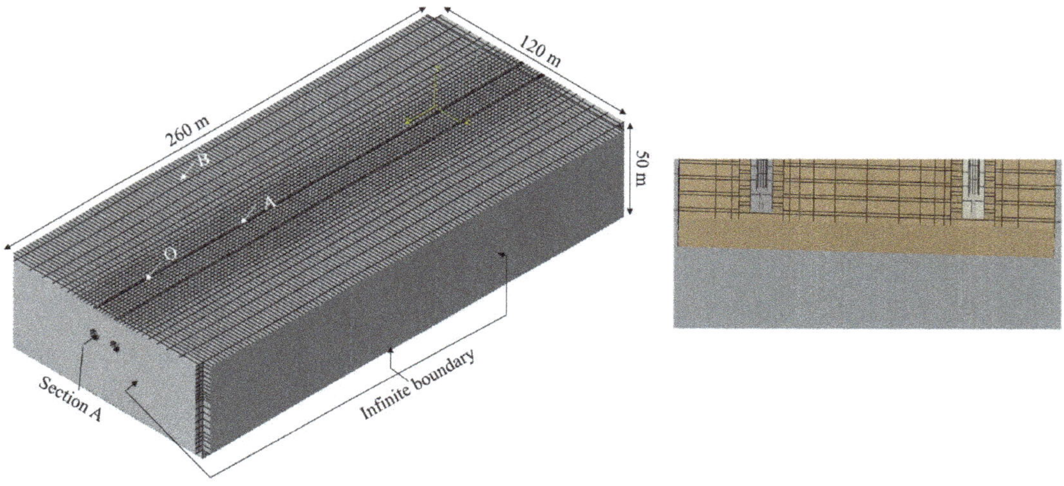

Figure 2. Model 3D mesh. (**a**) Mesh and boundary conditions view; (**b**) section A.

2.1.3. Boundary Condition and Material Damping

The boundary conditions imposed had some influence on the accuracy of the FE analysis results, thus it was important to choose appropriate and adequate boundary conditions. The boundary of a FE model is usually artificial; the wave generated by a moving subway train may propagate in the area of the structure until it is reflected at the artificial boundary, eventually contaminating the propagation of the wave in the domain. Therefore, previous studies have recommended the construction of non-reflecting or absorbing boundaries. As a result, an infinite element was chosen to eliminate the boundary effect of the tunnel model, absorb the wave energy, and reduce the degree of freedom in the far field. These elements are known to produce a quiet boundary without significant loss of precision for dynamic analyses [32].

The damping mechanism is a factor that affects the dynamic behavior of materials. It is related to the system itself, the viscosity of the surrounding medium, the energy dissipation of the subsurface, etc. Due to its complexity, it is difficult to accurately determine the damping matrix in FE analysis. The stress-energy factor method, modal damping method, stiffness factor method, and Rayleigh damping method are used to calculate the damping. In this FE analysis, the Rayleigh damping method was used, assuming that the damping matrix is a linear combination of the mass matrix and the stiffness damping proportion. The corresponding equation is defined in Equation (2).

$$C = \alpha[M] + \beta[K] \tag{2}$$

where α represents the mass damping coefficient; and β represents the stiffness damping coefficient. Therefore, the damping coefficients α and β depending on the structural natural frequency are assessed following Equation (3).

$$2\omega\xi = \alpha + \beta\omega^2 \tag{3}$$

With

$$\begin{cases} \alpha = 2\omega_1\omega_2\frac{\omega_1\xi-\omega_2\xi}{\omega_1^2-\omega_2^2} \\ \beta = 2\frac{\omega_1\xi-\omega_2\xi}{\omega_1^2-\omega_2^2} \end{cases} \tag{4}$$

The computational process of Rayleigh damping parameters is determined by a modal analysis performed with the numerical model implemented in Abaqus. The 50 natural frequencies were extracted (Table A1) to determine the natural angular frequency ω ($\omega = 2\pi f$). Therefore, the natural angular frequency ω_1 was determined by defining the fundamental natural frequency f_1. After defining the fundamental frequency, the second angular frequency ω_2 was determined based on the highest natural frequency selected from the other order vibration modes. It follows that the natural angular frequencies ω_1 and ω_2 determined for the calculation of the Rayleigh damping parameters are 17.31 and 19.89 rad/s, respectively. In this study, the damping rate ξ for the structure was chosen between 2% and 4%. With damping rate ξ between 2% and 4%, the Rayleigh damping coefficients α and β of the structure were calculated using Equation (4). The proportional damping value α is 0.69529 and the stiffness proportional damping value β is 0.0032.

2.2. Track Materials Characterization and Soil Profile

Assuming an elastic layer system, all railroad materials are linearly elastic. The rubber pads and the fasteners between the slab and the rail are considered as linear elastic components and were modeled using spring-damper elements with a damping coefficient of 45 kN s/m and a stiffness value of 45 kN/mm [33]. The material properties of the track and the specification values were taken from the results of previous work and are listed in Table 1.

Following some previous experiments [34,35], a soil profile corresponding to the typical conditions for soft soils in the Shanghai region was selected for this study. The soil profile studied at a depth of 100 m rests on an underlying rigid soil with an average shear wave velocity of 500 m/s. The rigid soil, which is at a greater depth, was not considered in this study to shorten the simulation time. In this way, the infinite boundary condition was applied below the last soil layer, as shown in Figure 2. The variation of shear wave velocities, density, cohesion C, and friction angle φ for the selected representative soil profile used for the numerical analysis is given in Table 1.

For the typical dynamic properties of Shanghai clay and sand, the variations of shear modulus $G/G_{\max}$ and damping ratio D with shear strain level γ are shown in Figure 3, based on the results of numerous resonant columns and cyclic triaxial tests [36]. It is worth noting that the groundwater effect was not considered in the dynamic analysis of the soil system. The shear modulus at low strain of the soil profile was derived from the dynamic properties as follows:

$$G_{\max} = \rho V_s^2 \tag{5}$$

where $G_{\max}$ represents the ground small strain-shear modulus, V_s represents the ground shear wave velocity, and ρ represents the ground density.

Table 1. Mechanical properties of the ground layers.

Materials Designation	Unit Weight (KN/m³)	Initial Void Ratio	Compression Modulus (MPa)	Poisson Ratio	Cohesion under CU Test (kPa)	Friction Angle under CU Test (°)	Permeability Coefficient	Shear Velocity t (m/s)
Rail	78	-	210×10^3	0.3	-	-	-	-
Track slab	25	-	32.4×10^3	0.2	-	-	-	-
				Soil type				
Artificial fill	17.8	0.96	5	0.33	10	13		110
Silty clay	18.6	0.86	9	0.32	14	12.5	1×10^{-7}	130
Alluvial clay	16.9	0.92	2.12	0.35	12	10.5	4×10^{-8}	140
Clay	17.5	0.82	3.13	0.33	16	12	2×10^{-7}	160
Sandy silt	18	0.78	5.44	0.28	12	18	8×10^{-7}	190
Sand	18.7	0.68	12.82	0.26	4	30	5×10^{-5}	250

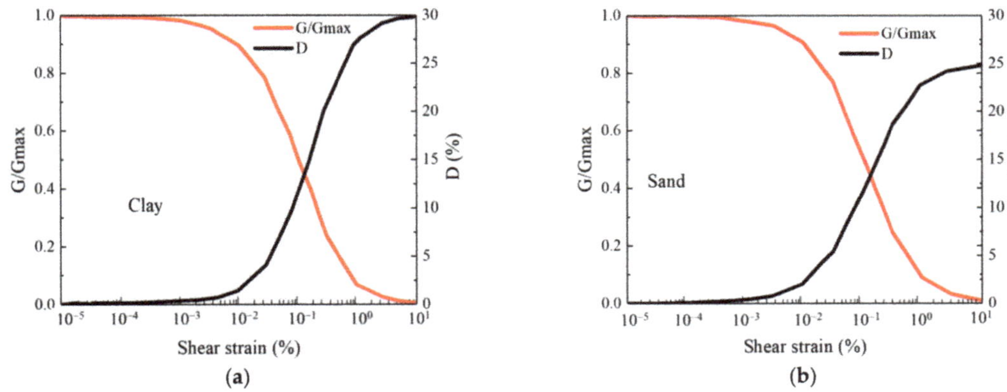

Figure 3. Typical G- γ -D curves for Shanghai soils. (**a**) Clays; (**b**) loose sands.

2.3. Characteristics of the Tunnel Structure

The tunnels of the Shanghai metro system are mainly constructed using the shield tunnelling method with a typical single-tube structure, i.e., a circular tunnel and a double-tube tunnel (DOT) with burial depths ranging from 9 to 33 m (Figure 4) [37]. The present work focuses on the dynamic response of the soil when a subway train passes through a subway tunnel. The subway studied here is a double tunnel with a typical circular cross-section. The two tunnels, which are 12.4 m apart and have an outer diameter of 6.2 m, were excavated using the shield tunnelling method, similar to the majority of tunnels in Shanghai. Therefore, after the tunnel was driven, the first lining (primary support) was made of steel ribs and fiber-reinforced shotcrete. Subsequently, the invert of the final lining was cast in plain concrete C35 with a thickness of 0.35 m.

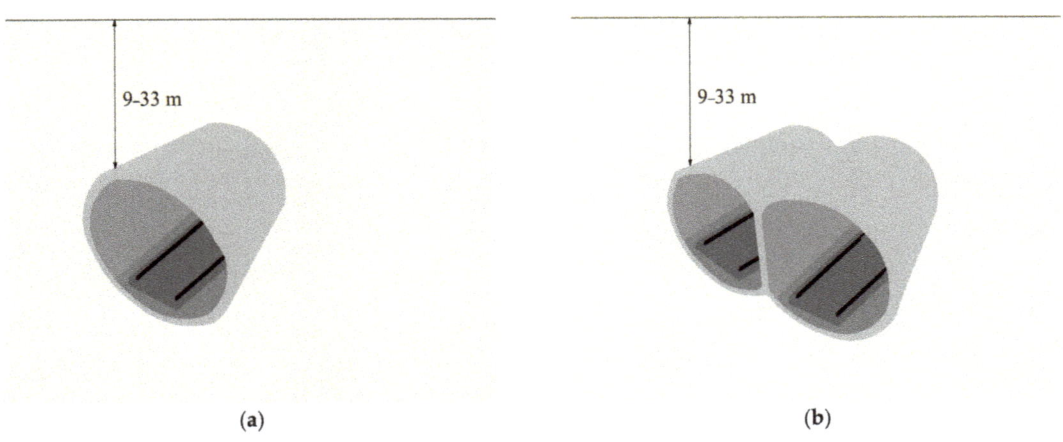

Figure 4. Type of tunnel often adopted in Shanghai. (**a**) Circular tunnel; (**b**) double-tube tunnel (DOT).

This study focuses on the effects of the final lining of the tunnel, which serves as a protective shield in the propagation of vibrations in the soil. Therefore, the resistive contribution of the first stage lining was neglected to reduce the simulation time due to the size of the 3D model. Moreover, this is a commonly accepted hypothesis since the stiffness of the shotcrete outer shell, which is in direct contact with the soil, is often neglected due to the fact that it may have undergone a physicochemical reaction that alters its original mechanical properties [38]. The geometrical and mechanical properties of the investigated tunnel sections are listed in Table 2.

Table 2. Physical parameters of the studied tunnel lining.

Density (g/cm^3)	Elastic Modulus (GPa)	Poisson Ratio	Dilatation Angle (°)	Viscosity Parameter	Eccentricity	K
2.4	3.55	0.2	35	0	0.1	0.67

Concrete damage plasticity "CDP" is one of the most popular concrete models used to simulate concrete behavior in Abaqus. For a complete definition of the CDP model in Abaqus, the following mandatory parameters should be entered: f_{b0}/f_{c0}, a ratio between the compressive strength in the biaxial state and the compressive strength in the uniaxial state, which is set to 1.16 as the default value in Abaqus, and the ratio K of the second stress invariant on the tensile meridian to the compressive meridian for the yield function, the dilatation angle, and the viscosity parameter [39].

2.4. Subway Dynamic Load Model

A six-car type A subway train manufactured by CNR Changchun Railway Vehicles Co., Changchun, China, was used for the field tests (Table 3). The individual components of the subway cars are considered as rigid parts connected by damper-spring elements. In this study, it was assumed that each subway car is properly proportioned and can be divided into four parts. The connection of the different car parts is shown in Figure 5. To specify the non-uniform distributed subway train load (DLOAD), the dynamic force due to the car movement on the rail was developed as a transient local dynamic load via a subroutine that allows the user to specify the magnitude change in the distributed load as a time functional form (TIME *), coordinates (COORDS *), the number of load integration points applied, and the number of domain elements [40]. The simplification of the calculation methods has led to the assumption that the wheels of the subway and the rails are always in contact, which indicates that there is no relative vertical displacement during the operation.

Table 3. Metro-train parameters of type A subway.

Mass of carriage (kg)	50,878	Inertia of Bogie/(kg·m^2)	3605
Mass of Bogie (kg)	2721	Stiffness of primary suspension spring (N/m)	2.14×10^6
Mass of Wheel Axle (kg)	1900	Damping of primary suspension spring (N·s/m)	4.9×10^4
Inertia of Carriage/(kg·m^2)	2.446×10^6	Stiffness of secondary suspension spring (N/m)	2.5×10^6
Distance of Wheel Axle in a Bogie (m)	2.50	Damping of primary suspension spring (N·s/m)	1.96×10^5
Distance of Bogies in a Carriage (m)	15.7	Radius of Wheel/m	0.42

The irregularity of the rail is regularly modeled as a simple or composite concave cosine wave [41]. Therefore, the presence of a cosine track irregularity (Z_w) on top of the rail is assumed with an amplitude A and wavenumber k_x (with λ, the wavelength). The track irregularity is defined as follows:

$$z_w(t) = A(1 - \cos(k_x t)) \tag{6}$$

The model of the subway train, which consists of six cars for the test, is created considering the quarter car model. Each car consists of two bogies with two sets of wheels per bogie. The contact force between the car wheelset and the rail, defined in [42], was expressed as follows, taking into account the track irregularities:

$$p(y,t) = \sum_{n=1}^{6}\sum_{i=1}^{4} p_{ni}(y - vt) \tag{7}$$

where y is the distance between the subway-train axle and a reference point at the subway-train head; v, t, and i are the metro-train speed, the time, and the axles numbering in a metro-carriage, respectively; p_{ni} $(y - vt)$ is the ith subway-train wheels pair in the n_{th} subway-train wheelset-rail contact force:

$$\begin{aligned}\sum_{i=1}^{4} p_{ni}(y-vt) = \quad & p_{n1}\delta\left(y - vt + \sum_{q=0}^{n-1} L_q + L_0\right) + p_{n2}\delta\left(y - vt + c_n + \sum_{s=0}^{n-1} L_q + L_0\right) \\ & + p_{n3}\delta\left(y - vt + c_n + d_n + \sum_{s=0}^{n-1} L_q + L_0\right) \\ & + p_{n4}\delta\left(y - vt + 2c_n + d_n + \sum_{s=0}^{n-1} L_q + L_0\right)\end{aligned} \tag{8}$$

where L_0 is the distance between a reference point in the operating direction and the subway-train head; L_q is the $(n - 1)$th length of a subway-carriage; c_n and d_n are the distance between two subway-train wheel axles in the n_{th} subway-carriage and the distance between two adjacent subway-train axles in n_{th} carriage bogies; δ represents a Dirac Delta function.

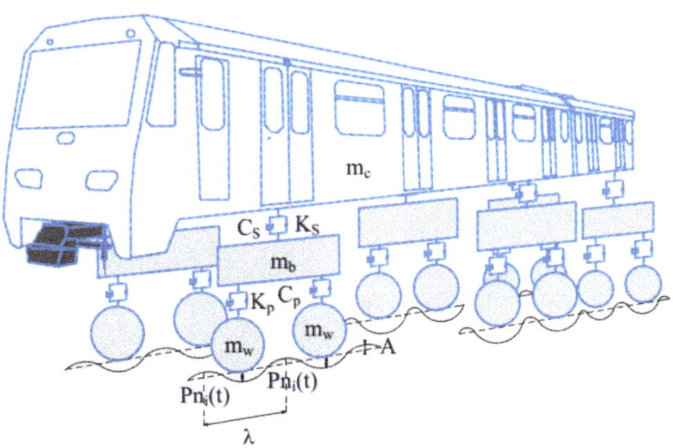

Figure 5. Metro-train geometry.

Substituting Equation (8) into the quarter-subway carriage model, the following expression was deducted:

$$\mathrm{P_{ni}}(\mathrm{k_x}, \omega) = \mathrm{W_{ni1}}\delta(\omega - \mathrm{k_x}\nu) + \mathrm{W_{ni2}}\delta(\omega - \omega_\mathrm{r} - \mathrm{k_x}\nu) + \mathrm{W_{ni3}}\delta(\omega + \omega_\mathrm{r} - \mathrm{k_x}\nu) \tag{9}$$

where δ is a Dirac Delta function, W_{ni} represents the sub-item of the ith carriage wheel–rail contact force in the n_{th} car, and ω_r represents the excitation frequency due to rail surface irregularity, which can be defined by $\omega_r = 2\pi v/\lambda$; λ is the rail wavelength.

The i_{th} transient dynamic of the subway wheel at a position x, obtained with Fortran and included in the implicit analysis, was defined by the combination of Equations (6) and (9). The equivalent parameters of the subway model are summarized in Table 3.

Depending on the type of section studied, two scenarios were considered, upward and downward. Therefore, the scenario where two subway trains run simultaneously in the upward and downward direction was also developed using the Fortran subroutine.

3. Numerical Model Validation

3.1. Comparative Analysis between Numerical Model Result and Field Measurements

The validation of the model was conducted in two stages, which are described below.

Stage 1

In the first phase, field measurements were conducted on top of a subway tunnel in the section of Line 8 between the Shanghai Hongkou District Youth Sports Centre and Xinghua Community. During the vibration measurement campaign, several points were selected that were triggered by subway vehicles (see Figure 6). The accelerometers used are LPMS-B2 series sensors with a frequency range of 0 to 400 Hz, a measurable acceleration field of 0.1 to 10 g, a latency of 20 ms, an accuracy of <0.5° (static) and <2° RMS (dynamic), an operating voltage of 5.5 VDC, an output voltage range of 0.5 to 4.5 V, and an operating temperature of −40 to 80 °C.

Figure 6. Distribution of monitoring point.

During the field measurement, the sensor frequency was set to 200 Hz while the metro train was traveling at approximately 80 km/h. Points D11 and D12 (see Figure 5) were selected for model validation due to their proximity to the tunnel. The other points were not considered due to their location, which could be influenced by external traffic.

The dynamic accelerations at points D11 and D12 (top of the ground) are shown in Figures 7 and 8. In both figures, the periodic occurrence of a wave crest series can be seen. The amplification of these waves in a given period corresponds to the dynamic effect of each subway bogie when it comes into contact with the respective location. Moreover, a great similarity was observed between the data measured in the field and the results calculated with the FE model. In Figure 7, the dynamic acceleration RMS measured in the field is 1.15×10^{-2} m/s^2 and the calculated dynamic acceleration RMS is 1.24×10^{-2} m/s^2. The specified difference is 7.26% on average. In Figure 8, the dynamic acceleration RMS measured in the field was 5.32×10^{-3} m/s^2, and the calculated dynamic acceleration RMS was 4.79×10^{-3} m/s^2. Here, the reported difference is 9.96% on average.

In Figure 9a,b, the frequency spectrum at points D11 and D12 were depicted. In both figures, a great similarity was observed between the data measured in the field and the results calculated with the FE model. In Figure 9a, the acceleration amplitude RMS measured in the field is 8.54×10^{-4} m/s^2/Hz and the calculated acceleration amplitude RMS is 7.61×10^{-4} m/s^2/Hz. The specified difference is 10.88% on average. In Figure 9b, the acceleration amplitude RMS measured in the field was 3.23×10^{-4} m/s^2/Hz, and the calculated acceleration amplitude RMS was 2.87×10^{-4} m/s^2/Hz. Here, the reported difference is 11.14% on average.

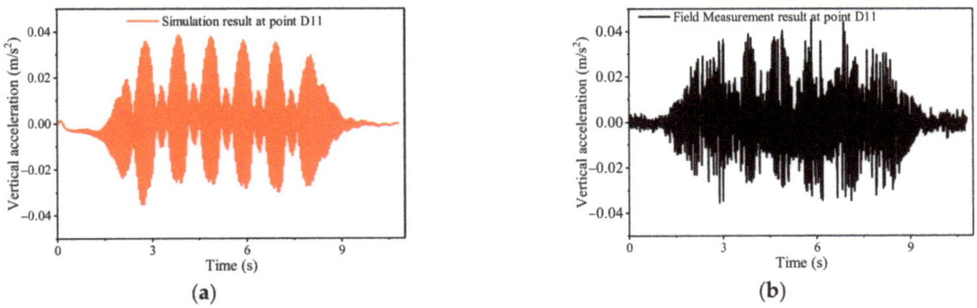

Figure 7. Vertical acceleration at point D11. (**a**) Numerical model result; (**b**) Field measurement result.

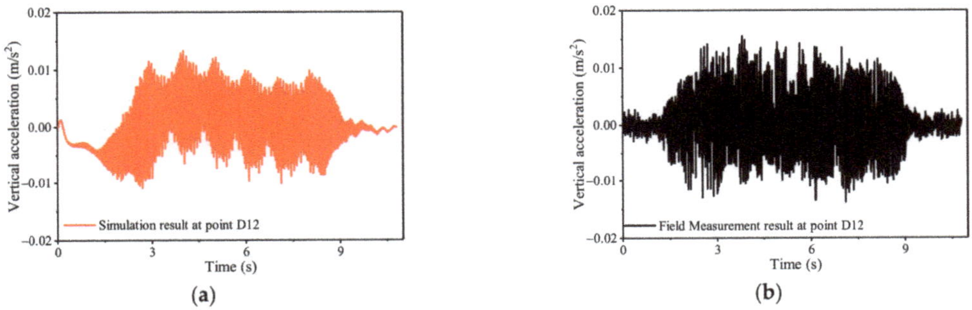

Figure 8. Vertical acceleration at point D12. (**a**) Numerical model result; (**b**) Field measurement result.

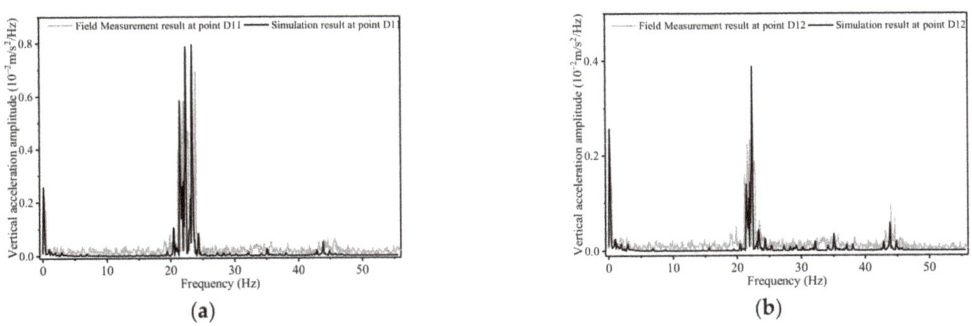

Figure 9. Vertical acceleration amplitude at points D11 and D12. (**a**) Numerical model result; (**b**) Field measurement result.

Stage 2

This second stage of numerical model validation is based on the in situ measurement of the dynamic response of a double tunnel system in Shanghai presented by S. Zhou et al. [43]. At this stage, only the time history was considered, since this is the only data available from Zhou's work to verify the results of the numerical model in the context of this work. A six-car type A subway train is also used on this subway line, which passes through the test site at a speed of about 54 km/h. To record the accelerations induced by the subway train, two measurement points were placed on the tunnel cross section as shown in Figure 10.

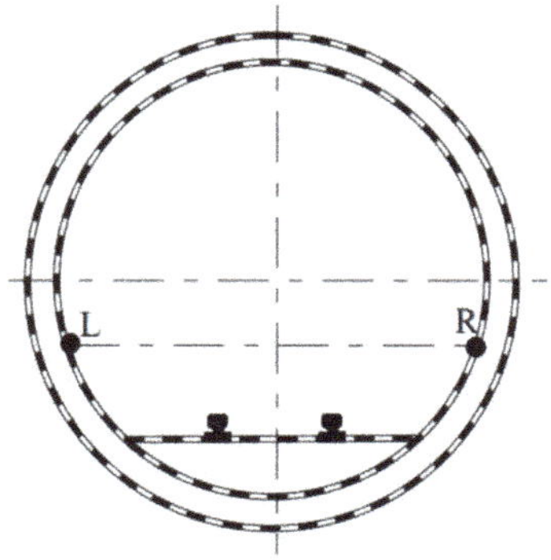

Figure 10. Schematic diagram of the measuring points in the tunnel.

The dynamic accelerations at points R and L are shown in Figures 11 and 12. In all figures, the periodic occurrence of a wave crest series can be seen. The amplification of these waves in a given period corresponds to the dynamic effect of each metro bogie when it is in contact with a given location. Moreover, a high agreement was observed between the data measured in the field and the results calculated with the FE model. In Figure 11, the maximum dynamic acceleration measured in the field is 0.338 m/s^2 and the calculated maximum dynamic acceleration is 0.366 m/s^2. The specified difference is 7.65% on average. In Figure 12, the maximum dynamic acceleration measured in the field was 0.341 m/s^2, and the calculated maximum dynamic acceleration was 0.360 m/s^2. Here, the reported difference is 5.28% on average.

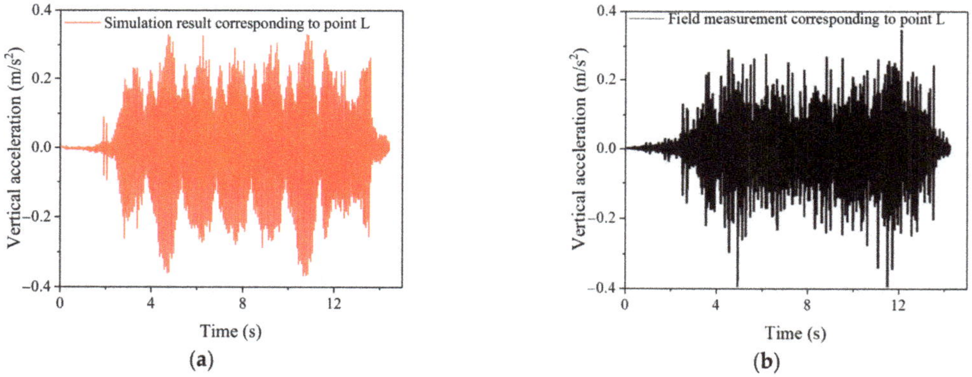

Figure 11. Vertical acceleration at point L. (**a**) Vertical acceleration from numerical model; (**b**) vertical acceleration from field test measurement Reprinted from Ref. [43]. 2019, Zhou Shunhua.

Combining the results observed in Stages 1 and 2, the causes of the observed differences could be due to some of the conditions described below:

(1) The intended speed fluctuated since it was difficult to maintain the speed of the metro train as stable during the operation.

(2) The sensors are subjected to vibrations when the subway train approaches and remain subjected to these vibrations even after the wheelset passes. As a result, the response of the sensors is affected. This illustrates the difference between the inclination of the spikes of the 3D FE model and the vibrations of the field measurement.

(3) The controlled wheel forces of the subway train were assumed to be uniformly distributed over a contact area between each wheelset and the rails.

(4) Rayleigh damping parameters are based on a modal analysis of the whole system, while in the field each material responds according to its damping.

(5) The external traffic may affect the recorded data slightly.

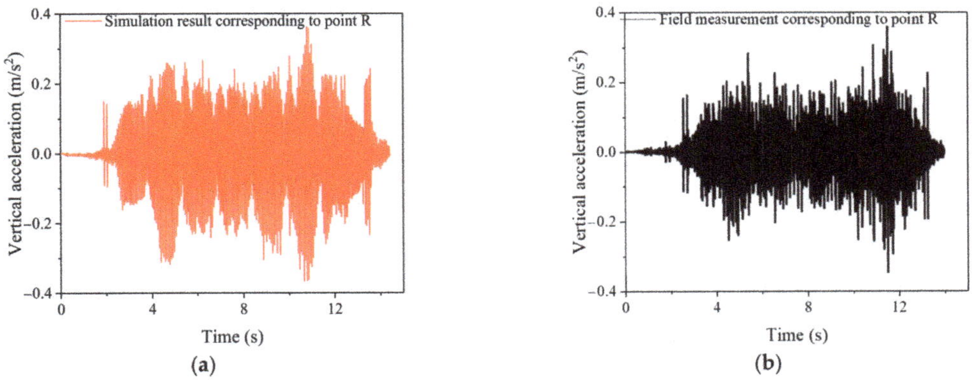

Figure 12. Vertical acceleration at point R. (**a**) Vertical acceleration from numerical model; (**b**) vertical acceleration from field test measurement Reprinted from Ref. [43]. 2019, Zhou Shunhua.

3.2. Correlation Analysis between the Numerical Results and Field Data Test

The comparative analysis between the field test data and the numerical results shows that the 3D FE model implemented in Abaqus accurately predicts the dynamic response of the metro system. Nevertheless, there is a slight margin of error in terms of the shape of the curve and the amplitude. To investigate the impact of these observed differences on the accuracy of the model in predicting the dynamic response of the structure, a correlation study was performed considering all variables. The measured vibration accelerations from the calculation with the numerical model and from the field tests were processed. The correlation between two variables u and v is obtained by calculating a coefficient Y_{UV}.

$$Y_{uv} = \frac{\sum(u_i - \overline{u})\sum(v_i - \overline{v})}{\sqrt{\sum(u_i - \overline{u})^2}\sqrt{\sum(v_i - \overline{v})^2}} \tag{10}$$

where $\overline{u} = \frac{1}{n}\sum_i^N u_i$ denotes the mean of u; and $\overline{v} = \frac{1}{n}\sum_i^N v_i$ denotes the mean of v.

The calculation of the Pearson coefficient of correlation leads to a result in the interval (−1;1); the sign indicates the direction of the relationship. However, r = 0 indicates that there is no linear relationship. The degree of correlation between the results calculated with the FE model and the measured field data is given in Table 4.

Table 4. Correlation matrix of field data and numerical model.

		FE Model				
		Vibration Acceleration at Point D11	**Vibration Acceleration at Point D12**	**Vibration Acceleration at Point L**	**Vibration Acceleration at Point R**	**Sig (Two-Tailed)**
Field Test	Vibration acceleration at point D11	0.985 **				0.000
	Vibration acceleration at point D12		0.991 **			0.000
	Vibration acceleration at point L			0.987 **		0.000
	Vibration acceleration at point R				0.982 **	0.000

**: Correlation is significant at the 0.001 level (two-tailed).

The presentation of the correlation matrix of the vibration accelerations measured during the field tests and the results of the FE model shows that the correlation factor varies between 0.982 and 0.991 depending on the location. Therefore, the dynamic response of the metro system calculated by the FE model and the vibrations measured during the field test at different locations have a very significant correlation at the level of 0.001. As a result, the difference between the results of the FE model and the field data in terms of the shape and magnitude of the curve does not affect the ability of the FE model to accurately predict the dynamic response of the subway system during the subway operation.

4. Influence of Subway Operation Direction on the Ground Vibration

Since the purpose of this study is to analyze the environmental ground vibration caused by the moving subway train, it is important to study the unfavorable case to better evaluate the dynamic behavior and the different solutions to improve the dynamic behavior of the ground. During the subway operation, the crossing of two subway trains occurs frequently. During this overlapping period, a change in vibration is noticeable. In this section, the ground vibration is evaluated when the subway train travels in one direction and in two directions (upward and downward). The vibrations caused by the subway train were investigated in two cases. Case 1 is described by the operation of the subway train in a single upward direction at 60 km/h. Case 2 is described by the operation of two subway trains in upward and downward direction moving at the same speed of 60 km/h.

Figures 13 and 14 show the time history of the vertical acceleration and amplitude spectrum at the ground surface (point O) for both cases studied. This point was chosen to avoid a complete crossing of the two subway trains during the calculation by the numerical model. As can be seen, the time history of acceleration in case 1 differs significantly from case 2. More precisely, the maximum acceleration for case 1 and case 2 is 0.417×10^{-2} m/s^2 and 2.313×10^{-2} m/s^2, respectively. The maximum acceleration for case 1, which is very small, is about 18% of the maximum acceleration for case 2. Moreover, the two amplitude spectral curves are similar in terms of the periodic occurrence of amplitude peaks in a certain frequency range of (0–3.75 Hz), (15.2–19.36 Hz), and (26.45–35.67 Hz). However, it should be noted that also in this section the maximum acceleration amplitude for case 1 is 21.6% of the maximum acceleration amplitude for case 2.

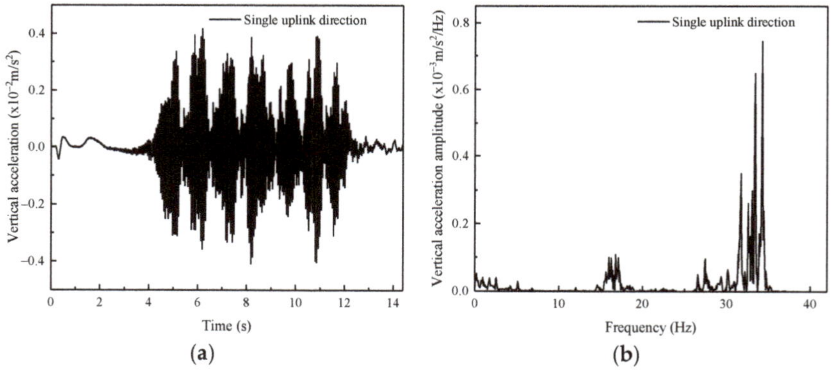

Figure 13. Vertical vibration at point O during the uplink operation. (**a**) Time history of dynamic acceleration; (**b**) acceleration amplitude-frequency.

From this analysis, it is clear that consideration of the ground vibrations caused by the two moving subway trains in the upward and downward directions is critical to evaluate the vibrations of the surrounding buildings and the effectiveness of the improvement methods. This analysis highlights the importance of considering the vibration effect caused by two trains traveling simultaneously in the upward and downward directions.

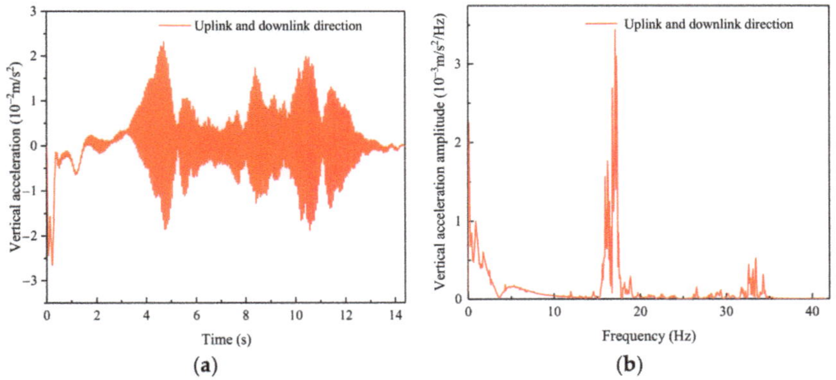

Figure 14. Vertical vibration at point O during the uplink and downlink operations. (**a**) Time history of dynamic acceleration; (**b**) acceleration amplitude-frequency.

5. Numerical Study of the Performance of the CFRP-Reinforced Tunnel

In this section, the effect of carbon fiber-reinforced polymers (CFRP) on the ground vibration caused by the moving subway train is investigated. In this case, the CFRP bars are used as reinforcement for the concrete of the tunnel lining.

5.1. Constitutive Model of CFRP Rebar-Reinforced Tunnel Lining Concrete

In typical applications, CFRP reinforcement consists of straight bars manufactured using a proven industrial pultrusion technology. In underground tunnels, reinforcement with a curved configuration is required, and the pultrusion process cannot be used. For this purpose, a modified pultrusion process called "forming" has been developed to produce curved bars with a constant and large radius of curvature. The behavior of prefabricated concrete segments with GFRP reinforcement produced by the tensile training technology is reported in [18,44]. In this study, the reinforcement cage consisted of continuous curved bars coupled in the longitudinal direction (intrados/extrados).

Starting from a traditional steel reinforcement cage (SR), which served as a reference, the CFRP reinforcement cage was designed in Abaqus. The reinforcement consists of curved bars of Ø12 longitudinal frames with a spacing of 11 cm. The transverse reinforcement consists of Ø12 bars of straight frames closed with stirrups spaced 11 cm apart, as shown in Figure 15. Elastic isotropic behavior to failure was assumed for the CFRP, which is considered a solid and homogeneous element. The Poisson's ratio was assumed to be 0.3. The reinforcement cage was embedded in the tunnel lining with a concrete cover of 50 mm. The typical mechanical properties of the CFRP are given in Table 5. The high modulus polyacrylonitrile carbon was used for this study. The embedded element option was chosen to connect the CFRP reinforcement to the concrete. The reinforcement was used as the embedded two-noded deformable truss (T3D2R) element. It is assumed that the CFRP cage is well anchored to the concrete to act as an effective shear or tension reinforcement element.

5.2. Dynamic Behavior of Soil during Operation of Subways

In this section, the influence of CFRP on dynamic soil response is evaluated. For this purpose, the characteristic distribution of soil vibrations and the influence of subway train speed on CFRP performance are analyzed.

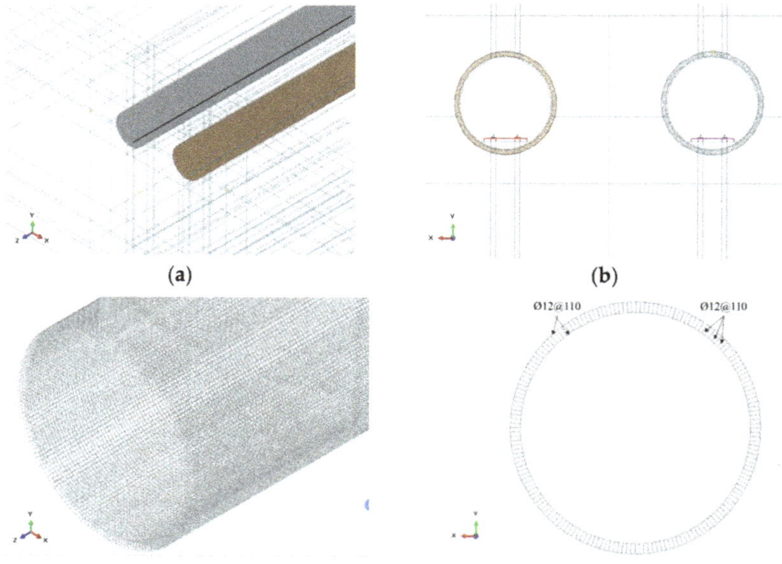

Figure 15. CFRP reinforcement cage model. (**a**) 3D view of the whole model with reinforcement; (**b**) 2D view of the model with reinforcement; (**c**) 3D view of the reinforcement cage model; (**d**) description of the rebar disposition.

Table 5. Typical mechanical property of carbon fiber-reinforced polymer (CFRP).

	Carbon Fiber			
	Polyacrylic Nitril Carbon		**Pitch Carbon**	
	High Strength	**High Modulus**	**Ordinary**	**High Modulus**
Density (kg/m^3)	1.7–1.8	1.9	1.65	2.0
Tensile strength (MPa)	3.4×10^3	3.2×10^3	0.9×10^3	3.2×10^3
Young's modulus (GPa)	228	517	38	620.4
Elongation (%)	1.55	0.6	2.3	0.95
Coefficient of thermal expansion (10^{-6}/°C)	−0.4	−0.65	−0.4	−0.8

5.2.1. Distribution Characteristics of Ground Vibrations Due to a Moving Subway Train

To represent the vibration characteristics caused by the subway train on the ground surface, a total of 4477 points are selected to evaluate the vibrations. The distance between two points follows the mesh lattice of the model, in order that two points located in the area of load application are very close to each other. The distance between two points increases the farther the points are from the load application area.

The color map of the ground vibrations in the time domain is obtained by linear interpolation, as shown in Figures 16 and 17. The vibrations at the top of the soil for the standard tunnel and the vibrations at the top of the soil for the structure reinforced with CFRP bars are shown in Figures 16 and 17, respectively. For both color plots, a large acceleration RMS (root-mean-square) is observed at the intersection of the two subway trains. The acceleration RMS decreases as the calculated point is close to the edges of the ground surface in the model. In Figure 16, three vibration amplified regions were observed in the crossing area of the two subway trains. Once the structure is reinforced with CFRP reinforcement (Figure 17), the vibration amplifications observed in the crossing area decrease. Moreover, when the tunnel structure is reinforced with CFRP reinforcement, an average reduction of 6.4% in acceleration RMS is observed over the entire ground surface. However, it should be noted that the reduction rate varies from one point to another, as shown in Figure 18a.

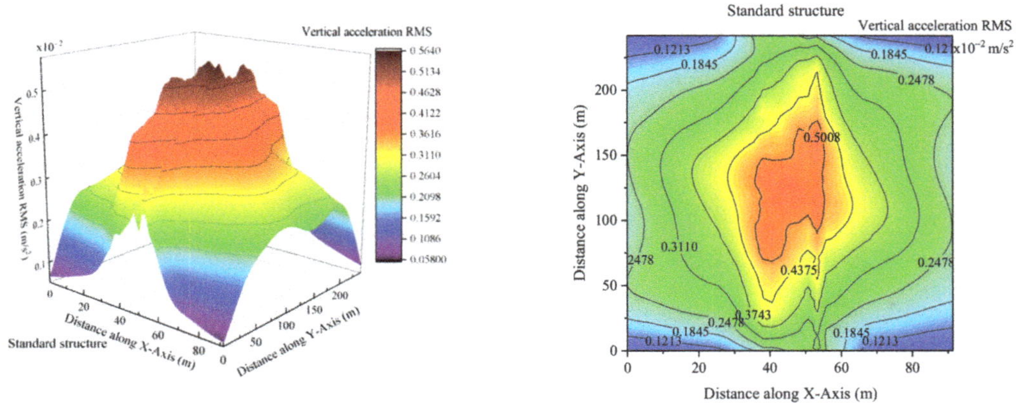

(a) (b)

Figure 16. Color chart of the vibration RMS for the standard structure. (**a**) 3D diagram; (**b**) vertical view.

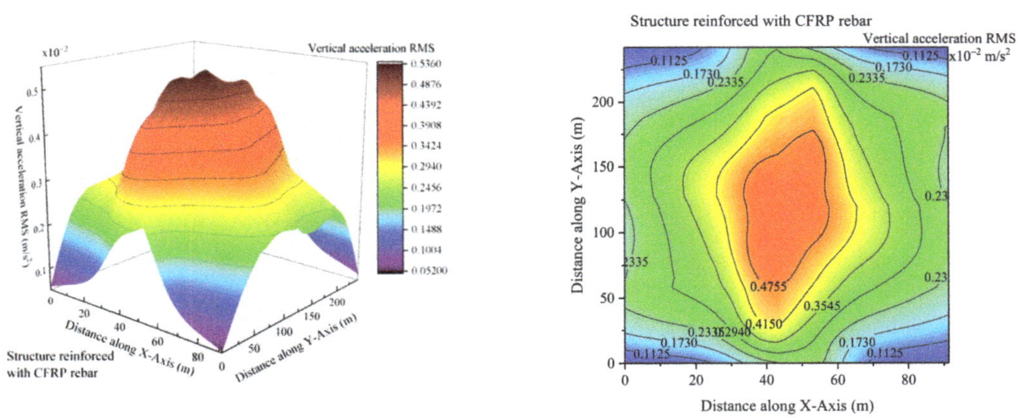

(a) (b)

Figure 17. Color chart of the vibration RMS for the structure reinforced with CFRP rebar. (**a**) 3D diagram; (**b**) vertical view.

Figure 18a shows the calculated acceleration RMS along the X-axis, perpendicular to the subway train travel direction. As can be seen, the maximum calculated acceleration RMS are located above each tunnel's center and decrease with the increasing distance. The calculation points corresponding to the two peaks are each located above the center line of each lane of the tunnel. Therefore, the combination of the effect of the two metro-train wheel lines located on either side of the track central line generates an increase in vibration at the track center of each tunnel. Therefore, vibration peaks are observed at the points above the center line of each tunnel track.

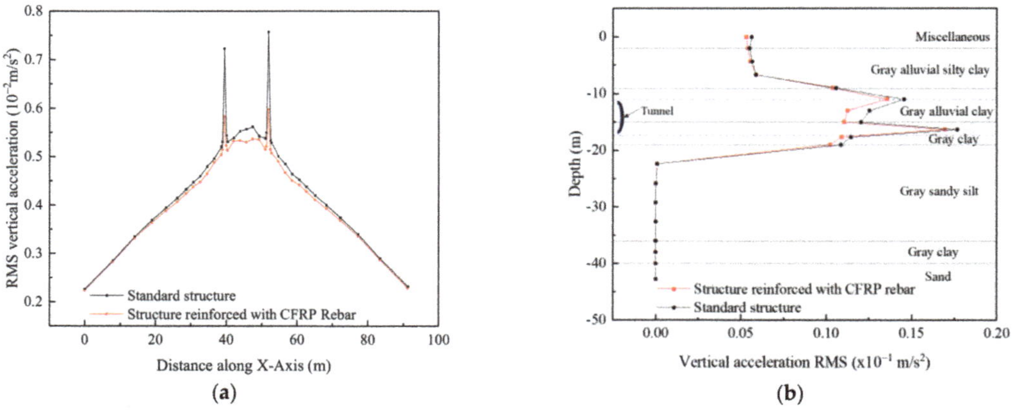

Figure 18. Vibration acceleration RMS. (**a**) Vibration RMS along model X-axis; (**b**) vibration RMS according to the depth.

For example, the acceleration RMS over the center (point A) of the standard tunnel and the tunnel reinforced with CFRP reinforcement is 7.57×10^{-3} m/s^2 and 5.97×10^{-3} m/s^2, respectively. At the edge of the ground surface (point B, e.g., 0 m, as described in the graph), the acceleration for the standard tunnel and the tunnel reinforced with CFRP reinforcement is 2.26×10^{-3} m/s^2 and 2.23×10^{-3} m/s^2, respectively. When the structure is reinforced with CFRP reinforcement, the acceleration RMS decreases by 21.14% above the center of the tunnel and by 6.63% on average in the area between the two tunnels. The reduction rate of the acceleration RMS varies from 3.89% to 1.33% as one moves away from the center of the tunnel.

The study of the distribution of acceleration RMS between the two tunnels as a function of depth is shown in Figure 18b. The acceleration RMS first increases with depth until it reaches a peak near the tunnel, then the acceleration RMS starts to decrease with the increasing depth. A peak was recorded at each pole line of the tunnel. In this case, amplified vibrations occur near the tunnel during subway operation. This phenomenon was also observed in the field measurement experiment by Qiang et al. [35]. When the tunnel lining is reinforced with CFRP reinforcement, a reduction in acceleration RMS in the soil is observed. The acceleration RMS is significant near the tunnel and decreases with the increasing or decreasing depth. The maximum acceleration RMS near the tunnel is 1.77×10^{-2} m/s^2 and 1.69×10^{-2} m/s^2 for the standard structure and the reinforced structure, respectively; which results in a reduced rate of 4.52%. Above the ground surface, the acceleration RMS reduction rate is about 5.51%.

Using Ø12 CFRP rods to reinforce the concrete of the tunnel lining helps in reducing the vibration distribution caused by the subway train. The reduction rate varies from 1.33% to 21.14% at the top of the ground and 3.82% on average at the bottom, considering the depth at the center of the two tunnels.

Tunnels with plain concrete have been built all over the world taking into account the load that the tunnel will carry. This work shows that tunnel lining can also play a key role in the propagation of the induced vibration. Therefore, its design method has an impact on the vibrations in the system. This is due to the results obtained when the reinforcement is made with bars and, in particular, with fiber-reinforced composites (FRP) due to the many other advantages. For example, reinforcing the tunnel lining with CFRP bars increases the dynamic capacity of the soil to respond to the vibrations caused by the passage of a subway train in the tunnel. Therefore, the installation of CFRP bars in the tunnel structure improves the dynamic properties of the soil by reducing the vibration in the time domain during train operation. This phenomenon is an advantage for the foundations of the surrounding buildings, which are less affected by the various dynamic loads caused by the subway train.

5.2.2. The Influence of the Speed of the Metro Train on the Ground Vibrations

Analysis at a Point Located on the Ground Surface above the Tunnel (Point A)

In this section, the frequency spectrum obtained with the fast Fourier transform is analyzed during subway operation. The objective of this analysis is, first, to investigate the dynamic behavior of the soil during subway train operation at different speeds in the tunnel and, second, to determine the effect of speed on the performance of the tunnel concrete reinforced with CFRP bars.

Figure 19 shows the acceleration spectrum at the ground surface above a tunnel centerline (point A). As can be seen, the number and size of the periodic amplitude cycles vary with speed. At 40 km/h, three periodic cycles are calculated with peak amplitudes averaging 2.22×10^{-1} m/s^2/Hz. The number of significant periodic amplitude cycles decreases with the increasing speed with an increase in the maximum peak in the frequency domain. At 40 km/h, the subway train motion is considered quasi-static. Therefore, the relatively long time of wheel-rail contact generates low ground vibrations.

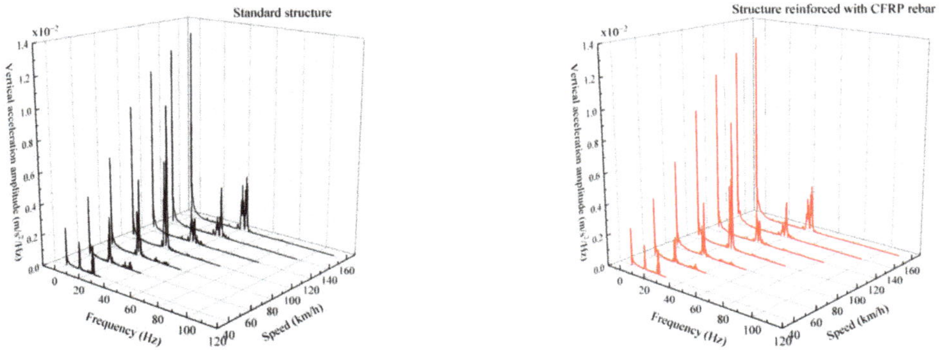

(a) (b)

Figure 19. Spectrum of the vibration amplitude at point A. (**a**) Acceleration amplitude-frequency for the standard structure; (**b**) acceleration amplitude-frequency for the structure reinforced with CFRP rebars.

When the subway train reaches a speed of 100 km/h, the peak values of the second periodic series of acceleration amplitudes are larger than the peak values of the first periodic series of amplitudes; subsequently, the peak values of the second cycle of acceleration amplitude decrease with the increasing speed. This phenomenon can be attributed to the dynamic amplification in the frequency domain when the speed of the subway train reaches a certain value.

When the tunnel is reinforced with CFRP reinforcement bars, the acceleration amplitude at the ground surface above the tunnel centerline decreases regardless of the speed of the subway train. At a speed of 100 km/h, the value of the second cycle of acceleration amplitude peaks decreases significantly and becomes smaller than the peak value of the first periodic cycle of the amplitude. From this analysis, it appears that reinforcing the concrete of the tunnel lining with CFRP bars increases the stiffness of the ground-tunnel system, which would contribute to the gradual elimination of the vibrations amplified in the ground when the train reaches a certain speed.

Figure 20 shows the acceleration amplitude RMS at the ground surface at various subway speeds. As can be seen, dynamic amplification occurs in the frequency domain when the subway train reaches a speed of 100 km/h. It can be seen in Table 1 that this velocity coincides with the shear wave velocity of a soil layer. The influence of the concrete reinforcement of the tunnel lining with CFRP reinforcement bars on the ground vibrations varies from speed to speed. The reduction rate of the acceleration amplitude RMS increases with the increasing speed up to 100 km/h. At this speed, where amplification dynamics in the frequency domain can be observed, the reduction rate of acceleration amplitude decreases, and then increases again with the increasing speed. Therefore, the reduction rate of vibrations at a subway speed of 40 and 80 km/h is 12.5% and 22.1%, respectively.

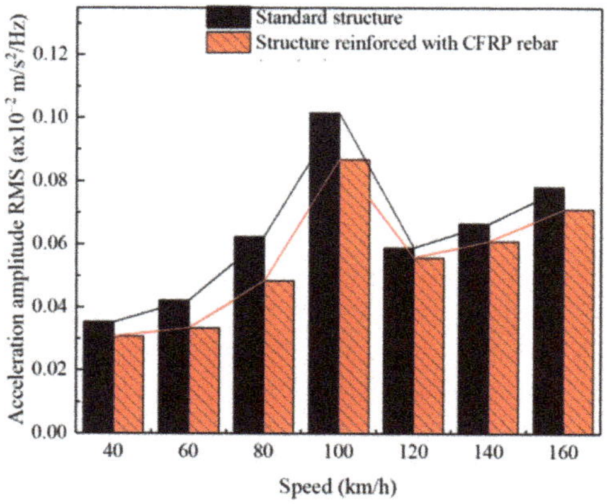

Figure 20. Acceleration amplitude RMS diagram at point A.

The installation of Ø12 CFRP bars in the concrete of the tunnel lining reduces the dynamic reinforcement to a certain extent. Using Ø12 CFRP bars helps in reducing the acceleration amplitude in the ground to a certain degree in order that the ground vibrations are less amplified when the speed changes from 80 to 120 km/h. This rate is determined by the gap between the acceleration amplitude RMS when the speed of the subway changes during operation, as described in Table 6. From this, it can be seen that the acceleration amplitudes RMS are close to each other when the structure is reinforced with CFRP bars. This value may vary depending on the type of CFRP used.

Table 6. Acceleration amplitude RMS at point A.

	Acceleration Amplitude RMS			**Gap between Acceleration Amplitude RMS**	
Speed (km/h)	80	100	120	80 to 100	100 to 120
Standard tunnel	6.23×10^{-4}	1.02×10^{-3}	5.88×10^{-4}	3.97×10^{-4}	4.32×10^{-4}
Tunnel With CFRP rebar Ø12	4.85×10^{-4}	8.68×10^{-4}	5.57×10^{-4}	3.83×10^{-4}	3.11×10^{-4}

Analysis at a Point on the Ground Surface 40 m from the Tunnel (Point B)

Figure 21 shows the amplitude spectral curves of acceleration at 40 m from the tunnel center at various subway speeds. As can be seen, the acceleration amplitude at a large distance from the tunnel center (point A) is small, with the peaks of the acceleration amplitude occurring in a single periodic cycle. However, the peaks of the acceleration amplitude increase as the speed of the subway train increases. No amplification dynamics are observed in the speed range used for the study, but the acceleration amplitude RMS increases significantly when the subway train reaches 100 km/h (Figure 22).

A slight reduction in acceleration amplitude is observed when the tunnel is reinforced with Ø12 CFRP reinforcement bars. When the metro train travels at 60, 80, and 100 km/h, the acceleration amplitude RMS is 9.312×10^{-5} m/s^2/Hz, 1.190×10^{-4} m/s^2/Hz, and 2.048×10^{-4} m/s^2/Hz for the standard structure, respectively. For the reinforced structure, the acceleration amplitude RMS is 9.180×10^{-5} m/s^2/Hz, 1.168×10^{-4} m/s^2/Hz, and 2.003×10^{-4} m/s^2/Hz, respectively, at train speeds of 60, 80, and 100 km/h. Therefore, the reduction in acceleration amplitude is 1.42%, 1.85%, and 2.2% when the subway train speed is 60, 80, and 100 km/h, respectively. It follows that the degree of vibration reduction at a point far from the center of the tunnel increases as the speed of the subway increases.

Therefore, the reinforcement of the tunnel with CFRP can contribute to the damping of vibrations in the surrounding buildings by reducing the ground vibrations caused by the passage of the metro train in the tunnel.

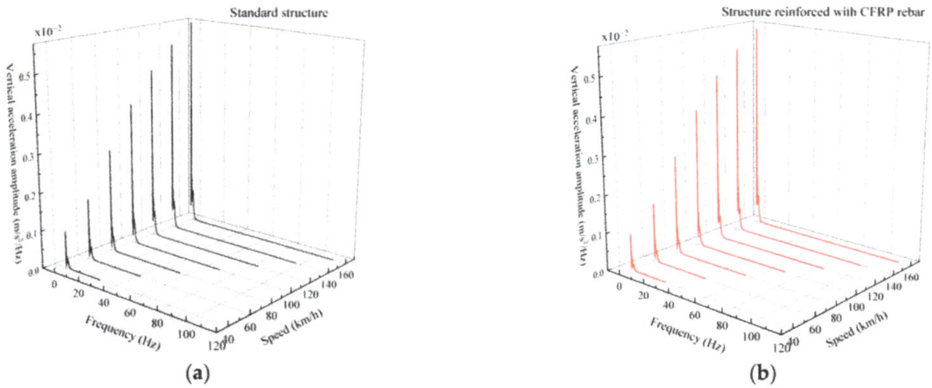

Figure 21. Spectrum of the vibration amplitude at point B. (**a**) Acceleration amplitude-frequency for the standard structure; (**b**) acceleration amplitude-frequency for the structure reinforced with CFRP rebars.

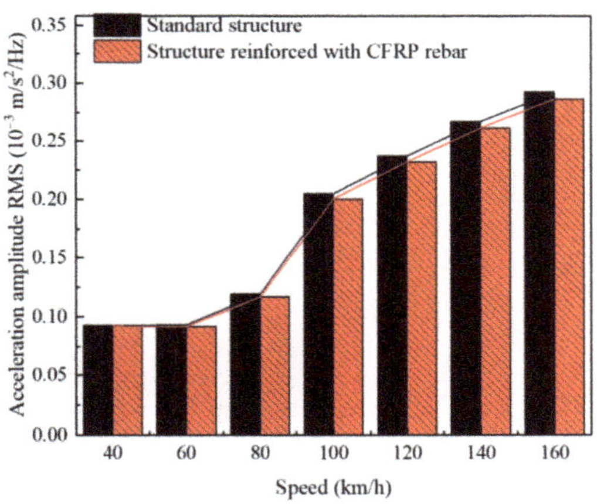

Figure 22. Acceleration amplitude RMS diagram at point B.

6. Statistical Analysis of CFRP Performance on Soil Dynamic Response

6.1. Experimental Setup

The vibrations caused by a moving train become weaker with the increasing distance. Nevertheless, the surrounding buildings are affected by the vibrations while the subway train is moving. In this section, the performance of CFRP in reducing the ground vibration at short and long distances from the tunnel is investigated. Therefore, the influence of some categorical factors (as described in Table 7) and their interaction on the reduction rate of ground vibration due to the travel of the metro train is investigated. The various types of CFRP used are listed in Table 5. At the end of this investigation, a simple and reliable prediction model was proposed.

Table 7. Factor and level selected in this analysis.

			Level		
Parameter Factors	Unit	Code	1	2	3
CFRP content (bar diameter)	mm	A	Ø10	Ø12	Ø14
CFRP type	-	B	PNCHS	PNCHM	PCHM
Distance from tunnel center	m	C	0	11	40
Response variable		Reduction rate of the ground-borne vibration			

PNCHS: Polyacrylic nitril carbon high strength; PNCHM: Polyacrylic nitril carbon high modulus; PCHM: Pitch carbon high modulus.

For this purpose, Minitab software was used to analyze the ground vibrations when the tunnel is reinforced with CFRP rebars by applying an analysis of variance to the designed orthogonal array [45]. A full fractional factorial design was used, mainly the standard L27 (3^{13-10}) orthogonal array. The orthogonal array consisted of three factors at three levels with 27 runs, which was chosen for its ability to fully capture the interaction between the independent variables [46]. Three levels were defined for each factor, as shown in Table 7. The first row (A) is associated with the CFRP content defined by the bar diameter, the second row (B) indicates the type of CFRP used, and the third row (C) represents the distance from the tunnel center (point A). The three levels of each factor are defined by "1, 2 or 3" in the standard orthogonal arrangement L27 (3^{13-10}) [47,48]. In this way, the ground vibration reduction rate defined in Equation (11) was determined from the calculated data for each numerical model run. The data were selected randomly to ensure that the model met certain statistical assumptions and to minimize the effects of factors not included in the experimental design.

$$r_{GV} = \left(1 - \frac{D_{\text{withCFRP}}}{D_{\text{standard}}}\right) \times 100 \tag{11}$$

where D_{withCFRP} is the ground acceleration amplitude RMS when the tunnel lining concrete is reinforced with CFRP rebar and D_{standard} is the ground acceleration amplitude RMS for the standard tunnel.

6.2. Analysis of Factors Influencing the Reduction Rate of Ground Vibrations

The analysis of variance consists of various variables, which are described as follows:

- the degrees of freedom (DF), which correspond to the information content of the experimental design;
- adjusted sums of squares (Adj SS), which correspond to the measures of variation for the different components of the experimental design;
- adjusted mean squares (Adj MS), which measure the proportion of variation explained by a term or experimental design;
- sequential sums of squares (Seq SS), which correspond to measures of variation for different components of the experimental design;
- contribution, which indicates the percentage contribution of each source in the table ANOVA under Sequential Sums of Squares Total (Seq SS);
- F-value, which appears for each term in the analysis of variance table; and
- *p*-Value is a probability that measures the evidence against the null hypothesis.

The variables, all of which are related, contribute to the definition of the *p*-value. Further details on the methods, principles, and formulas related to the statistical index used in the present study can be found in [49]. The analysis of variance presented in Tables 8 and 9 show that all terms in the linear and two-way interaction models had a *p*-value below the 5% significance level (β = 0.05).

Table 8. Analysis of variance for speed of 100 km/h.

Source	DF	Seq SS	Contribution	Adj SS	Adj MS	F-Value	*p*-Value
Model	18	1231.56	98.30%	1231.56	68.420	25.68	0.000
Linear	6	962.67	76.84%	962.67	160.446	60.23	0.000
CFRP content	2	106.44	8.50%	106.44	53.219	19.98	0.001
CFRP type	2	147.88	11.8%	147.88	73.941	27.76	0.000
Distance from tunnel center	2	708.35	56.34%	708.35	354.177	132.95	0.000
Two-way interactions	12	268.88	21.46%	268.88	22.407	8.41	0.003
CFRP content * CFRP type	4	18.47	1.47%	18.47	4.617	1.73	0.235
CFRP bar * Distance from tunnel center	4	111.09	8.87%	111.09	27.772	10.43	0.003
CFRP type * Distance from tunnel center	4	139.32	11.12%	139.32	34.831	13.06	0.001
Error	8	21.31	1.70%	21.31	2.664		
Total	26	1252.87	100%				

"*" define the coupling of two variables.

Table 9. Analysis of variance for speed of 60 km/h.

Source	DF	Seq SS	Contribution	Adj SS	Adj MS	F-Value	*p*-Value
Model	18	2022.26	99.81%	2022.26	112.348	204.68	0.000
Linear	6	1848.41	91.23%	1712.08	285.347	519.85	0.000
CFRP content	2	48.58	2.40%	46.66	23.329	42.50	0.000
CFRP type	2	172.53	8.52%	109.19	54.597	99.47	0.000
Distance from tunnel center	2	1627.30	80.32%	1321.98	660.988	1204.21	0.000
Two-way interactions	12	173.85	8.58%	173.85	14.487	26.39	0.000
CFRP content * CFRP type	4	11.67	0.58%	3.69	0.923	1.68	0.257
CFRP content * Distance from tunnel center	4	59.45	2.93%	49.09	12.273	22.36	0.000
CFRP type * Distance from tunnel center	4	102.73	5.07%	102.73	25.684	46.79	0.000
Error	7	3.84	0.19%	3.84	0.549		
Total	25	2026.10	100%				

"*" define the coupling of two variables.

The results of the analysis of variance presented in Tables 8 and 9 for the speeds of 100 and 60 km/h, respectively, show that all three factors of the linear model have a *p*-value of less than 5%, and thus have a very significant influence on the distribution of the ground vibration characteristics. For the two-way interaction model, the analysis shows that the *p*-value of the interaction effect between the CFRP content and the CFRP type is greater than 5%, and thus insignificant for the reduction in ground vibrations.

The main effect and interaction diagrams of the studied factors affecting ground vibrations are shown in Figures 23 and 24. Only analyses with a significant *p*-value are considered in this section.

In the mean effect plots in Figure 23a–c, each point represents the mean rate of reduction in ground vibrations for a corresponding factor. The horizontal reference lines indicate the mean response for all test series. If the points of a particular factor connected by a line are close to the horizontal reference line, the factor influence is insignificant. On the other hand, the steeper the line connecting a factor point, the greater its influence. As can be seen from the diagram of the mean values, the mean values of the responses differ from one factor to another, although the shape of the curve for the speeds of 60 and 100 km/h is almost identical.

Moreover, the influence of distance (distance between the tunnel and the calculation point) on the distribution of vibration characteristics is dominant due to the steep slope of the line connecting the points (Figure 23c). Therefore, the distance greatly influences the reduction rate of ground vibrations. However, it is observed that the performance of the CFRP reinforcements increases depending on the type or diameter of the bar used (Figure 23a,b).

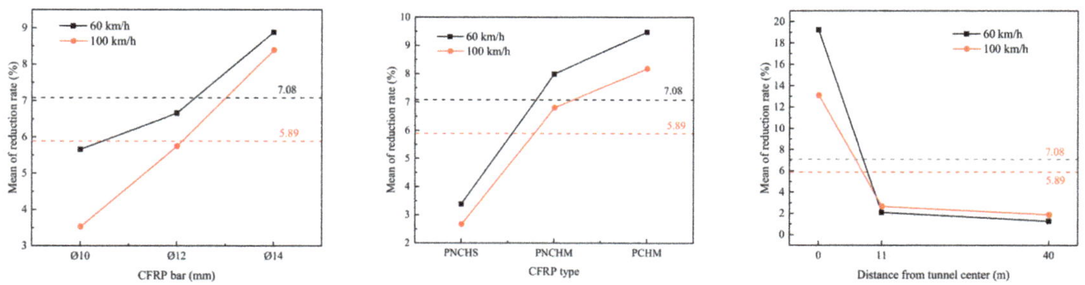

(a) (b) (c)

Figure 23. Reduction rate for each selected factor. (a) CFRP content; (b) CFRP type; (c) distance from the tunnel center.

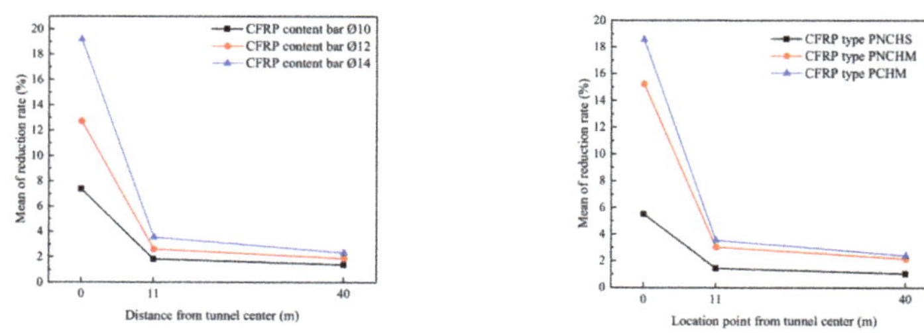

(a) (b)

Figure 24. Interaction plot matrix of the reduction rate of the ground vibration at the speed of 100 km/h. (a) CFRP content * Distance; (b) CFRP Type * Distance.

Figure 24a,b shows the average response of CFRP reinforcement's influence on the soil's dynamic response during the passage of a subway train at 100 km/h in the tunnel. In this section, the degree of vibration reduction at the ground surface was analyzed considering the interaction between the CFRP bar used and the distance. The interaction diagrams confirm the analysis performed above. The degree of vibration reduction is very large at 0 m, regardless of the content and type of CFRP, and decreases drastically at 40 m distance from the tunnel. The vibration reduction rate at the ground surface strongly depends on the proximity to the tunnel. Near the tunnel, the vibration reduction rate is very high and decreases drastically with the increasing distance from the tunnel. Therefore, the distance has a significant influence on the performance of the CFRP reinforcement.

To better assess the effects of each factor studied, a Pareto diagram of the standardized effect at a speed of 100 km/h was also created (see Figure 25). In this Pareto diagram, the bars representing factors C, B, A, BC, and AC cross the reference line 2.31. These factors are statistically significant at the 0.05 level with the terms of the presented model. According to the magnitude of each term described in the Pareto diagram, the distance between the calculation points and the tunnel significantly influences the reduction in ground vibrations. Since the objective of this section is to study the performance of CFRP reinforcement, the factor "CFRP type" contributes more to the reduction in ground vibration than the factor "CFRP content". As can be seen from the Pareto diagram, the BC term is larger than the AC term, indicating that the CFRP type factor is influential over a large distance from the tunnel.

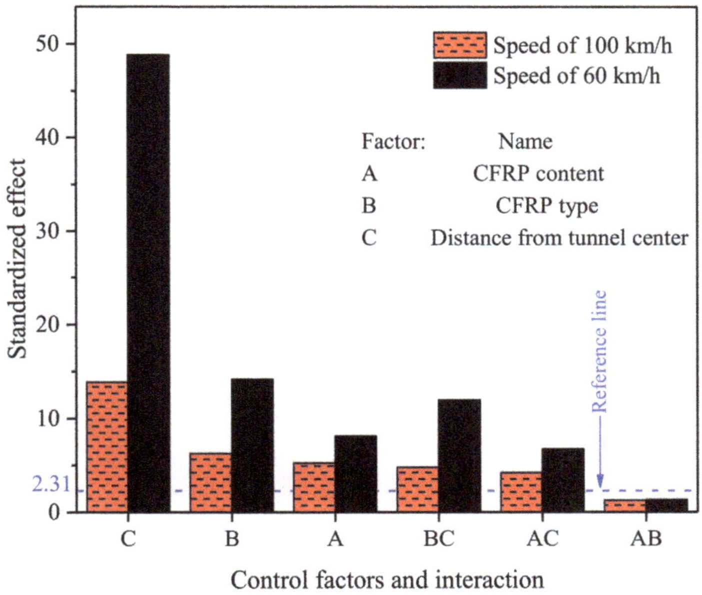

Figure 25. Pareto diagram of the standardized effect for vibration propagation reduction.

Spacing has a significant impact on the effect of CFRP in reducing ground vibration. CFRP reinforcement is most effective in the area of load application. Nevertheless, a slight reduction is observed at a distance far from the tunnel. According to the statistical analyses, the factor "CFRP content" in the tunnel lining has a more negligible influence on the reduction in ground vibrations than the factor "CFRP type". According to the Pareto diagram and the variance analysis table, the combination of both factors has very little influence on the vibration reduction rate. Therefore, the combination of the two factors in a tunnel structure would be a loss. It is recommended to give priority to the "CFRP type" factor. Therefore, CFRP reinforcement with high stiffness is better. However, if it is difficult to find CFRP with high stiffness, the second factor "CFRP content" can be considered.

6.3. A Predictive Model of the Vibration Propagation Reduction at the Ground Surface

To determine the dynamic response of the ground surface independent of the reinforcement properties (CFRP type and CFRP content) and the distance from the tunnel center, the relationship between the reduction rate of ground vibration and the above factors was established using the multiple linear regression equation (Equation (12)) shown below. In establishing the equation, the interaction CFRP-type CFRP-content term, which is statistically insignificant, was excluded.

$$y = \chi + \delta_1 Z_1 + \delta_2 Z_2 + \ldots + \delta_k Z_k \tag{12}$$

where y represents the dependent or response variable (ground vibration reduction rate); χ intercept is the response value when all the independent variables are zero; $\delta_1, \delta_2, \ldots, \delta_k$ represent coefficients or parameters, which reflect the contributions of each predictor in predicting the response; $Z_1, Z_2, \ldots, Z_k$ are independent variables or values of the influential factors and their coupling.

A multivariate regression analysis was performed individually for the train traveling at different speeds to create a specific predictive model to accurately estimate the dynamic response of the soil with the CFRP reinforcement according to the previously defined orthogonal arrangement. The prediction models were developed for a tunnel buried in soft soil with the subway train traveling at 60 and 100 km/h. The regression parameters provided a reasonably accurate estimate of the ground vibration reduction rate, which is shown in Table 10 following Equation (12). In this table, y_{RGV} is the ground vibration reduction rate, and the influencing factors are specifically the CFRP content, the CFRP type, and the distance from the tunnel center. The CFRP content distance and the CFRP type distance are respectively denoted as C_{RGV}, T_{RGV}, D_{RGV}, $C_{RGV}D_{RGV}$, and $T_{RGV}D_{RGV}$. As for the CFRP type, only the elastic modulus was considered in the calculations. The standard errors (SE) of each estimated parameter were also presented, along with their 95% confidence interval (CI). From the general factor regression analysis, all the prediction models had regression coefficients (R^2) close to 0.983, which indicates that these equations have a relatively high degree of fit. Therefore, the reduction rate of the vibration at the ground surface according to the CFRP material can be calculated by these formulations.

Table 10. Summary of prediction model.

Prediction Model for Reduction Rate of the Vibration at the Ground Upper Surface $y_{RGV} = \chi + \delta_1 C_{RSD} + \delta_2 T_{RGV} + \delta_3 D_{RGV} + \delta_4 C_{RGV} D_{RGV} + \delta_5 T_{RGV} D_{RGV}$						
Parameter	χ	δ_1	δ_2	δ_3	δ_4	δ_5
			Train speed of 100 km/h			
Estimate	−27.2	2.160	0.02439	0.714	−0.0556	−0.000607
SE Estimate	10.0	0.777	0.00762	0.419	0.0324	0.000318
95% CI	(−48.1, −6.4)	(0.545, 3.775)	(0.00853, 0.04024)	(−0.156, 1.584)	(−0.1230, 0.0118)	(−0.001269, 0.000055)
			Train speed of 60 km/h			
Estimate	−15.2	1.44	0.0241	0.396	−0.0376	−0.000633
SE Estimate	13.6	1.05	0.0103	0.568	0.0440	0.000432
95% CI	(−43.5, 13.0)	(−0.75, 3.63)	(0.0026, 0.0457)	(−0.785, 1.576)	(−0.1291, 0.0538)	(−0.001531, 0.000265)

7. Conclusions

Ground vibrations caused by the operation of a subway train in a tunnel were investigated by implementing a nonlinear 3D finite element model in Abaqus software [25]. The reliability of the numerical model, which was developed in accordance with the dimensions and conditions of the Shanghai subway section, was verified by comparing the simulation results with field test data from previous work. The load of the subway wheels was integrated into the numerical model as a transient moving load via a subroutine DLOAD developed in Fortran. Once the created model was successfully validated after calibrations, the ground vibrations during the operation in a tunnel reinforced with CFRP bars were investigated, and a series of parameter studies were performed. In addition, statistical analyses were performed to investigate the relationship between the dynamic response of the soil in a tunnel lining reinforced with CFRP bars and influencing factors, such as the CFRP bars used, the CFRP type, and the location of the calculation point. Finally, a prediction model for the dynamic response of the soil in a tunnel lining reinforced with CFRP bars was proposed. The following conclusions were drawn from this study:

- The developed 3D FE model was suitable for accurately simulating the dynamic behavior of the soil during the subway operation in a tunnel, since the error calculated by the comparative analysis was small.
- The effects of two subway trains running simultaneously in the downward and upward directions should be taken into account in the design of buildings in the vicinity of the subway line since the crossing of subway trains is unpredictable, while the vibrations generated by these two subway trains are almost twice as high as those generated by a single subway train.
- Reinforcing the concrete of the tunnel lining with CFRP rebars increases the dynamic capacity of the soil to respond to the loads caused by the passage of a subway train in a tunnel. A reduction in vibration is observed at the surface of the soil as well as at a depth of the soil, with the rate of reduction varying from one point to another. Therefore, reinforcing the tunnel lining with CFRP bars mitigates to some extent the discomfort experienced by surrounding buildings during the subway operation.

- The analysis of the effect of the subway's travel speed on the ground vibration shows that the maximum peak of the vibration amplitude increases with speed. After calculating the vibration amplitude RMS in the frequency domain, dynamic amplification was observed at the ground surface above the tunnel when the speed of the subway train reached 100 km/h. The dynamic amplification disappears with the increasing distance from the tunnel. Nevertheless, the difference between the vibration amplitude RMS at speeds from 80 to 100 km/h is huge. The speed of 100 km/h, which coincides with the shear wave velocity of a soil layer, illustrates the influence of the system mechanical properties on the soil's dynamic response.
- The reinforcement of the tunnel with CFRP bars reduces vibrations in the ground regardless of the operating speed of the subway. In general, the reduction rate increases as the speed of the metro train increases. The effect of CFRP bars in the structure decreases when the train reaches the critical speed, which coincides with the shear wave velocity of a soil layer. Nevertheless, the vibrations are less amplified when the distance between the dynamic acceleration and the speed of the subway train decreases from 80 to 120 km/h.

In addition, a statistical analysis of the dynamic response of the soil was performed when the tunnel lining is reinforced with CFRP bars, considering three factors (CFRP rod diameter, CFRP type, and distance from the tunnel center). The following results were obtained:

- The distance has a significant impact on the effect of CFRP reinforcement in reducing ground vibration. According to the results, the CFRP reinforcement is more effective near the tunnel. According to the statistical analyses, the factor "CFRP content" in the tunnel lining has a more negligible influence on the reduction in ground vibration than the factor "CFRP type". The combination of the two factors has little influence on the vibration reduction rate. Therefore, it would be a waste to combine the effect of both factors in a tunnel design. As a result, the use of a CFRP reinforcement with high stiffness is recommended. However, if it is difficult to find CFRP with high stiffness, the CFRP content in the tunnel lining can be considered.
- The proposed multiple linear regression model predicted with reasonable accuracy the reduction in ground vibration as a function of the type and content of CFRP used at any distance from the tunnel center buried in a soft deposit.

Author Contributions: Conceptualization, I.V.L.C. and J.C.; methodology, I.V.L.C. and J.C.; software, I.V.L.C.; validation, I.V.L.C. and J.C.; formal analysis, I.V.L.C. and J.C.; investigation, I.V.L.C. and J.C.; resources, J.C.; data curation, I.V.L.C. and J.C.; writing—original draft preparation, I.V.L.C.; writing—review and editing, I.V.L.C. and J.C.; visualization, I.V.L.C.; supervision, J.C.; project administration, J.C.; funding acquisition, J.C. All authors have read and agreed to the published version of the manuscript.

Funding: This research was funded by the National Natural Science Foundation of China (Grant No. 52178151); Shanghai TCM Chronic Disease Prevention and Health Service Innovation Center (Grant No. ZYJKFW201811009); and State Key Laboratory for Disaster Reduction of Civil Engineering (Grant No. SLDRCE19-B-22).

Data Availability Statement: Not applicable.

Acknowledgments: The authors acknowledge the support from the National Natural Science Foundation of China; Shanghai TCM Chronic Disease Prevention and Health Service Innovation Center; and State Key Laboratory for Disaster Reduction of Civil Engineering.

Conflicts of Interest: The authors declare no conflict of interest.

Appendix A

Table A1. Natural frequency obtained from the modal analysis (cycle/second).

Mode	Frequency	Mode	Frequency	Mode	Frequency	Mode	Frequency
1	2.75497	14	3.00939	27	3.06791	40	3.141291
2	2.75498	15	3.01641	28	3.07973	41	3.15007
3	2.86599	16	3.01931	29	3.08658	42	3.15021
4	2.86510	17	3.02062	30	3.08669	43	3.15334
5	2.98193	18	3.03278	31	3.09806	44	3.15411
6	2.98341	19	3.03346	32	3.10431	45	3.16050
7	2.98193	20	3.03457	33	3.10664	46	3.16303
8	2.98193	21	3.04378	34	3.10670	47	3.16306
9	2.98742	22	3.04887	35	3.10746	48	3.16457
10	2.99916	23	3.04978	36	3.11628	49	3.16610
11	3.00061	24	3.05046	37	3.11866	50	3.16624
12	3.00172	25	3.05361	38	3.12790		
13	3.00194	26	3.06765	39	3.12791		

References

1. Xia, H.; Cao, Y. Problem of railway traffic induced vibrations of environments. *J. Railw. Sci. Eng.* **2004**, *1*, 44–51.
2. Guo, T.; Cao, Z.; Zhang, Z.; Li, A. Numerical simulation of floor vibrations of a metro depot under moving subway trains. *J. Vib. Control* **2018**, *24*, 4353–4366. [CrossRef]
3. Lopes, P.; Ruiz, J.F.; Costa, P.A.; Rodríguez, L.M.; Cardoso, A.S. Vibrations inside buildings due to subway railway traffic. Experimental validation of a comprehensive prediction model. *Sci. Total Environ.* **2016**, *568*, 1333–1343. [PubMed]
4. Connolly, D.P.; Marecki, G.P.; Kouroussis, G.; Thalassinakis, I.; Woodward, P.K. The growth of railway ground vibration problems—A review. *Sci. Total Environ.* **2016**, *568*, 1276–1282. [CrossRef] [PubMed]
5. López-Mendoza, D.; Romero, A.; Connolly, D.P.; Galvín, P. Scoping assessment of building vibration induced by railway traffic. *Soil Dyn. Earthq. Eng.* **2017**, *93*, 147–161. [CrossRef]
6. Lang, J. Ground-borne vibrations caused by trams, and control measures. *J. Sound Vib.* **1988**, *120*, 407–412. [CrossRef]
7. Xu, P.; Wu, Y.; Huang, L.; Zhang, K. Study on the Progressive Deterioration of Tunnel Lining Structures in Cold Regions Experiencing Freeze–Thaw Cycles. *Appl. Sci.* **2021**, *11*, 5903. [CrossRef]
8. Li, R.; Gao, X.; Kong, C. Study on the aging deterioration curve of tunnel lining in water rich area. In *IOP Conference Series: Materials Science and Engineering*; IOP Publishing: Bristol, UK, 2020.
9. Han, W.; Jiang, Y.; Gao, Y.; Koga, D. Study on design of tunnel lining reinforced by combination of pcm shotcrete and FRP grid technique. In *IOP Conference Series: Earth and Environmental Science*; IOP Publishing: Bristol, UK, 2020.
10. Jiang, Y.; Wang, X.; Li, B.; Higashi, Y.; Taniguchi, K.; Ishida, K. Estimation of reinforcing effects of FRP-PCM method on degraded tunnel linings. *Soils Found.* **2017**, *57*, 327–340. [CrossRef]
11. Hollaway, L. A review of the present and future utilisation of FRP composites in the civil infrastructure with reference to their important in-service properties. *Constr. Build. Mater.* **2010**, *24*, 2419–2445. [CrossRef]
12. Weaver, A. Composites: World markets and opportunities. *Mater. Today* **1999**, *2*, 3–6. [CrossRef]
13. Han, W.; Jiang, Y.; Zhang, X.; Koga, D.; Gao, Y. Quantitative assessment on the reinforcing behavior of the CFRP-PCM method on tunnel linings. *Geomech. Eng.* **2021**, *25*, 123–134.
14. Li, W.; Gang, W.; Dong, Z.; Wang, S. Experimental study on a pre-damaged scaled tunnel model strengthened with CFRP grids. *J. Southeast Univ.* **2017**, *33*, 196–202.
15. Dolan, C.W. FRP Prestressing in the USA. *Concr. Int.* **1999**, *21*, 21–24.
16. Fukuyama, H. FRP composites in Japan. *Concr. Int.* **1999**, *21*, 29–32.
17. Bakis, C.E.; Bank, L.C.; Brown, V.; Cosenza, E.; Davalos, J.; Lesko, J.; Machida, A.; Rizkalla, S.; Triantafillou, T. Fiber-reinforced polymer composites for construction-state-of-the-art review. *J. Compos. Constr.* **2002**, *6*, 73–87. [CrossRef]
18. Caratelli, A.; Meda, A.; Rinaldi, Z.; Spagnuolo, S. Precast tunnel segments with GFRP reinforcement. *Tunn. Undergr. Space Technol.* **2016**, *60*, 10–20. [CrossRef]
19. Laníková, I.; Stepanek, P.; Venclovský, J. Optimization of a tunnel lining reinforced with FRP. In *Key Engineering Materials*; Trans Tech Publication: Zurich, Switzerland, 2016.
20. Caratelli, A.; Meda, A.; Rinaldi, Z.; Spagnuolo, S.; Maddaluno, G. Optimization of GFRP reinforcement in precast segments for metro tunnel lining. *Compos. Struct.* **2017**, *181*, 336–346. [CrossRef]
21. Spagnuolo, S.; Meda, A.; Rinaldi, Z.; Nanni, A. Curvilinear GFRP bars for tunnel segments applications. *Compos. Part B Eng.* **2018**, *141*, 137–147. [CrossRef]
22. Zhao, C.; Tang, Z.; Wang, P.; Feng, J.; Zhou, J.; Kong, X.; Geng, H.; Jin, F.; Fan, H. Blast responses of shallow-buried prefabricated modular concrete tunnels reinforced by BFRP-steel bars. *Undergr. Space* **2022**, *7*, 184–198. [CrossRef]

23. Abbood, I.S.; Aldeen Odaa, S.; Hasan, K.F.; Jasim, M.A. Properties evaluation of fiber reinforced polymers and their constituent materials used in structures—A review. *Mater. Today Proc.* **2021**, *43*, 1003–1008.
24. Karsli, N.G.; Ozkan, C.; Aytac, A.; Deniz, V. Effects of sizing materials on the properties of carbon fiber-reinforced polyamide 6, 6 composites. *Polym. Compos.* **2013**, *34*, 1583–1590. [CrossRef]
25. Abaqus, F. *Dassault Systemes Simulia Corporation*; FEA Abaqus: Providence, RI, USA, 2014.
26. Bathe, K.-J. *Finite Element Procedures*; Prentice Hall: Upper Saddle River, NJ, USA, 2006.
27. Lee, I.-W.; Kim, D.-O.; Jung, G.-H. Natural frequency and mode shape sensitivities of damped systems: Part I, distinct natural frequencies. *J. Sound Vib.* **1999**, *223*, 399–412. [CrossRef]
28. Zhai, W.; Xia, H.; Cai, C.; Gao, M.; Li, X.; Guo, X.; Zhang, N.; Wang, K. High-speed train–track–bridge dynamic interactions–Part I: Theoretical model and numerical simulation. *Int. J. Rail Transp.* **2013**, *1*, 3–24. [CrossRef]
29. Chango, I.V.L.; Assogba, O.C.; Yan, M. Estimating static and dynamic stress distribution in a railway embankment reinforced by geogrid and supported by pile system. *Transp. Infrastruct. Geotechnol.* **2022**, *2022*, 1–28. [CrossRef]
30. Assogba, O.C.; Sun, Z.; Tan, Y.; Nonde, L.; Bin, Z. Finite-element simulation of instrumented asphalt pavement response under moving vehicular load. *Int. J. Geomech.* **2020**, *20*, 04020006. [CrossRef]
31. Zienkiewicz, O.C.; Taylor, R.L.; Zhu, J.Z. *The Finite Element Method: Its Basis and Fundamentals*; Elsevier: Amsterdam, The Netherlands, 2005.
32. Li, S.; Guo, Z.; Yang, Y. Dynamic viscoelastic response of an instrumented asphalt pavement under various axles with non-uniform stress distribution. *Road Mater. Pavement Des.* **2016**, *17*, 446–465. [CrossRef]
33. Chango, I.V.L.; Assogba, G.C.; Yan, M.; Xianzhang, L.; Mitobaba, J.G. Impact Assessment of Asphalt Concrete in Geogrid-Reinforced-Pile-Supported Embankment During High-Speed Train Traffic. *Balt. J. Road Bridge Eng.* **2022**, *17*, 135–163. [CrossRef]
34. Huang, Z.-K.; Pitilakis, K.; Tsinidis, G.; Argyroudis, S.; Zhang, D.-M. Seismic vulnerability of circular tunnels in soft soil deposits: The case of Shanghai metropolitan system. *Tunn. Undergr. Space Technol.* **2020**, *98*, 103341. [CrossRef]
35. Huang, Q.; Li, P.; Zhang, D.; Huang, H.; Zhang, F. Field Measurement and Numerical Simulation of Train-Induced Vibration from a Metro Tunnel in Soft Deposits. *Adv. Civ. Eng.* **2021**, *2021*, 6688746. [CrossRef]
36. Yang, L.; Wang, G.; Liu, Q.; Lei, G. A study on the dynamic properties of soft soil in Shanghai. In *Soil and Rock Behavior and Modeling*; American Society of Civil Engineers: Reston, VA, USA, 2006; pp. 466–473.
37. Zhang, D.-M.; Huang, Z.-K.; Wang, R.-L.; Yan, J.-Y.; Zhang, J. Grouting-based treatment of tunnel settlement: Practice in Shanghai. *Tunn. Undergr. Space Technol.* **2018**, *80*, 181–196. [CrossRef]
38. Oreste, P. Analysis of structural interaction in tunnels using the covergence–confinement approach. *Tunn. Undergr. Space Technol.* **2003**, *18*, 347–363. [CrossRef]
39. Michał, S.; Andrzej, W. Calibration of the CDP model parameters in Abaqus. In Proceedings of the 2015 Wourld Congress on Advances in Structural Engineering and Mechanics (ASEM15), Incheon, Korea, 7 September 2015.
40. Chango, I.V.L.; Yan, M.; Ling, X.; Liang, T.; Assogba, O.C. Dynamic Response Analysis of Geogrid Reinforced Embankment Supported by CFG Pile Structure During a High-Speed Train Operation. *Lat. Am. J. Solids Struct.* **2019**, *16*. [CrossRef]
41. Gao, J.; Zhai, W. Dynamic effect and safety limits of rail weld irregularity on high-speed railways. *Sci. Sin. Technol.* **2014**, *44*, 697–706.
42. Bian, X.; Jiang, H.; Chang, C.; Hu, J.; Chen, Y. Track and ground vibrations generated by high-speed train running on ballastless railway with excitation of vertical track irregularities. *Soil Dyn. Earthq. Eng.* **2015**, *76*, 29–43. [CrossRef]
43. Zhou, S. *Dynamics of Rail Transit Tunnel Systems*; Academic Press: Cambridge, MA, USA, 2019.
44. Spagnuolo, S.; Meda, A.; Rinaldi, Z.; Nanni, A. Precast concrete tunnel segments with GFRP reinforcement. *J. Compos. Constr.* **2017**, *21*, 04017020. [CrossRef]
45. Tanaydin, S. *Robust Design and Analysis for Quality Engineering*; Taylor & Francis: Abingdon, UK, 1998.
46. Ghani, J.A.; Choudhury, I.; Hassan, H. Application of Taguchi method in the optimization of end milling parameters. *J. Mater. Process. Technol.* **2004**, *145*, 84–92. [CrossRef]
47. Assogba, O.C.; Tan, Y.; Dong, W.; Lv, H.; Anato, N.J. Field evaluation and statistical analysis of the dynamic response of semi-rigid pavement under full-scale moving truck load. *Road Mater. Pavement Des.* **2021**, *2021*, 1–29. [CrossRef]
48. Anato, N.J.; Assogba, O.C.; Tang, A.; Diakité, Y.; Cho Mya, D. Numerical and statistical investigation of the performance of closed-cell aluminium foam as a seismic isolation layer for tunnel linings. *Eur. J. Environ. Civ. Eng.* **2021**, 1–25. [CrossRef]
49. Bryman, A.; Cramer, D. *Quantitative Data Analysis with Minitab: A Guide for Social Scientists*; Routledge: Oxfordshire, UK, 2003.

MDPI
St. Alban-Anlage 66
4052 Basel
Switzerland
www.mdpi.com

Buildings Editorial Office
E-mail: buildings@mdpi.com
www.mdpi.com/journal/buildings

www.ingramcontent.com/pod-product-compliance
Lightning Source LLC
Chambersburg PA
CBHW061847210926
51850CB00006B/991
9783036587066